THE LAND HAS CHANGED

AFRICA: MISSING VOICES SERIES

Donald I. Ray, general editor

ISSN 1703-1826

University of Calgary Press has a long history of publishing academic works on Africa. *Africa: Missing Voices* illuminates issues and topics concerning Africa that have been ignored or are missing from current global debates. This series will fill a gap in African scholarship by addressing concerns that have been long overlooked in political, social, and historical discussions about this continent.

UNIVERSITY OF
CALGARY
PRESS

THE LAND HAS CHANGED

History, Society and Gender in Colonial Eastern Nigeria

CHIMA J. KORIEH

© 2010 Chima J. Korieh

University of Calgary Press

2500 University Drive NW

Calgary, Alberta

Canada T2N 1N4

www.uofcpress.com

LIBRARY AND ARCHIVES CANADA CATALOGUING IN PUBLICATION

Korieh, Chima J. (Chima Jacob), 1962-

The land has changed : history, society and gender in colonial Eastern Nigeria / Chima J. Korieh.

Includes bibliographical references and index.

ISBN 978-1-55238-268-4

1. Igbo (African people)–Agriculture–Nigeria, Eastern–History. 2. Agriculture–Nigeria, Eastern–History. 3. Agriculture–Social aspects–Nigeria, Eastern–History. 4. Agriculture and state–Nigeria, Eastern–History. 5. Igbo (African people)–Nigeria, Eastern–Social conditions. 6. Women, Igbo–Nigeria, Eastern–Social conditions. 7. Igbo (African people)–Nigeria, Eastern–History. 8. Great Britain–Colonies–Nigeria. I. Title.

HD1021.K67 2009   306.3'490966946   C2009-906542-8

The University of Calgary Press acknowledges the support of the Alberta Foundation for the Arts for our publications. We acknowledge the financial support of the Government of Canada through the Book Publishing Industry Development Program (BPIDP) for our publishing activities. We acknowledge the financial support of the Canada Council for the Arts for our publishing program.

Alberta Foundation for the Arts   Canada   Canada Council for the Arts / Conseil des Arts du Canada

Cover Photo: "Sorting the Palm Nuts for Making Oil"
Photographer: Jeanne Tabachnick

Cover design by Melina Cusano
Page design and typesetting by Melina Cusano

In memory of my father, Linus Onyegbule Korieh
For my daughter, Akunna Chinaza Korieh

TABLE OF CONTENTS

LIST OF ILLUSTRATIONS

LIST OF TABLES

FOREWORD

I consider it a distinct honour and privilege to have been invited to write a foreword for this book, *The Land Has Changed: History, Society and Gender in Colonial Eastern Nigeria*, which is a major and significant contribution to Igbo studies. In this volume, the author provides a vivid account of the patterns of socioeconomic change in the former Eastern Region of Nigeria, but focusing primary attention on present Imo and Abia states. The book deals with the transformations of agriculture during the colonial and postcolonial eras. It is certainly a study whose time has come, insofar as this genre, relating specifically to the Igbo, has been in the doldrums for too long. In other words, studies dealing with colonial agrarian policies and their impact on the indigenous peoples seem thus far to have been neglected by both indigenous and foreign scholars. Korieh's book, therefore, fills an important gap in the existing studies dealing with British colonial innovations or changes relating to agriculture in Nigeria. As a remedy, Korieh examines and analyzes the transformations in agriculture in southeastern Nigeria with a special focus on the colonial and postcolonial periods.

In *The Land Has Changed*, a title borrowed from an agonized interviewee, the author confronts the problems of modernization as perceived by the British and challenges the notion that British colonial agricultural policies redounded to the benefit of the colonized. Rather, the transformations laid the foundation for decades of resistance, the decline of agriculture, and the onset of perennial hunger and poverty. In the author's words, "The book centres on the British attempt to transform a colonial society (continued in the early postcolonial period) through the modernization of agriculture and the experiences, actions, and perceptions of the local population whom colonial officials often characterized as backward and unchanging." In the final analysis, we are concerned here with, first, the dynamic processes of colonial and postcolonial socioeconomic engineering with their insidious consequences. Secondly, we witness the patterns of local/Igbo reactions towards the radical changes in the political economy.

Because of the accident of history, Nigeria (including Igboland) came under British direct rule from about 1900 to 1960. With the advent of the British, therefore, things literally fell apart with special reference to the political economy (agriculture). For example, Nigeria was brought inexorably under the vortex of the Western capitalist system, and ipso facto was engulfed in the capitalist economy. This meant that agricultural production *for export* became the foundation of the new economy. In a sense, agricultural products for the British industries became a British colonial economic preoccupation, and agricultural products now became the chief contributors to Nigeria's gross domestic product. Put differently, Nigeria, as well as most parts of West Africa, entered into what A. G. Hopkins described as the "open" phase of colonial economic development, meaning the entry of the economy into the phase of the Western capitalist economic model of colonialism – exploitation. Of course, this implied changes in the patterns of production and commodity exchange (trade). The characteristics include the export of a range of agricultural and mineral products in exchange for a variety of British manufactures, chiefly consumer goods, the domination of some sectors of the economy by expatriates, and perhaps most importantly, the assertion of considerable influence and control over Nigerian economic policy by the British.[1] These changes invariably heralded the age of British imperial control, or unequal relationship, and thus the basis of colonial exploitation and other forms of oppression.

It might be pertinent at this juncture to examine the impact of the transformations in agriculture in the context of culture and life. We should emphasize the fact that agriculture (farming) was essentially the principal occupation of the Igbo of southeastern Nigeria from time immemorial. In particular, agriculture was the core of the people's source of livelihood, and agriculture dominated the Igbo/Nigerian economy up to the 1960s. Virtually every adult household (family) was engaged in farming and, accordingly, food production (particularly yams, the king of foodstuffs) and other crops like cassava, corn/maize, and a variety of vegetables formed the basic staples of food. Not surprisingly, most people had enough food to eat, and quite correspondingly populations grew. Historically, therefore, Igboland was the land of plenty, where food production was ample and abundant, and people thus prided themselves as being blessed with surplus food and wealth. Is it any wonder that the elders still speak nostalgically of the "good old days," in reference to when food was plentiful and poverty and starvation had no place in the land? In fact, studies show that agriculture was central to the evolution of the highly advanced Igbo civilization.[2]

Admittedly, there were times of food scarcity/shortages and hence temporary periods of hunger, known popularly as the period of *Onwo*. As Victor Uchendu explains: "By June, all the yams have been planted. A period of food shortage called *onwo* sets in…. *Onwo* is caused by lack of a well-developed system of storing surplus yams until the new crop is harvested. The chief staples then become cassava, cocoyams, and, in some parts of southern Igbo, the three-leaved yam locally known as eno."[3] These food crops are popularly referred to as "women's" crops, while yam, the king crop, is known as "men's" crop.

Production for exchange, of course, was an important component of the Igbo political economy. "Field studies in the eastern region [of Nigeria] have shown that the region possesses a highly developed system of distributive trade. This system is not recent in origin but appears to have been developed over centuries of slow and unimpeded growth."[4] Indeed, food production and trade significantly contributed to the dynamic process of population growth, therefore making Igboland one of the most densely populated regions of the African continent, if not in the world. Yes, Robert Stevenson affirms, "There can be little question that the I[g]bo region in south eastern Nigeria constitutes one of the greatest nodes of rural population density in all of sub-Saharan Africa."[5] Stevenson adds that, "the very nature of the density and the large area involved bespeak a rather respectable antiquity."[6]

The point of emphasis here is that in Igboland food was plentiful, and poverty and starvation had no place in the land. But sadly, as Korieh's study illustrates, this supposedly pristine and prosperous country (land) was "spoiled by man" – arguably the British. The proverb/idiom, "A Country is spoiled by men," or alternatively, "It is man who 'spoils' the country," cryptically captures the Igbo ideological perception of change. Culturally and philosophically, the Igbo accept the principle/notion of change. In essence, they are fully aware that change is inevitable and can come from within or from without. But change, they also say, must be useful; otherwise, it must be rejected out of hand. Thus, in consonance with their cultural disposition that "life [hazards] must be faced and its problems overcome," British colonial problems via agricultural changes had to be overcome. As illustrated in this book and in other studies, the British were the fundamental sources of trouble, with particular regard to the socioeconomic and political innovations that upset the apple cart. The female victims of agricultural change, for instance, resolutely challenged the oppressive regime, as evidenced in the Women's War of 1929. In effect, British innovations triggered a variety of reactions, which were at times quite tragically violent in nature.

Also reacting specifically to the transformations in agriculture and the restrictions placed on trade, the aggrieved women expressed their discontent as follows:

> Our grievances are that the land has changed and we are all dying. [Specifically,] since the white men came, our [palm] oil does not fetch money.

> Imagine our suffering [as we are] restricted from exporting garri [cassava flour] to the north especially when the small profit accruing from the trade is being exhausted….

> [And,] consider the lives of a family, which may perish as a result of the measures, which have been taken to restrict the garri trade.[7]

The introduction of the gender equation – voices of women – in this volume is critically important. It helps to call into question the assumption that African/Igbo women were simply the silent victims of colonial oppression. But this and other studies clearly show that, because women, after all, were (and still are) the backbone of the extractive (farming) sector of the economy, they could hardly remain aloof over colonial intrusions into their sources of livelihood. Consequently, any study/studies of structural change(s) in colonial/postcolonial agriculture, which excludes women's voices, must certainly be declared as truncated. The author, therefore, makes a major contribution in Africanist scholarship and historiography by approaching his subject from a holistic perspective.

Using primarily archival and oral sources (interviews), Korieh navigates the complex and, at times, conflicting perceptions/notions of change, as they relate to agriculture. In particular, he critically examines the transition from an essentially subsistence agriculture-based economy to a capitalist export-oriented economy. And he further focuses attention on how colonialism and the international economic environment created "poor terms of trade" for Igbo/Nigeria's primary agricultural exports, and the onset and consequences of the dependency syndrome on developed economies.

Korieh notes, however, that from the 1970s Nigeria's socioeconomic woes began with the collapse or decline of agriculture, even though ambitious large-scale agricultural development projects had been attempted, e.g., Operation Feed the Nation, the importation of fertilizers, etc. But "Despite the heavy financial investments … agricultural productivity remains abysmal." Several reasons are suggested for the colossal failures, including corruption and the crisis of governance, and the role of external factors – international agencies such as the World Bank and the International Monetary Fund. In the final analysis, Korieh attributes the ominous decline of agriculture to the shift from agriculture as the base of Nigeria's economy to reliance or dependence on petroleum as the principal source of revenue from the 1970s. His words:

> The discovery of oil at Oloibiri in 1956 and the production [of oil] that began in 1958 made Nigeria the largest oil producer in Africa and the eleventh in the world. The 1970s coincided with the rise in the world oil price, and Nigeria reaped instant riches from petroleum. The emergence of petroleum economy [thus] ushered in an important new phase in Nigeria's economy. Agriculture, which earned significant revenue for the [Eastern] region and the country before the oil boom, suffered from low investment, leaving

the rural population vulnerable. Rural regions witnessed a major crisis, with social, infrastructural, and economic collapse in the 1980s.[8]

The Land Has Changed: History, Society and Gender in Colonial Eastern Nigeria, based on carefully researched archival and oral sources, provides us with a wealth of knowledge about changes in the Nigerian economy during the colonial and postcolonial eras. Although this volume presents a rather depressing picture of contemporary Nigeria faced with problems of dire hunger and poverty because it seems no longer self-sufficient in food production and now depends on the importation of food, it offers some ray of hope. Economic recovery under more patriotic and committed leaders might still bring Nigeria back to its lost reputation as the giant of Africa.

Felix K. Ekechi
Professor Emeritus
Kent State University

ACKNOWLEDGMENTS

This book has been long in the making. The impetus for the book stems from my study of gender relations focusing on widowhood practices among the Igbo of Eastern Nigeria in 1996. During that research, I found that various aspects of rural life, including social and cultural practices, were interlinked with the economic life of the people. The research also raised important questions about the economy and rural livelihoods in a transforming society. Although I was interested in the cultural aspects of widowhood, it was apparent that widowhood practices had been influenced by external factors, including colonialism, Westernization, and new economic structures that had their origins in the increased commercialization of agriculture in the early parts of the twentieth century. The increased importance of palm produce (oil and kernels) for the European market in particular led to significant changes among the Igbo, including the commoditization of land and changes in gender relations of production. The changes that emerged and the impact of colonial construction of new economic and social formations necessary to achieve increased export production, as well as the responses of rural people in a predominantly agrarian economy, piqued my interest. How did official policies and attitudes influence the nature of agricultural change among the Igbo of southeastern Nigeria? I sought to answer this question by exploring official agricultural policy and its influences through a multi-layered study of the Igbo of southeastern Nigeria. I was convinced that official perceptions about the local agricultural environment, the mode of production, and the role of colonial subjects – male and female – represented a narrow understanding of the local production system of the Igbo and how the culture affected economic life. The structure of the local production pattern was more complex than officials conceptualized it. By examining the historically important role that agriculture played in the encounter between the Igbo and the British colonial government, I sought to clarify the ways state policies transformed agriculture and, in particular, how local people responded to the transformation of the local society.

I owe a huge debt of gratitude to several people who made this book possible. Let me start by expressing such gratitude to Professor Martin Klein. His support when I was a graduate student at the University of Toronto gave me the courage to pursue this dream. He read the early drafts of this book and offered useful insights. Marty and his wife Susan shared their home with me on many occasions and I remain grateful for their continued friendship. I would also like to express my thanks to Professor Michael Levin. The Levins offered me accommodation for a full semester on my return from field work in 2001. I thank them immensely for their assistance.

I would also like to thank the African Research Center, Leiden, Netherlands, for a visiting fellowship in 2007 that enabled me to complete the manuscript. The British Academy fellowship at Oxford University in 2008 gave me the opportunity to explore additional resources at Rhodes House and to revise the manuscript. I am grateful to my hosts: the African Studies Center and St. Anthony's College for supporting me during the period of my fellowship. Professor David Anderson, the Director of the African Studies Center and Dr. David Pratten, made the trip possible, and I thank them immensely.

Several people read my proposal and early drafts of the manuscript. I am particularly grateful to Professors U. D. Anyanwu, John Inikori, Toyin Falola, the late Ogbu Kalu, Dmitri van den Bersselaar, Dorothy Ukaegbu, and Faye Harrison, for their insightful comments and criticism. I owe a lot of debt to the staff of Enugu and Calabar archives. I have used both archives over several years and have developed enduring friendships. I thank the former director of the Enugu archives, Dr. U. Esse, and the current director, Chief Nsoro, for their assistance. Other staffs of both archives have been of immense assistance during my research and I thank them. At Marquette, the chair of the history department, James Marten, and my other colleagues have offered encouragement and support.

Finally I want to thank my family for their support. Akunna and Chidi deserve my sincere gratitude for putting up with my frequent absences while I conducted the research and wrote this book. I cannot repay such a debt. My father was an inspiration and I dedicate this work to him. My mother and siblings gave me more love and support than I could ask for. I especially want to thank my sister Adanna, her husband Mazi Onuegbu, and their children; my uncle, Rev. T.A.U. Iwuh, his wife Kate, and their children, for their hospitality during my many research visits to Nigeria. I am grateful to my research assistants, Mrs. Kate Emmanuel and Henry Onyema, for assistance during fieldwork in Nigeria. I benefited from the support of many other family members and friends in the process of writing this book: Bobby, Mama Charity Ihediwa, Uloaku and Niyi Adesun, Raph Njoku, Ebby Madera, Bright and Ijeoma Okoronkwo, Ifeanyi Ezeonu, Anayo Okorie and his family, Naomi Diptee, Ndu Life Njoku, Dr. Ugwuanya Nwosu and Patience, Bill Carrigan and Emily Blanck of Rowan University, and Christopher Obiukwu. I thank them sincerely for their support and enduring friendship.

I am very grateful to the staff at University of Calgary Press for their enthusiasm and for the insightful comments of my editor John King throughout the production of this book.

In the process of writing this book, I have come to respect the men and women I met and interacted with in the Igbo countryside. Their tenacity and resilience and their ability to cope amidst significant stress on their households and communities amaze me. I learnt valuable life lessons from them, especially the importance of listening to other voices – those often at the margins of society – in reconstructing the history of rural societies. The voices and criticism I received from readers helped to improve the manuscript, but I remain responsible for the imperfections that it contains and its contribution to the scholarship.

Map of Igboland

INTRODUCTION: PERSPECTIVES, SETTING, SOURCES

> Few attempts have been made to teach the native more mod-
> ern methods of farming, but in this direction the native has so
> far shown himself to be quite unteachable in this part of Nigeria.
> – *Edward Morris Falk, British District Officer, Aba, 1920*

> Our grievances are that the land has changed and we are all
> dying. It is a long time since the chiefs and the people who know
> book [Western-educated people] have been oppressing us. We are
> telling you that we have been oppressed. – *Female Witness, Aba
> Commission of Inquiry, 1930*

"The land has changed!" This is how my father began when I asked him in the year 2000 to relate his experiences of the nature of change since the colonial period in my village, a small rural community in central Igboland. Many rural dwellers shared similar sentiments. However, when African people tell stories about the colonial period, some talk about a period of plenty while others talk about a period of deprivation. Grace Chidomere, a rural farmer and small-scale trader, looked back on the colonial era with nostalgia, describing it as the "good old days."[1] Eighty-year-old Francis Eneremadu, who served as a tax assessment officer in the colonial period, remembers, with a degree of sadness, a period when "things were very good."[2] "People lived off what they earned from produce sales, trading, and farming," said Comfort Anabalam, a rural farmer in Mbaise.[3]

Not all, however, hold the view that the colonial period was a time of plenty. Some associated the colonial period with mere "survival" because it included several periods of crisis for the rural population. The depression of the late 1920s and early 1930s and the economic difficulties it engendered in low export prices for palm oil and kernels and high costs of imported goods generated a crisis in the local economy for which the rural and urban populations paid the price. Reacting to the colonial restrictions imposed on African traders during the Second World War, for example, A. Jamola, a trader in Aba, who had been refused a permit to trade in gari (a local staple produced from cassava), expressed the following sentiments in his petition to the district officer on 21 July 1943.

> Imagine our suffering, Sir, during this period for which I am restricted from exporting garri to the north especially when the small profit accruing from the trade is being exhausted. I can assure you, Sir, that I do appreciate your point of view in this rather difficult question, but at the same time I would very respectfully and humbly ask that in addition to your viewing the matter from the official stand-point, you may consider the lives of a family which may perish as a result of the measures which have been taken to restrict the garri trade.[4]

Jamola's request was an act of desperation, but such sentiments were widespread and reflected the diverse outcomes of colonial encounters from the native viewpoint. One woman's view of the society and economy during this period supports this perspective: "since the white men came, our oil does not fetch money. Our kernels do not fetch money. If we take goats or yams to market to sell, court messengers who wear a uniform take all these things from us."[5]

From the late nineteenth and early twentieth centuries onward, African societies underwent significant transformations. This is evident both in the regions where the attempt to transform rural agriculture was enforced through rural peasants and in colonies where commercial production by a white settler community left indelible inequality between the settlers and the African population. In both cases, production for export worked to link African societies to the industrialized societies of Europe and provided opportunities for some and a struggle for others. Yet the history of agriculture in the twentieth century was one of growth and prosperity that was followed by progressive decline from the 1970s in many areas. In many parts of Africa at the beginning of the century, the agricultural sector was demographically larger, wealthier, and more productive than most other sectors of the economy. The agricultural sector had become demographically smaller, poorer, and less productive than ever before by the end the century. Farmers at the end of the century had declining incomes and were worse off, on average, than those who did not farm. There is considerable agreement among scholars and other commentators that the rapid decline and disintegration of the rural economy accelerated under the pressure of state policies and modernization.

In Igboland, as in most of colonial Africa, agricultural production was an instrument and focus of colonial governance, and high productivity was the ambition of officials, since government revenue depended largely upon peasant taxes, export duties, and import duties. From its establishment in 1910, the Nigerian Department of Agriculture pursued a policy of encouraging indigenous development of the palm oil industry as a means of achieving economic development and drawing rural folk into the cash economy. Colonial policies and European demand for palm oil and kernels led to an expansion in the productive capacity of Igbo households and drew them closer to the capitalist market. Until recent times, the Igbo region was largely dependent on rural agriculture, and palm oil was the most important agricultural produce and an important source of accumulation.[6] Household members participated in the formal economy as producers and marketers of palm produce, and, for most of the colonial and early post-colonial period, they built substantial wealth until the agricultural decline that set in from the late 1960s. How is this change reflected in the history of the region and the memories of those who witnessed these changes?

This book addresses this historical problem. The context for rural change varied across much of Africa and the decline and crisis that have occurred were not generally related to state policies alone. Structural changes, over which the rural population as much as the state had little control, the people themselves, and their activism, were also powerful forces of change, as this book suggests for the Igbo society. Rather than concentrating on state policies alone, the book will instead focus on the complex processes through which an African society responded to state intervention during the colonial and early post-colonial period. While I do not attempt to describe local life in its entirety, I use agricultural change as a lens through which to view socio-economic change, political struggle, cultural change, and colonial hegemony. The book centres on the British attempt to transform a colonial society (continued in the early post-colonial period) through the modernization of agriculture and the experiences, actions, and perceptions of the local population whom colonial officials often characterized as backward and unchanging. Such perceptions of African societies' attitudes toward official ideas and about development and change would come to reflect the top-down approach in the attempt to modernize agriculture.

The focus is the central Igbo region in southeastern Nigeria, defined as the areas included in Owerri Province in the colonial period or contemporary Imo State and parts of Abia State. The area lies within the oil palm belt, which includes, roughly, the area west of the Cross River and south of the region between Onitsha and Afikpo. This was one of the most important centres of palm oil production in the colonial and early post-colonial periods, and it is, therefore, an important case study in the impact of agricultural change and local responses. The region is also characterized by high population density, significant out-migration, and relatively poor soil.[7] This was also an area that witnessed major political agitations during the colonial period, including the tax revolts of 1929, that were deeply rooted in the agrarian economy as described in chapter 4. In chapter 5, I also examine the link between the rural economy and the forms of political consciousness that emerged as this economy was threatened by the depression caused by the Second World War. These unique characteristics presented peculiar challenges to the population in this part of Igboland, but they also provide an opportunity to assess how local historical contexts mediate major societal transformations. The advantages of this regional approach include the opportunity to collect data in depth, to provide comparative perspectives on how different parts of Igboland responded to varying ecological niches, and to explain the varying responses to the transformation that came with colonialism.

The forms and patterns of change addressed in this book have occurred throughout Igboland, but it would be a mistake to assume that these transformations occurred uniformly and that there were no variations. Although different parts of Igboland experienced unique circumstances due to diversity in population, ecology, and socio-economic structures, this book stresses that there is some level of homogeneity and continuously draws upon the collective experiences of the territory historically known as southeastern Nigeria, where the Igbo remain the dominant population. Therefore, I have not confined myself to the central region alone.

While the book covers the period from the middle of the nineteenth century onward, its central concerns revolve around the periods of substantial growth in agriculture – the beginning of European colonization and the subsequent colonial period (1880s–1960), and the (1960s–1970s), which witnessed a gradual but significant decline in agricultural output. These were periods when the majority of the population made their living from agriculture and the trade in agricultural produce. My study illuminates the changing nature of agricultural production and the rural economy over different historical periods and shows how the rural population responded to these changes.

In most of colonial Africa, agricultural commodity production was seen as an essential engine of growth upon which the development of African farmers and societies depended.[8] Colonial intervention increased the pace and scope of agricultural change in several parts of sub-Saharan Africa, leading some commentators to characterize this period as an era of agricultural revolution.[9] The degree of social and economic changes had fundamental implications for agricultural sustainability and rural subsistence.[10] Development and economics-oriented studies have concluded that economic reforms driven by cash crop production in both the colonial and the post-colonial period led to agricultural decline and threatened rural survival.[11] They have blamed contemporary agricultural decline on colonialism, government mismanagement of resources, and the effects of a global economy.[12] These factors constitute the general explanations for the forms of agricultural change that have taken place in most parts of Africa, but they do not provide explanations for regional experiences.[13] The responses and actions of the local population, the social context in which they produced crops, demographic factors, environmental factors, and political changes undoubtedly contributed to the pace and forms of changes that occurred – for government policies did not function in a vacuum. The policies pursued by colonial officials, at times through coercion, on the one hand, and the response of local farmers to the opportunities created by the new economy, on the other, account for the pace, scope, and nature of the transformations that occurred in the twentieth century.

Scholars are still struggling with how to explain these changes throughout Africa and how to characterize African producers like the ones described in this study. Two major perspectives have been used to explain Africa's disappearing farming population.[14] At one end of the spectrum, scholars posit that Africa's agricultural decline was the result of the incorporation of African economies into the world capitalist system. At the other end of the spectrum are scholars and international agencies such as the World Bank, the Food and Agricultural Organization of the United Nations (FAO), and the International Monetary Fund who argue that the causes of agrarian breakdown are rooted in the internal economic policies of African governments since independence.[15]

The externalist argument has focused on the adverse effects of colonialism and an international economic environment that created poor terms of trade for Africa's primary agricultural exports and dependency on developed economies. This line of thought suggests that the dependency of African countries on the export of primary products has proven risky for both state and peasant revenue as the high cost of imports and very low prices for export produce lead to economic stagnation and poverty. Agricultural decline, therefore, is seen as a legacy of colonialism, which continues to support the economies of the developed countries to the detriment of primary producers in Africa.[16] Indeed, Marxist-informed analyses blame the plight of African societies on colonialism and neocolonialism. The plethora of work in this area is grounded on dependency theory. There is widespread agreement within the dependency school that the infrastructure in colonial Africa was built to facilitate the exploitation of local resources, particularly agricultural raw materials that were essential to the industries of Europe.[17] This development pattern disrupted indigenous economies and political structures and rendered them dependent on Europe and the developed world.[18]

The British, like other colonial powers, did not have a well-articulated plan for agricultural development. The inability to work out a coherent and long-term strategy for agricultural development, John Levi and Michael Havinden have argued, was chiefly due to the fact that "policy was unduly subject to passing fashions ... or to pressure from powerful individuals who pursued particular enthusiasms."[19] According to Michael Watts, "While the success of metropolitan capital depended upon expanded commodity production by households which subsidized the reproduction of their own labor power, the demands of capital and the effects of commodity production simultaneously undermined (and occasionally threatened) the survival of those upon whom it ultimately depended."[20] The export market spawned new and often "contradictory" forms of production in rural societies, while the form of capital accumulation that emerged stifled the survival of a viable local peasantry. The emphasis on export production turned rural peasants into producers of raw materials largely for the benefit of European traders and industries. In good years, farmers could earn a modest profit, but the structure of the colonial economy did not develop an independent and self-sustaining peasantry. Overall, the externalist argument remains as valid today as it was two decades ago. However, the dependency theory paradigm and its basic concepts of "incorporation" and "centre-periphery" relations ignore internal dynamics of change.

Proponents of the internalist argument agree that the policies pursued by African leaders since independence have contributed significantly to the decline in agriculture. International economic and financial institutions such as the International Monetary Fund and the World Bank mainly base their argument on the role of the state in directing the pace and structure of agricultural change in many African countries. As Sara Berry notes, these institutions have interpreted Africa's agrarian stagnation as "a crisis of production, arising in part from historical factors, but also exacerbated by African governments' penchant for excessive and ill-advised regulation of economic activities."[21] Their contention that African states have generally ignored agriculture informed the imposition of structural adjustment programs in Africa. However, the biases and prescriptions of these institutions have been deplored. Their development models often are based on Western experience and a development ideology that is largely prejudiced against African economic and social systems.

The internalist perspective also focuses on factors confronting many African societies, including high rates of population growth, the lack of adequate infrastructure and industrialization, political instability, the neglect of agricultural investment, and environmental forces such as drought. Yet, the argument that the post-colonial state in Africa has produced little positive result in agriculture remains the dominant one. Robert Clute notes in his overview of the role of agriculture in African development that the continent remains the only part of the world to experience a decline in both agricultural and food indices in recent years.[22] The greatest problem in African agricultural development, he argues, is "the pervasiveness of political elites with an urban bias."[23] The marked contrasts between the ideology professed by the African elite and the reality of their practical actions have led to "a growing gap between the urban elite and the rural masses."[24] The broader issues facing African agriculture go beyond the development ideology pursued by the African elite. Moreover, the focus on state intervention in the post-colonial period ignores the historical origins of state intervention in peasant agriculture dating back to the colonial period and the long-term impact of incorporating local economies into the capitalist market.

The dichotomy that emerged in the debate has left unanswered questions about the dynamics and the effects of change on studied societies as well as about how African societies have responded to the crisis in agriculture. Indeed, the gendered impact of social and economic change is crucial and understudied even though Ester Boserup's groundbreaking work, *Women's Role in Economic Development*, drew attention to the centrality of female labour in African farming systems.[25] Boserup addressed the most important concern of feminist scholars – the role of women in agricultural production – while glossing over gender analysis,[26] and important relational changes taking place as a result of state intervention in rural economies. Much of the scholarly literature about gender and socio-economic transformation in Africa has a long and problematic history in which gender or economic transformation is not analyzed in the context of studied societies. Furthermore, the fact that gender relations are automatically equated with women's subordination has meant that important issues concerning women and men and the ways in which government policies impinged on the lives of rural people have not been well analyzed. The result is that we still lack an understanding of agricultural change as one among a number of mutually constitutive factors such as state policy, demography, ecology, gender, the household, and other social and economic imperatives that order people's lives in their locale but are also influenced by people's relationships with the world beyond their own immediate societies.

The key to understanding these changes has varied widely across Africa. Researchers in western Africa, in particular, have articulated the process through which rural farmers were integrated into the complex economic structures that developed in colonial Africa as well as the complex structures that influenced this process and informed its outcomes.[27] Sara Berry's study of how cocoa production stimulated the development of capitalist social structures in rural Yorubaland, including the evolution of private property rights in land and a land market, is a good example of the market-driven transformation of pre-colonial social structures. Still, the absence of a distinct class or category of labour and the availability of land in Yorubaland gave the development of peasant agriculture its character.[28] Polly Hill's study of the development of rural capitalism in Ghana may also be noted.[29]

The Igbo context was different from the above examples in the process of incorporating the region into the colonial economy. For Igboland, J. Lagemann points to a "more equal distribution of agricultural resources and farm income as population pressure increases, and greater inequalities in the distribution of total income as non-farm employment becomes more important."[30] Unlike the situation in western Nigeria and Ghana, where cocoa was a new crop, the oil palm was indigenous to Igboland and had been developed as an export product before the advent of colonialism. Access to the oil palm would, generally speaking, be open to most households, based on existing primogeniture with regard to land. Large-scale capitalization and new labour and land arrangements were not features of peasant farming in the oil palm production zones of Igboland. Susan Martin's study of the Ngwa Igbo may be noted in this regard.[31]

While the cocoa-producing areas in Western Nigeria and Ghana, about which Berry and Hill have written, attracted migrants, the Igbo area continued to be a net exporter of labour. Lineage and other forms of labour and land use persisted into the late colonial and post-independence periods. The extensive quantity of oil and kernels produced in the region was, therefore, achieved by the mobilization of household labour and continued reliance on a land-tenure pattern based on lineage and relationships controlled by elders. Despite large-scale production, what emerged among the Igbo was a class of farmers and traders that was largely undifferentiated.

The trajectories of change were also shaped by developments that emerged from the period of British disengagement from Nigeria. The mid-1950s were particularly significant in this regard as Nigeria entered the era of internal self-rule in 1954. The Lyttelton Constitution of 1954 provided for regional governments (Eastern, Western, and Northern) with wide powers in political and economic affairs at the regional level. When independence was obtained from Britain on October 1, 1960, the government of the Eastern Region had the opportunity to fully implement its economic policy and ideology. The new elite, under the leadership of the pragmatic premier of the Eastern Region, Dr. M. I. Okpara, rejected the colonial political order, but they inevitably accepted the economic one bequeathed by the British. Agriculture was perceived as the source of economic development. The regional government focused on the establishment of community plantations and farm settlements for export production, although innovations were introduced to encourage food production and to draw farmers further into official agricultural programs.[32] However, the attempt to modernize indigenous agriculture revealed the paradox of an elitist state that aspired to "modernize" a population that viewed state intervention with skepticism.

A complex web of structural and political developments eroded the agricultural base before these policies could be fully tested. By 1966, the Nigerian federation was in a political crisis that led to a civil war between the predominantly Igbo-speaking people of Eastern Nigeria and the rest of the federation in 1967. Igbo society faced several challenges because of the civil war, which lasted until 1970.[33] Unprecedented famine followed, and agricultural crisis became a permanent feature of an area that was already a "food-reserve-deficit" area before the war.[34] The 1970s also witnessed major changes in the economic base of the Nigerian economy. The discovery of crude oil at Oloibiri in 1956 and the production that began in 1958 made Nigeria the largest oil producer in Africa and the eleventh in the world. The 1970s coincided with a rise in the world oil price, and Nigeria reaped instant riches from petroleum.[35] The emergence of the petroleum economy ushered in an important new phase in Nigeria's economy. Agriculture, which earned significant revenue for the region and the country before the oil boom, suffered from low investment, leaving the rural population vulnerable. Rural regions witnessed a major crisis, with social, infrastructural, and economic collapse in the 1980s.

The book pays attention to circumstances that are uniquely Igbo. An important one is the cultural ethos associated with farming in Igbo society, especially the cultural impacts of yam production – the dominant male crop – thereby revealing the gender and power relation that exists and the poorly understood connection between farming, culture, and Igbo identity. This book thus illuminates the shifting nature of rural identity in parts of Igboland, such as central Igboland, where the agricultural crisis has been more severe and where non-agricultural sources of income have become more important from the 1970s onward. It reveals that the foundations upon which Igbo masculinities were built have shifted, gradually after the crisis of the Nigerian civil war (1967–70), and more dramatically in the petroleum boom era of the late 1970s. What has followed has been a progressive reordering of social relations and reinterpretations of "traditional" gender ideology despite the persistence of a dominating idea of male power.

The overall pattern of change that emerged in the colonial and post-colonial period was inseparably linked to perceived gender roles, which lay at the heart of the political and economic discourse of both the colonial and post-colonial states.[36] Indeed, the discourse on the colonial impact on agriculture and gender has produced two important but related lines of critique. First, a substantive critique has focused on the adverse impact of the neglect of the role of women in agriculture stemming from the patriarchal ideology of officials. Second, feminist economists have drawn attention to the adverse consequences of hidden bias in the economic theories underpinning agricultural development.[37] Oyeronke Oyewumi rightly claims that "theories of colonialism relate a dialectical world of the colonizer and the colonized that are often presumed as male ... while it is not difficult to sustain the idea that the colonizer was predominantly male ... the idea that the colonized was uniformly male is less so."[38] While represented as gender-neutral, official policies, which disregarded women's work, were in fact not gender-blind. The character of agricultural change was therefore distinguished by the unanticipated impact of male-centred improvement schemes and the role that gender ideology played in shaping the processes of change. During the colonial period, officials excluded women from extension services, agricultural education, technological innovations, and agricultural loans. Thus, this study is also the story of the gendered nature of this encounter, particularly the colonial notion of the "male farmers" and its outcomes. The institutions created by colonialism contrasted with pre-colonial systems, transforming the roles men and women had previously played in their societies while creating new gender and class relations.[39] As systematic exploitation of colonized peoples, colonialism affected both men and women, but as a gendered process, it affected men and women in both similar and dissimilar ways. These trends continued in the post-colonial era, but the patterns and assumptions about the role and responsibilities of the genders within rural households have been challenged by economic and social changes.

The book emphasizes that economic and social changes were not the result of official policies and actions alone, even though the choices made by local farmers and traders and the rewards they received were determined by conditions created by the state. Therefore, this book attempts to present, through a particularly local dimension, the responses of rural households to official attempts to transform agriculture and other mechanism of state control. Since rural people were important agents of change and the instruments through which the state often implemented its agenda, they are the medium through which official policies and its outcome can be assessed.

Economic and political forces beyond their control often influenced the rural populations. But these populations also exerted an influence upon the state by employing various strategies, including rebellions or petitioning authorities to seek redress for grievances. These means, by which the local population influenced state behaviours, were employed in greater frequency during the Great Depression and during the Second World War in response to new market forces and state control that often imposed limits on rural people's ability to control their own lives. Rebellions place the colonized population at the centre, show them as historical actors with pliable cultures and communities, and provide a more accurate understanding of the processes of colonization, its impacts on particular groups and communities, and its role in social and economic change. Through their engagement in the expanding economy as producers and traders and the use of revolts and petitions, ordinary people accumulated political and economic capital that helped to shape societal transformations in the colonial period.

Petitions to colonial officials were used extensively from the 1930s as a means to seek remedy for grievances over a number of actions including taxation, court decisions, and policies that directly affected the rural economy (production and marketing of agricultural goods). The petitions sent to colonial officials are unparalleled as intimate and immediate records of life in colonial Nigeria, and as means for understanding African expression and the process of negotiating with a hegemonic colonial state. Furthermore, petitions highlight the important political effects of subtle forms of resistance and negotiation by rural men and women in the colonial context. Overall, they represent local perceptions and characterizations of societal transformations and show why ordinary men and women took matters into their own hands and in some cases influenced the outcomes of official policies.

Igbo farmers and traders adopted other strategies to win concessions when their resilience failed. They revolted and resisted when official policies and market forces beyond their control put their survival at risk or when colonial initiatives threatened them. These struggles, which took place in the rural areas, markets, urban areas, and colonial courts in southeastern Nigeria, helped shape the political and economic landscape of the colonial state, formulated in imperial offices in London and colonial offices in Nigeria. As instruments of social and political change, rural revolts particularly helped create political and economic space for ordinary people. The actions of rural people in Igboland had parallels elsewhere. However, the Igbos' actions took place in a particular cultural, historical, and economic context – within an economic setting that struggled with a mono-economy (dominated by palm produce), high population density, and severe land scarcity. It is, therefore, a history of an African society's experiences, contributions, collaborations, and resistance to colonialism and its consequences.

I have not attempted to discuss rural life in all its ramifications as it relates to colonial and post-colonial policies. The two general themes that stand out in this book – themes that highlight the role of the state in a colonial context as well as the rural population as agents of historical change – suggest that, despite the actions and top-down approach of official policies, the voices and actions of ordinary people proved critical in determining the forms of social and economic changes that followed state intervention in Igbo society. These particular themes (state policies and peasants responses) have been chosen both because they are timely topics of debate among scholars of colonialism and African development and because they allow for a historical narrative that fully explores the processes of economic and social transformation among the Igbo. A critical assessment of these themes also suggests that, although the forces of intervention were mostly beyond the control of rural people, these people often negotiated as individuals and as a group, offering competing claims for political and economic rights throughout the colonial period.

The salient features of the Igbo region in general and central Igboland in particular have been underscored in the literature. I will comment on them in order to outline the distinguishing features of this book. Some of the earlier commentaries included the work of Northcote W. Thomas. Thomas commented on the links among poor soil quality, the shortage of land, and subsistence insecurity in the densely populated parts of Eastern Nigeria at the beginning of the twentieth century.[40] Indeed, labour migration from barren lands within the Onitsha-Awka axis to more favoured regions by the late nineteenth century provides valuable information for periodizing land scarcity and the emergence of an agricultural crisis in parts of Igboland before the imposition of colonialism. Barry Floyd made similar observations and suggested that the soil in the region, which was never of high fertility even under the original high forest cover, was further impoverished by continued use under traditional farming methods.[41] Central Igboland shared in these constraints.

References to the population pressure found in missionary letters and travellers' journals suggest that high population numbers were already an economic problem by the end of the nineteenth century.[42] The "population is so great that if they hear we shall want carriers, they come in great numbers begging to be used, even during the farming season," reported a missionary in the Owerri region in 1866.[43] Land became so crowded in most of the region that fallow periods had already been shortened by the 1940s.[44] G.E.K. Ofomata's recent edited collection, *A Survey of the Igbo Nation*, devoted considerable space to the ecological character of Igboland and its impact on agriculture and livelihood.[45] These studies provide a useful framework for the study of rural subsistence and survival strategies in a changing agrarian landscape.

Perhaps the most important aspect of these works is their usefulness in locating the origins of agricultural crisis in several parts of Igboland in the late nineteenth century and its increasing spread in the early twentieth century.

By the beginning of the twentieth century, the overwhelming majority of the Igbo were still engaged in agriculture. Many combined subsistence production with the production and marketing of palm produce. Scholars and commentators have examined the changing nature of Igbo agriculture since the extensive commercialization of the economy began in the beginning of the twentieth century. *Trade and Imperialism* by Walter Ofonagoro, published in 1979, is an important historical work on British commerce and the establishment of colonial rule in Southern Nigeria. This volume is an introduction to the uneven and sometimes antagonistic process through which British colonial authority and commerce were extended to the southeastern Nigerian hinterland. With specific regard to agriculture, Eno J. Usoro's work on the palm oil industry remains an important study.[46] Although largely a quantitative analysis of the export sector, it provides a useful source for the study of the palm produce trade during the colonial and early post-colonial periods.

A recent work by Simon Ottenberg, *Farmers and Townspeople in a Changing Nigeria: Abakaliki during Colonial Times (1905–1960)*, is the story of successful rural farmers in the midst of an emerging town and the ethnic interrelationships, integration, and conflict between the town and the rural areas that occurred. The broad framework of this study supports my argument that the history of Africa's encounter with colonialism is also a story of African responses, resistance, accommodation, and innovation. The anthropological and historical approaches are welcome. The current work places local responses and innovation in a broader historical context and framework.

The gender dimension of agricultural and economic transformation among the Igbo has received significant attention from scholars. One of the most significant works relating to Igbo agrarian change is Susan Martin's monograph, *Palm Oil and Protest*. Her detailed analysis of the changing nature of the household economy among the Ngwa Igbo enables a broader understanding of how the commercialization of palm oil trade spurred the transformation of the rural agrarian economy and changes in gender relations of production. Martin's work on the Ngwa region shows that local initiatives played a significant role in the patterns of export growth, capital investment, and the food production sector, a major deviation from both the dependency and the vent-for-surplus theories, which had dominated the discourse on the development of the export sector. Yet, Martin's conclusions, particularly in regards to gender transformation and women's ability to participate in the new economy, are problematic since access to land and other production resources was not often under the control of lineage and household heads. Individual land tenure, which had become widespread in most parts of Igboland by the beginning of the nineteenth century, had more significant implications for gender and the household economy than Martin acknowledged. Men and women shared a common basic outlook toward the colonial economy, and the production patterns that emerged occasionally empowered women to claim more economic rights and engage in the formal economy. Indeed, Martin's earlier article, "Gender and Innovation: Farming, Cooking, and Palm Processing in the Ngwa Region of Southeastern Nigeria, 1900–1930,"[47] which analyzed the role of women in the agrarian change in this region, argues that women responded to the expanding oil palm economy by introducing innovations, developing efficient time management strategies, and adopting new crops such as cassava.

However, Martin's conclusion that men took the lion's share of the proceeds and amassed capital from palm oil production because they maintained political and economic control of the Ngwa lineage system of production ignores women's ability to work around these social institutions for their own benefit. My own study shows that the transformations that occurred empowered women to claim economic rights and engage in the formal economy. Indeed, Gloria Chuku's *Igbo Women and Economic Transformation in Southeastern Nigeria, 1900–1960* has outlined women's contributions to agriculture, commerce, and craft manufacture.[48] Chuku demonstrates that some women were able to seize the new opportunities offered by the colonial economy in the areas of export crops such as palm oil, trade, and commerce, despite the gendered ideology that tended to favour men. Nwando Achebe's *Farmers, Traders, Warriors, and Kings: Female Power and Authority in Northern Igboland, 1900–1960*, challenges the notion of women's invisibility in African societies.[49] As farmers, traders, potters, and weavers, some successful females translated their economic successes into political and social power and authority.

While these studies provide a useful framework to assess the implications of expanded commodity production for rural farmers and the way in which rural producers carried the burden of economic transformation, the present study of central Igboland underscores the centrality of rural men and women in the transformation of rural societies and extends the analysis by emphasizing how rural farmers adapted to state-induced transformations and market forces. By highlighting the role of rural people, this work underscores a fundamental fact, which is that one cannot truly understand the dynamics of change in an agrarian society without a detailed understanding of the history of its major actors – rural farmers and traders.

The book is important for several reasons. The book is, in many ways, a history of policies pursued by colonial (and, to some extent, post-colonial) governments in Nigeria, supplemented by voices "from below" and some degree of re-interpretation of sources, in order to bring in the perspective of local people. Most rural farmers did not leave individual sources such as diaries and memoirs, but they left their imprints in other places. They left records in the forms of petitions, letters, and memories. Drawing on an extensive array of previously unread colonial archival sources and oral accounts, this book reveals the "silent voices" in history, their resilience, adjustments, and adaptations in a changing society. Their voices and concerns, as reflected in petitions and supplications, challenge universalistic and essentialist categories in history. Capturing the "silent voices" in history, including those of women, local agency, and contestations of the dominant modernization ideology of the post-colonial elite, is at the core of this book. It puts the reader in direct contact with ordinary victims of colonial control, evoking a feeling of what it was like to live through the era. As such, it is also a social and cultural history of economic change and the rural farmers and traders who were important agents of change.

Second to the Atlantic slave trade, no story is more important in shaping the history of the region than the British attempt to transform agriculture. Yet official policies and perceptions about the local agricultural environment, the mode of production, and the role of colonial subjects – male and female – reflected a narrow understanding of African production systems.

Local production patterns were more complex than officials realized. Household production patterns and strategies were influenced by several factors, including market forces, kinship structures, and modes of resource allocation. The contradictions that emerged from the official construction of new economic and social formations and the responses of rural farmers have not piqued the interest of historians of colonial Eastern Nigeria – a region that was heavily dependent on local production in the colonial and immediate post-colonial periods. The significant role of local actors and local responses to the changes engendered by the expansion of commodity production and colonial intervention, and the forms of transformations that emerged, shaped the outcome of the colonial encounter and local economies.

This book challenges the largely deterministic notion that the social and economic transformation of colonized societies was the making of the state alone. It shows that rural people acted as important agents of change for themselves and their communities. Significant internal forces also often forced officials to interrogate European discourses and strategies. The ways in which the British exercised power over the rural population and their response also reveal these interactions as negotiated encounters between colonial officials and natives and challenge the simplistic notions of a hegemonic colonial state and a compromising native population.

The book explains the salience of agricultural change in central Igboland in terms of the interaction of state agency, structural, historical, and social factors, all induced by the opportunities to generate income from cash crops. The economic structures that emerged drew from the social and cultural backgrounds of both imperial Europe and African societies. Amid several forces, including imperial Britain, European traders, physical and environmental conditions, high population density, land tenure systems, war, famine, a rebellious peasant population, and a gender ideology largely indifferent to the local context, the transformation of the rural economy presented a daunting task for officials and local farmers alike. The members of the African population were impressive in their capacity to balance state demands with their own cultural ethos as farmers in a landscape that was changing rapidly.

The history of socio-economic change, as depicted in the case of the Igbo-speaking people of southeastern Nigeria, raises far more complex – if less easily defined – questions than the classic debates about the long-term effects of imperialism on colonized societies.[50] What was the response of local people in southeastern Nigeria? In reconstructing a narrative of the changes that occurred, I have paid attention to how state policies, market forces, and significant events in the twentieth century, including colonialism, the Great Depression, and the Second World War, combined with the actions and responses of the African population to shape the character and nature of change in this region.

A NOTE ON METHODOLOGY AND SOURCES

Throughout this work, I use various terms, including *farmers*, *traders*, and *peasants*, to refer to the groups of people whose lives, economic activities, and social activities are reflected here. Despite the difficulty of finding an acceptable definition for African peasants, the term peasant remains useful in examining the lives of people who refer to themselves as farmers or traders or both. Although the classification of Africa's agrarian social structures, especially the application of the term *peasantry* as a social category, remains problematic, Igbo producers of the colonial and early post-colonial periods embody the basic characteristics of peasants outlined by T. Shanin, namely the pursuit of an agricultural livelihood for subsistence and commercial purposes; reliance on family labour as a unit of production; subordination to state authority and market forces through which peasants' surpluses are extracted, and membership of a community that defines peasants' outlook, attitude, and worldview.[51]

While one can debate endlessly whether the Igbo and other groups in southeastern Nigeria fit this categorization, the majority of Igbo farmers were involved in an agricultural export economy, using household labour, to meet their survival needs and the demands (taxes, rents, and other fees) that arose from their incorporation into state institutions. In the colonial and early post-colonial periods, there were certainly those who moved from farming to non-farming occupations and vice versa. Whether as traders, farmers, or both, all were affected by the social and economic processes of these periods. In employing the term *peasants*, this book considers the term as a fluid and unstable one and uses it mainly for descriptive purposes. Although the term peasant is employed here mainly descriptively, it should be said that the actions of the state and the forms of hegemony it exercised over the rural population highlight the extractive power of the state over the rural population – a basic characteristic of a peasant-state relationship.

The development of a peasant population, however, must be understood in the context of earlier commercial developments. The Igbo region of the hinterland of the Bight of Biafra was a major source of slaves for the Atlantic slave trade in the seventeenth and eighteenth centuries. The trade, mostly facilitated by the Aro and their trading outposts throughout Igboland, linked the region to the economy of the Atlantic world two centuries before colonialism. Commercial food production in yams and palm oil developed along with the slave trade to feed the market that the slave trade created. The abolition of the slave trade and the development of a trade in palm oil drew the region closer to the European market in the nineteenth century. The region represented Europe's most important centre of the production of palm oil – a product that shaped the Igbo economy from the end of the Atlantic slave trade onward. On their own initiative, the Igbo produced palm oil to feed the expanding demand for tropical raw materials that developed with the Industrial Revolution in Europe. The imperialism of the late nineteenth century made greater integration of the Igbo economy into the European economy inevitable.

Thus, the transformation of the rural Igbo producers can roughly be associated with two historical periods. The first stage represents the period when the Igbo were drawn into the international economy through the nineteenth-century commercial transition from the slave trade to the palm produce trade. The produce trade was an important factor in the transformation of Igbo agriculture. Yet in this period, many members of the Igbo population could not be regarded as peasants simply because they produced palm oil for export. This was particularly true of central Igboland before the colonial period. Igbo farmers enjoyed the freedom to produce on their own account using their labour without much external interference. The next stage coincided with the imposition of colonial rule, when Igbo farmers became increasingly commercially oriented due to the influence of the colonial state and foreign traders. Igbo farmers were drawn into more commercial agricultural production from the beginning of colonialism – the coercive instrument for the extraction of peasant surpluses. The farmers in this period were different from earlier subsistence producers. They produced more for the market. Their increasing incorporation into the colonial and increasing European market that now included other traders made them more vulnerable to state control and to the dictates of the world economy beyond their control – at least in controlling the price of produce.

Regarding sources, I have relied on both oral sources and colonial documents to make sense of life in the colonial and early post-colonial state, the attempts by the state to transform the rural economy, and the ways the local population responded and shaped the outcomes. As Paul Thompson explains, "Oral history can transform the content of history by shifting the focus and opening new areas of inquiry, by challenging some of the assumptions and accepted judgments of historians, by bringing recognition to substantial groups of people who had been ignored."[52] In our context, oral sources stress African voices and explain agricultural change as rural people lived it; indeed, they provide, essentially, history "from below."[53] The use of oral history rectifies the omission of rural voices, which are often ignored in official historical documents and have mostly been omitted in past studies of agricultural change.

This work is based on a people with whom I share a common language and culture. Situating myself in relation to the research context enabled me to deal with issues of representation and interpretation of the experiences of the informants and the social context.[54] I interviewed four members of my family: my father, mother, brother, and uncle. For my family, access to historical knowledge was seen as part of the process of acculturation and education about the past. This situation confronts the researcher with a different set of problems – the difference between individual and group history.[55] I was perceived as an "insider" and provided insights into matters that might be denied to an "outsider." While these claims contradict those made by critics of insider research, it does not diminish the advantage the insider has in understanding the complexity and nature of his/her own society. There is less inclination by the insider to construct opaque stereotypes of a society. An insider perspective also has an impact on epistemology because the insider may have a better understanding of the social structures that generated a particular set of oral data.[56] Indeed, Ndaywel Nziem has written of the African researcher as one who has a double role as "observer and actor."[57]

However, I have been keenly aware of what Richard Wright called the "unscriptable factor" – "perspective."[58] Faced with the task of maintaining what Obioma Nnaemeka called "a balanced distance between alienation and over-identification," I was conscious of the problems of both interpretation and mediation,[59] since both the insider and the "outsider" perspectives present the possibility of distortion and preconception of social reality. I was conscious of the fact that I too may be looked upon as an outsider, albeit in a very different context. However, while the historian's own sense of personal identity may place limitations on his/her work, as Raphael Samuel rightly argues, at the same time, that sense of personal identity gives it thrust and direction.[60]

Some informants talked freely about general trends and more reluctantly about some aspects of individual lives. Some of my informants were glad to talk to me but reluctant to discuss certain aspects of their lives for fear that what they said might be discussed with other people. Many informants made personal references in the third person. Some people generalized personal experiences while discussing traumatic events or poverty. I have tried to conceptualize these issues by moving between different levels of analysis: from the social and cultural norms of the Igbo, to consideration of the privacy of my informant, to the collective experience of the Igbo. As an insider, it was possible for me to transfer some of the unspoken words and actions to the spoken because a phrase sometimes represents a very long story or a deep sense of emotion. However, the generalization of personal experiences by informants shows how personal reality can translate into group identity and group reality. Interestingly, people often used the word "we" when relating emotional issues such as poverty and food insecurity, as a way of depersonalizing their suffering.

Oral historians must also deal with the problem of memory, on which oral history is heavily dependent but which is human and fallible.[61] The conditional acceptance of oral history is recognized. This study, however, seeks to present a balanced view by combining oral data with other primary sources and with secondary data. But as Hoopes succinctly put it, "all historical documents, including both oral and written, reflect the particular subjective minds of their creators."[62] Yet the oral sources used here "give a "feel" for the "facts" that "can be provided only by one who lived with them."[63] Therefore, I urged informants to recount their own experiences and to remember specific events in the lives of their parents and families.

I have used a wide range of other sources, some of which have not been previously used. I consulted various documents left by imperial Britain in the National Archives of Nigeria, Enugu, the Public Record Office (National Archives), London, and Rhodes House Library at the University of Oxford. The archival materials contain concrete, dateable information and data, but limitations abound. Statistical data and information on agricultural production during the colonial period may not be the most reliable indicator of aggregate economic performance. Colonial data often did not provide an accurate level of agricultural production or indicate the level of peasant production. Often, colonial reports reflected the economic and political objectives of the colonial officials rather than the economic realities of the colonized societies. Colonial officials often relied upon data from experimental farms, which operated under optimum management conditions, to estimate food and export crop yields for the provinces and the country at large. In most cases, the optimum conditions differed immensely from the conditions that farmers faced in their natural environments. Such evidence of agricultural performance and conditions is problematic and unsatisfactory when used to reach general conclusions.

Many bits and pieces of useful information about rural life are missing from these reports. The effects of the shift to an export-oriented agriculture, the short- and long-term implications of colonial agricultural policies on farmers, the implications for subsistence production, and the general quality of life of rural producers are not easily discernible. As in many official colonial records, the voices of the local people are absent, although their actions speak for them. Official texts also obscure the life of the rural farmer, especially the contribution of women to agricultural production. The expression the "genuine farmer" was used in the colonial reports to refer to male farmers. Reading between the lines of the reports reveals the gendered nature of colonial policies. The colonial assumption that the farmer was male silenced the contributions of women and children. However, reading past the "colonial approach" reveals that this was an unrealistic assumption.

Colonial reports contain a mass of information on economic statistics and agricultural conditions in the colonies, but there is little information on the method of data collection. Generally, colonial reports tended to be too optimistic in their estimation and projection of production in the colonies. Failures were often ignored or left unexplained. This is understandable because the jobs of colonial officials depended upon positive justifications of their activities in the colonies. Positive and idealistic reports also guaranteed continued interest in the colonies by the British government. The reports are largely couched within an ideological framework informed by the economic motives of the colonial enterprise. For these reasons, I have considered the circumstances under which colonial reports were generated. Despite the problems with colonial sources, however, they document the evidence of agricultural crisis and the social conditions during the colonial period. They present a means of putting faces and voices behind official statistics.

I encountered a unique type of source – petitions – while researching this book. The hundreds of petitions written by ordinary people and local political activists provide a different perspective on the colonial encounter, agricultural change, and rural responses. The lives of ordinary people during the colonial period come alive through these letters, petitions, and supplications. The personal – and often intimate – letters of Africans create a unique portrait of the rough-and-tumble times of the colonial period. Most of the letters and petitions from ordinary people began to appear from the period of the Great Depression, but they became very common during the Second World War. The ability of Africans to write or to hire professional letter writers gave them the opportunity to respond to power and establish a dialogue with colonial powers – a dialogue that received some level of respect. Overall, these petitions reveal the impact of colonial policies on peasants and their attitude towards these policies.

Still, the use of colonial documents to examine changes in colonial societies and relationships between the colonizer and the colonized calls for some caution regarding their limitations. Indeed, questions abound. Were events and data recorded factually or were they reflections of the perspectives of officials in the colonies? Did colonial officers make history to portray themselves and government programs in a good light? What kinds of documents were preserved, and what types were destroyed, as frequently happened? The lack of local voices and their stories make it easy to promote an officially sanctioned point of view. As Thomas Spear has argued, every action, interaction, and dialogue within a colonial context ought to be assessed from the "perspective of the different actors to understand the particular meaning each gave to it."[64] This caution over perspective informed my assessment of colonial sources and my attempt to provide some insight into the working of the minds of ordinary people in this period.

The Land Has Changed employs a diachronic and multidisciplinary approach and juxtaposes peasant production, peasant resistance, gender ideology, and the culture and identity of the African farmer with the broader scholarship on colonialism. The approach here has been one of blending between two fields of historical inquiry – social history and economic history – using oral, secondary, and archival sources. In following this approach, the book links the past and the present and integrates individual and personal experiences to the explanation of the agricultural crisis in Igboland. This has created the opportunity to put text and people in context and allowed a range of individuals and circumstances to be understood.

STRUCTURE OF THE BOOK

The book is arranged thematically, but each chapter addresses a specific historical issue from the beginning of colonial rule onward, thereby providing a level of chronological sequence. Chapters 1 and 2 discuss the structure of Igbo society, politics, and economy on the eve of colonialism and the attempt by the British to restructure them in order to achieve the goals of imperialism. Chapter 3 examines how the mission to transform African agriculture and achieve uplift was implemented through the dominant Western gendered ideology of the male farmer – an ideology deeply embedded in colonial patriarchal thinking. Chapters 4, 5, and 6 take up the issues related to African responses to colonial policy and in relation to the major structural changes that defined the twentieth century, including the Great Depression and the Second World War. These events had a direct impact on the local society, often drawing rural peasants into the colonial economy, and at times generating considerable hardship to which rural peasants responded by rebellions. Chapter 7 focuses on the agricultural policy of the Eastern Region after independence – its espousal of a development ideology that continued to espouse agriculture as the instrument for rural transformation. The chapter details how the agricultural programs of the early post-colonial period were stifled by political and structural factors including the Nigerian civil war and the emergence of petroleum exportation as the major source of income for Nigeria. Chapter 8 is a reflection on agricultural crisis, responses to it, rural coping strategies, the resilience of rural people, and the changing social relationships engendered by the crisis of the 1970s and 1980s.

The aim of the book is to write a history of Igbo agrarian change and societal transformation as ordinary people responded to colonial policies and the structural changes that came with colonialism. It was the actions, responses, and, at times, the resistance of these ordinary men and women that helped shape both the colonial society and the society the Igbo inherited at the end of colonialism.

CHAPTER ONE

"WE HAVE ALWAYS BEEN FARMERS": SOCIETY AND ECONOMY AT THE CLOSE OF THE NINETEENTH CENTURY

This chapter examines the socio-economic constitution of Igbo society, especially the ways in which the intersection of agriculture with individual and group identity illuminates historical patterns and processes of change. It outlines the key elements of the economy in relation to agriculture and trade before the twentieth century. In so doing, it links the political and social landscape of the Igbo to its political economy and sources of identity and provides the background needed to assess the developments in all these areas from the beginning of the twentieth century onward.

The Igbo-speaking people are one of the largest single ethnicities in Nigeria. The Igbo have been characterized in the extant literature as decentralized or as forming what anthropologists termed "stateless societies." These early classifications, as advanced by M. Fortes and E. E. Evans-Pritchard, were based on the relevance and centrality of the lineage system of social organization in the regulation of political relations within and outside particular groups.[1] Under this system, the responsibility for leadership is said to rest on village councils. Yet scholars who have identified a much more complex political organization in both the centralized societies and those founded on the segmentary-lineage principle have challenged such a broad dichotomy as applied to pre-colonial African political systems. As M. G. Smith noted, all political systems have a shared characteristic of competition at various levels – a competitive dimension that cuts across lineage groups for control of decision-making, creating the opportunity for the development of a hierarchical structure even among elementary societies.[2]

Although most parts of Igboland exhibit a decentralized political structure, they embody a much more complex political system. Indeed, scholars have identified models that range from kinship and lineage networks to societies where titled persons and age-sets have controlled the instruments of power and authority.[3] Other parts of Igboland, such as Oguta, Onitsha, and Umueri, have monarchical political systems. At all levels of government, however, the Igbo have practised a form of direct participatory and representative democracy.[4] As Raphael Njoku argues, "the distribution of power, authority, and the processes of local administration are the same" in both the monarchies and the village republics. He notes further that the "elders ruled [sic] by representation, participation and negotiation," and the notion of difference is found primarily in "the king's ceremonial objects and titles."[5]

Most important to note at this juncture, however, is the interconnectedness between the political structures of the Igbo and sociocultural and economic life and how this interplay relates to agriculture. The role of indigenous political actors and their influence on matters of production, reproduction, access to productive resources, and control of labour for production provide the contexts within which the Igbo economy can be understood. This role relates to their control of the distribution of resources and the impact on gender relations of production. This relational approach is important in understanding the forms of economic change that have occurred and the ways in which they reveal the trends in the historical patterns of production as outlined in subsequent chapters.

The reconstruction of the origin and antiquity of plant domestication and agriculture among the Igbo, as among their counterparts in many parts of western Africa, is largely based on conjecture.[6] Like most forest dwellers, the Igbo shifting cultivators produced such staple root crops as yams (*ji*; *Discorea spp.*), cocoyam (*ede*), and various kinds of cultivated bananas (*unere*). In addition, trees such as the kola tree (*oji*), the oil palm (*nkwu*), and the raffia palm (*ngwo*) provided additional sources of food, beverage, and income.[7] The long history of plant domestication is evident in the large variety of food crops and edible plants in the Igbo subregion of West Africa.[8] The variety of crops and plants that attest to the antiquity of agriculture in the Igbo region include African rice (*oriza glaberrina; osikapa obara obara*), Guinea corn (*sorghum vulgare; oka okiri*), bulrush millet (*oka mileti*), hungry rice (*osikapa ocha*), Bambara groundnuts (*voandzela subterranea* or *ahueekere otu anya*), cowpea (*vigna unguiculata, akidi*), benniseed (*sesamum indicum*), okra (*hibiscus esculenta*), kaffir or Hausa potato (*pectchranthus esculentus*), fluted pumpkin (*telfaria occidentalis, ugu*), gourd (*legenaria isiceraria, agbo nwanru*), and several other plant species. However, root crops provided the largest percentage of calories in the Igbo diet, followed by oils and fats from oil palm and cereals.[9]

It is possible to paint a clearer picture of agriculture in modern times. As elsewhere on the West African coast, contact with the Europeans from the fifteenth century was a major transformative factor for indigenous agriculture. The arrival of the Portuguese in Benin and later on the southwestern lagoon coast and in the creeks of the Niger and Cross river basins resulted in the introduction of important food plants of central and South American origin. The arrival of maize (*oka*), cassava (*akpu*), groundnuts (*ahu ekere*), sweet potatoes (*ji nwa nnu*), tomatoes, tobacco (*utagba*), and several varieties of citrus fruits helped to diversify the region's food plants. By the seventeenth century, maize in particular had become an important secondary crop in the forest region.[10] The new food crops and fruits led to the clearing of forests and changes in production relations.

The introduction of Southeast Asian crops, including certain species of yam, cocoyam, rice, and banana, provided new varieties of food of high nutritive value that made it possible to support large populations.[11] Several species of fruits, including oranges (*oroma*), lime (*oroma nkirisi*), tangerine, grapefruit, and mangoes, and also sugar cane were part of this exchange. It is difficult to date precisely when these crops were introduced, but it is speculated that they came into the Igbo region either by northern routes via Egypt and the Sudan, or across equatorial Africa, or by the sea route round the Cape of Good Hope.[12] The new patterns of agriculture that emerged with the introduction of these crops entailed a more efficient use of human and material resources in Igboland in ways that J. E. Flint characterized as "perhaps the most efficient in Africa."[13]

Asian and American crops, especially cocoyams (*ede*), had a long-term impact on Igbo society, especially on gender ideology. Cocoyams came to be known as a woman's crop and women performed the routine work of planting, weeding, and harvesting the crop. Control of cocoyam production gave women a distinct identity as they took related titles such as *eze-ede* (cocoyam king) as a mark of success in farming, just as men took the yam-related title of *eze ji* (yam king).[14] Yams and cocoyams are the only crops that attained this status in Igbo cosmology, although they are never regarded as equals. The significance of both crops as expressions of male and female identity was confirmed by Nwanyiafo Obasi when she asked, "What is man without yams and what is woman without cocoyams?"[15] In all, considerable dietary adjustments occurred among the Igbo with the adaptation of the so-called women's crops.

Indigenous protein-rich foods were in short supply even in normal times. The few sheep, goats, and chicken that were kept and wild game provided an irregular, but quite significant, source of protein. Cattle, which were of the dwarf tsetse-fly resistant variety (*ehi*) and poor in meat yields, were kept mainly for prestige and ritual purposes.[16] Thus the Igbo could not by all accounts be regarded as animal husbandpersons for most of their subsistence, aside from occasional income, did not come from keeping animals. Dried fish were generally bought from coastal areas until the colonial period when the Igbo people became the main consumers of the tons of Norwegian stockfish, locally known as *okporoko*, that were imported into Nigeria.

Although agriculture dominated both the subsistence and exchange economies, the Igbo pursued other productive activities and moved between occupations as the demands for special goods or skills warranted. Such movement of goods and skills produced additional specializations like salt production and fishing, and also led to the formation of guilds and the rise of ritual specialists. Other people facilitated trade as hosts and guides.[17] The role of the last mentioned was particularly important in a region that lacked a strong centralized authority to guarantee safe passage from one village to the other.[18]

The Igbo economy had become part of a larger regional economy, that of the Bight of Biafra, by the seventeenth century.[19] Although Europeans did not visit Igboland until the late nineteenth century, early European travellers to the coast of West Africa and the Bight of Biafra recorded life and livelihood among the Igbo in connection with the regional economy of the lower Nigeria basin. Duarte Pacheco Pereira, a Portuguese geographer visiting in the sixteenth century, observed a vigorous trade in salt between the hinterland and coastal societies of southeastern Nigeria.[20] The Igbo brought forest and agricultural products to the coast in exchange for salt and dried fish.[21] Writing in 1699 on Kalabari/Igbo relations, John Grazilhier notes: "The land about the town [Kalabari] being very barren, the inhabitants fetch all their subsistence from the country lying to the northward of them, called the Hackbous [Igbo] Blacks." Grazilhier reported further: "In their territories there are two market-days every week, for slaves and provisions, which Calabar Blacks keep very regularly, to supply themselves both with provisions and slaves, palm-oil, palm-wine, etc. there being great plenty of the last."[22]

The commercial relations that had developed prior to European contact expanded with the development of the slave trade in the eighteenth century. The accounts of captives of Igbo origin in the New World provide a rare and unusual glimpse of the agricultural and exchange economy of the Igbo in the two centuries before European colonization. An important source is the autobiography of Olaudah Equiano, who was born in an Igbo village in ca. 1745 and sold into slavery at the age of eleven.[23] Equiano recorded the social and economic life of eighteenth-century Igbo society in *The Interesting Narrative of the Life of Olaudah Equiano, or Gustavas Vassa, the African*, published in 1788. Agriculture, he wrote, "is our chief employment; and everyone, even the children and women, are engaged in it. Thus we are all habituated to labour from our earliest years."[24] Equiano's description of Igbo society and economy in the eighteenth century is supported by the accounts given by other enslaved Africans of Igbo origin. Archibald Monteith, born about 1799 in Igboland and taken to Jamaica as a slave during the Atlantic slave trade, affirmed in his memoir that yams, potatoes, and Indian corn were an integral part of Igbo agriculture in the eighteenth century.[25]

Nineteenth-century European accounts offer more perspectives on the Igbo agricultural past. In his journey to Bonny in 1840, Hermann Koler, a German doctor, took notice that the Igbo exported agricultural produce and metal goods to Bonny. He described Igboland as rich in natural products – maize, rice, yams, oil palms, dyewoods, cotton, and so forth, and as "quite indispensable" in the provision of staple foodstuffs for Bonny.[26] W. E.

Carew, who visited Ndoki in 1866, was impressed by the abundance of agricultural produce among other provisions in the market: "It abounds in corn, palm wine, rum, fish, deer's flesh, dogs' flesh, cat, fowls, tobacco, yam, eggs, spices, pineapple, palm oil, bananas, cassava, cloths, gun powder, pipes, and things which I could not number."[27] During an expedition to Akwete in 1897, A. B. Harcourt described seemingly extreme fertile land and "yam plantations extending for miles in all directions."[28] In a journey to Nsugbe and Nteje in 1897, S. R. Smith observed what would obviously be classified as women's crops: "I noticed, growing in the Ntegi farms, pepper, cotton, black rice, cassava (*Manihot esculenta Crantz*), and several other edible useful plants."[29]

The prodigy and efficiency of the Ezza farmers of northern Igboland, especially "their efficient application of manure made from leaves, crop remains, animal dung and night soil," impressed European visitors in the early twentieth century.[30] Sir W. Egerton described the Ezza as "thrifty and excellent farmers."[31] Miss Holbrook, who visited Umudioka near Ogidi in 1904, recognized the complementary role played by both sexes in agriculture: "The people on the whole are very industrious, the women especially, one seldom finds them at leisure, very often they are out helping the men plant or tend the yams, or planting cassava on their own account."[32]

Much of this suggests an economy that had developed far beyond subsistence levels and that met the food and exchange needs of the population before the colonial encounter. By the sixteenth century, settlement on the fringes of the Guinean rainforest and on the Sub-Guinean environment had produced communities of town dwellers increasingly shifting to rotational bush fallow cultivation instead of earlier forms of shifting cultivation.[33] This led to a consistent use of already cleared forest rather than virgin land and accounts for the high population concentration in the Igbo region.

Gender ideology played important social and economic functions in regulating production relations, the land tenure system, and the labour process. Traditional Igbo farming was based on the complementary participation of men, women, and children. In this circumstance, farming defined both male and female identity and instilled the value of hard work in both adult and young Igbo persons. Equiano recalled, "Everyone contributes something to the common stock; and, as we are unacquainted with idleness, we have no beggars."[34] It was the height of social snobbery for an Igbo person, as Victor

Uchendu concluded, to be referred to as *ori mgbe ahia loro*, – that is, one who depended on the market for subsistence.[35] This cultural ethos defined Igbo identity and attitudes to farming until recent times.

ENVIRONMENT, DEMOGRAPHY, AND ECONOMY

> In the past our soil was fertile because people were fewer and there was less cultivation. In fact, land was cultivated every four years. Now land is scarce because of too many people – *Luke Osunwoke, Interview, 5 January 2000.*

The physical and environmental conditions, including climate, rainfall patterns, and soil formations influenced agricultural practices among the Igbo. Most of Igboland falls within the Guinean and Sub-Guinean environment. The Guinea region is characterized by an annual rainfall in excess of fifty inches, less than three months of dry season, and a mean monthly humidity of 90 per cent or more throughout the year. For the regions in the Sub-Guinean vegetation zone (northern Igboland), the rainfall ranges between forty and sixty inches in the year with a dry season lasting between three and four months.[36] The pattern of rainfall produced two distinct patterns of vegetation. The southern part of the region was characterized by heavy rainfall, producing a dense rainforest, which thinned out northwards into a savannah. However, many centuries of human habitation and activities have turned the whole region into a secondary forest, with only pockets of forest remaining. Although the region is fortunate to receive a well-distributed annual rainfall that can support a variety of crops, the soil has never been very fertile.[37] Little variations in the timing and quantity of rain in a growing season can upset the agricultural cycle, food production, and food security. Stories of *unwu* (famine) abound in local folklore.

The demographic characteristic of southeastern Nigeria, particularly in the areas inhabited by the Igbo, was a major determinant of the economic life of the people and the social structures that emerged from it. The Igbo occupy a little over half of the land area of southeastern Nigeria but comprise over 60 per cent of the total population. The need for crop land in a rapidly expanding population environment led to the clearing of the original vegetation and the emergence of grassland dotted with oil palms and other useful trees.[38] Morgan observes that "the Guinean environment of the I[g]bo and Ibibio-land may have been comparatively easy to clear since the soil consisted mainly of deep, well-drained sands."[39] By the middle of the twentieth century, the vegetation in many parts of Igboland was already composed of palm groves ranging from 100 to 200 trees per acre in some areas. The palm groves have largely survived human activities because of the importance of the edible oil derived from the oil palm and its increased importance as a source of cash after the abolition of the Atlantic slave trade in the nineteenth century.[40]

Colonial estimates and ecological characteristics are proof of both high population and extensive use of the land. When the population density was about 236 persons per square mile, the colonial Resident for Onitsha observed in 1929 that land was quite limited in proportion to the population.[41] By the 1940s, a population density of 1,000 persons per square kilometre was recorded in parts of Igboland[42] and by the 1960s the density in the Igbo areas was about four times the Nigerian average.[43] The high population density in many parts of Igboland often resulted in the over-exploitation of the soil and consequent erosion and destruction of the top soil. According to J. R. Mackie, the director of the agricultural department from the 1930s:

> The Agricultural problems are extremely difficult. Much of the soil is a very acid sand [sic] the Onitsha and Owerri Provinces carry a very dense population which is far greater than the soil can support.... Once they have become exhausted beyond a certain point recovery under fallow takes a very long time."[44]

As people were boxed into a diminishing land space, their survival became dependent on their ability to remaster an already exhausted environment.

The key to Igbo population densities and to the distinctive crowded landscape of compounds and oil palms lies in the social organization and the character of the environment and the agricultural economy. Food production, especially the production of yams, may account for the unusually high concentration of population in central Igboland despite the large-scale enslavement of people from this region during the Atlantic slave trade. Although such an increase in population often leads to the emergence of complex societies, this did not occur here in terms of the formation of large centralized organizations. A simplistic explanation will include the absence of endemic warfare and conflict that would have resulted in the conquest and incorporation of other groups to reach the size of chiefdoms, states, or empires. Even the Aro and the Nri did not have the capacity to embark on such wars of conquest, so their influence over the rest of Igboland was more ritualistic than political. So, most of the Igbo remained organized in small groups and lineage organizations. The desire to live near one's own cropland, away from the authority of others, appears to have overridden all other considerations. This was a dominant factor in the evolution of the Igbo landscape.[45] Morgan and Pugh have argued that the "attractive environment of the homeland, and the lack of any but local contacts, produced an immense crowding of people."[46] The unique residential pattern of the Igbo has been explained by the perception that the Igbo think in human, not in geographical, terms.[47] Voluntary shifting cultivation would have been impossible to practice under conditions of high population density.[48] Local overcrowding increased the rate of agricultural intensification such that "probably nowhere else in West Africa is there so low a proportion of waste [land] and so high a proportion of cropland."[49]

Igbo farmers had a deep understanding of their environment, the way to manipulate it to suit their needs, and the consequences of human actions on productivity.[50] The Igbo, for example, rated the fertility of the soil by its colour and its distance from the homestead.[51] So, the difference in colour between *ala nkwuru* (compound land) and *ala mbara* (distant land) was an indicator of fertility and determined what crops or species of crop could be planted on the land. By the end of the nineteenth century, ecological and demographic challenges had forced farmers to devise new strategies to manage their environment. They adopted crop rotation and shortened their fallow periods.

The agricultural system used was rotational bush fallow (popularly known as swidden or shifting cultivation). Land was cleared and cultivated until its fertility decreased, and then the plot was left fallow to regenerate its fertility. During the fallow period, plant cover and litter protects the soil from the impact of high intensity rain, and the roots help bind the soil, increase water filtration, and reduce run off and soil erosion.[52] In addition to providing the soil with nutrients, the fallow system provides supplementary food, animal feed, staking materials for yam vines, firewood, and herbal medicine.[53]

While the historical origins of these practices are difficult to establish, they probably arose in response to the limitations imposed by an increasingly less mobile population dating back more than two centuries.[54] To sustain their agricultural economy, the Igbo applied a set of conscious, interrelated practices aimed at making food production possible on their land through and beyond the foreseeable future. Igbo farmers made use of organic fertilizer, cover crops, and mulching to increase the fertility of the soil and to protect crops. The Ezza Igbo, for example, who were the greatest yam farmers, made compost in open circular pits, even using human excrement.[55] Their experience in farming and their efficient application of compost from leaves, crop remains, animal dung, and night soil surprised but impressed early colonial officials, who in general formed an unfavourable impression of the capabilities of African farmers. They interpreted unoccupied lands as unused or spare territory, which Africans were incapable of developing due to lack of skill or initiative.[56] Hence, techniques such as shifting cultivation were seen as wasteful. But they failed to grasp the principles underlying shifting cultivation.[57]

INDIGENOUS MODE OF PRODUCTION

The work of Claude Meillassoux and other French Marxist anthropologists has been very instrumental in explaining the mode of production in traditional Africa.[58] Yet, as an example of the economy of the so-called stateless society, pre-colonial Igboland did not have an egalitarian mode of production, where all classes and groups had equal access to the means of production. The patriarchal and gerontologically based mode of production under the control of male elders precluded class and gender equity.[59] The specific method adopted in subsistence production, the control of the productive forces, and the relations of production gave elders control over communal land, the labour of younger members of the lineage, and the labour of wives and children.[60]

Access to land was linked to the membership of a kin-group. Until recent times, land ownership was predominantly communal and held in common by an entire village, kindred group, or extended family.[61] Such lands were apportioned out to members during each agricultural season.[62] As land held in reserve for the benefit of the whole group, communal land could not be alienated without the consent of the group as a whole. This tenure system survived into the twentieth century in some parts of Igboland. A district officer wrote in 1929 that land in Owerri was "under communal control and ownership."[63] European officials observed that land was controlled by the lineage heads who allocated it to relatives based on their need in Aba and Orlu areas at the beginning of the twentieth century.[64] W. B. Morgan observed among the Ngwa in the 1950s that "community land is still divided by the decision of the elders."[65] Individual land tenure developed rapidly in the twentieth century as a result of population pressure and intense commoditization of land during the colonial period.

Age, gender, and marital status were important elements that structured access to resources and assigned roles in production. The principle of seniority guided relations between males and females within the lineage group, between co-wives, and among male and female children.[66] The right to hold land was governed by inheritance rules and formed part of an elaborate kinship structure based on patriarchy in which only men could inherit land.[67]

The social organization of production and familial relationships, especially the institution of marriage, were linked. The marriage system, as a well-defined exogamous residential system, deprived women of the right to inherit land, based on the idea that they would marry and leave the group. So a woman's usufruct right to land was relationally derived either through a husband or through children.[68] This customary land practice gave women room to manoeuvre within a very strong patriarchal inheritance system.

Labour was organized at the level of the nuclear family (*umunne*) and the extended family (*umunna*). An extended family, consisting of a group of close patrilineal relatives of about three or four generations, was the most important socio-economic unit. The household constituted the first resort for labour recruitment. In such a household, a man and his wife constituted the major organizers of farm labour, including the labour of their children. Through their labour contribution, dependents gained access to land, the most important means of production.[69] The importance of the family as a production unit perhaps explains why the Igbo valued large families. Mbagwu Korieh, who until the Nigeria-Biafra Civil War in 1970 was a very successful yam farmer, stressed the importance of human labour in Igbo agriculture: "The size of a man's household, the labour he could demand from relatives and clients, determined his success as a farmer."[70] The more wives and children a man had, the greater his ability to increase production and his prestige.

Often members of an extended family helped each other by providing labour. Lineage heads had access to additional labour through the *oru orie* system. In this arrangement, the entire membership of the kindred provided labour to the head of the lineage (usually the oldest male in the lineage or *ofo* holder). This was usually on the *orie* day in some parts of Igboland.[71] It was an important traditional practice, legitimized by a gerontologically based ideology in which the kindred head symbolized the ancestors. As the head of the kindred and custodian of their *ofo* (symbol of authority and justice), he was compensated for these services.[72] The system gave an advantage to the kindred head since he could draw on a large pool of labour.[73] The formulation of *oru orie* labour arrangements in Igbo society had its root in a strong, traditional kinship structure where the provision of labour guaranteed the younger members sustenance, shelter, and other social needs in exchange for labour.

Another method of raising labour among the Igbo was an appeal to friends (*irio oru*). Under this system, the farmer appealed to friends or kin for help during the farming season. Men often made appeals during yam planting and harvesting. Women often made use of this labour system during the planting and harvesting of their crops. Through this work arrangement, women as well as men relied on social networks to execute farm work. For women in particular, the work arrangement gave them substantial access to labour. The system benefited both the host farmer and the labourer because the host always provided lavish entertainment on the farm and in the house at the end of the day's work. An Igbo adage puts it rightly: *akpuzie onye oru, ya abia ozo* (If a labourer/worker is well fed, he/she comes again). Workers in this arrangement were often given parting gifts of yams and other food items.

Mutual exchange of labour (*owe oru*) was an equally important method of labour recruitment. A work group was the basic system adopted in an increasingly labour-intensive agrarian system to meet its labour needs. This system of labour organization and management was organized along the lines of age grades, friendship circles, or social or finance clubs (*isusu*). These groups worked in turn for their members. For women in particular, this was a very popular method of providing labour for cocoyam and cassava cultivation.[74] Women married into the same kindred group (*ndom alu alu*) often took turns helping each other with farm work. Cash payments were not usually made, but the host provided food and drinks.[75] Work groups may have concentrated on clearing the forest (young men's groups in particular); others may have concentrated on tilling the soil or other phases in the farming cycle up to the harvest. But the most serious groups participated in the three most important phases in the farming cycle: clearing the forest, planting the crop, and weeding. The system was beneficial to all concerned because it solved fundamental labour shortage problems, especially for widows and others who did not have a large, family-based labour force.

There was evidence of the wide use of unfree labor in the form of both slaves and pawns (*ohu*).[76] Pawns were usually debtors who provided labour or other services to a creditor for a specified period or until the loan was repaid.[77] In some instances, debtors who pawned themselves lived with the creditor until they worked off the debt. The debtor was not usually paid for his/her labour throughout the period and still had to pay the loan in full.[78] This system was important because it guaranteed constant labour throughout the year. The existence of the *ohu* system, in an otherwise egalitarian society, indicates the existence of poverty and social and economic differentiation.

Labour was often organized along gender lines and many African societies distinguished between the labour of men and women. Yet, early European perceptions of Africa gave the impression that women did most of the farm work while men enjoyed much leisure time. A German trader, David van Nyendael, described women in late-seventeenth-century Benin as having so much employment that they "ought not to sit still."[79] The women of Whydah, Dahomey, in the same period, "Till[ed] the ground, for their husbands,"[80] and the Capuchin priest, Denis de Carli, reported that women in west-central Africa worked the fields while men made little or no contribution to agriculture.[81] Walter Rodney identified the same trends in the societies of the Upper Guinea Coast.[82] Though not an expert on the Igbo, French historian Catherine Coquery-Vidrovitch has noted that "it was women who worked in the fields" among the Igbo.[83] Anthony Hopkins cites Bamenda (Cameroon) as an example of a pre-colonial society in which women were very important in farming, while among the Yoruba they were much more prominent in trade.[84]

A gendered division of labour was the underlying principle that structured production relations among the Igbo, but there was ample flexibility in the tasks performed by men and women.[85] From Equiano's description, the most important source of energy (human labour) among the Igbo remained relatively unchanged into the contemporary era. Equiano wrote in his autobiography: "Our tillage is exercised in a large plain or common, some hours walk from our dwellings, and all the neighbours resort thither in a body. They use no beasts or husbandry, and their only instruments are hoes, axes, shovels, and beaks, or pointed iron, to dig with."[86]

Tilling the ground for yams, planting them, and staking them were predominantly male tasks. Women and children carried out important farm operations such as clearing the land at the beginning of the farming season, weeding, planting, and tending certain crops such as melon (*egusi*), maize, cocoyams, and cassava. Men were involved in clearing virgin forests. Yet the division of labour was fashioned to suit the economic and demographic realities. Convenience was important in dictating how the family did its work. Basden notes that women took their full share of tasks that were regarded as men's work, especially turning the soil and mounding up the yam beds.[87] Chief Onyema states, "No taboo barred women from participating in all agricultural practices or activities."[88] Even though the cultivation of yams is regarded as a man's affair, a division of labour still existed. Luke Osunwoke of Umuorlu, explains: "In the production of yam there was a division of labour. Men prepared the mounds and also measured the soil for the yam stands. Women weeded the land after the crop had been planted."[89] J. Harris concluded that the division of labour in farm work was anything but rigid and few strict rules were enforced.[90] Yet, women have been known to play less significant roles in agriculture, concentrating on trade and craft manufacture in some Igbo groups. Women in the Nsukka area played a less crucial role in agriculture, concentrating rather on cloth weaving, pottery production, and petty trading, while men generally dominated farming.[91]

ETHOS OF THE IGBO AGRICULTURAL ECONOMY

> There can be little doubt that a crop as precious and as demand-
> ing as the yam would in time acquire this "status of a god of life."
> – *M.J.C. Echeruo, Ahiajoku Lecture, 1979*

At the core of Igbo farming was the cultivation of yams. It was the most important symbol of prestige and other forms of identity were directly or indirectly tied to its production and consumption. Until recent times, yams also
shaped important elements of Igbo cosmology and food and dietary habits.
Alexander Falconbridge, a British slave ship surgeon, who visited the Bight
of Biafra during the second half of the eighteenth century, described yams as
"the favourite food of the Eboe [Igbo]."[92] D. G. Coursey referred to the Igbo as
"the most enthusiastic yam cultivators in the world."[93] According to Reverend
G. T. Basden, yam "stands for [the Igbo] as the potato does for the typical
Irishman.... A shortage of yam supply is a cause of genuine distress, for no
substitute gives the same sense of satisfaction."[94] H. H. Dobinson described
yams as the chief product of Asaba in 1891.[95] Linus Anabalam, a rural farmer
in Mbaise, said, "A man who bought yams from the market in the olden days
was ridiculed and regarded as lazy."[96]

Yam, a highly ritualized crop, was synonymous with the agricultural
ethos of the Igbo people. Emmanuel Nlenanya Onwu notes: "The discovery
of yam cultivation formed not only the economic base of Igbo civilization but
it also carried tremendous religious import. It was of such great importance
that it was given ritual and symbolic expressions in many areas of Igbo life."[97]
The most significant expression of the rituals was the celebration of Ahiajoku
(the yam spirit) – a thanksgiving dedicated to the earth goddess (Ala) before
the harvesting of new yams. The new yam festival was not just an offering
of thanksgiving to the gods; it signified community life and social cohesion.
Although celebrated in recent times largely as a symbolic activity and devoid
of its ritual components, the new yam festival remains a memorial to the most
important crop in Igbo life.

Associated with the Igbo penchant for the yam are cultural habits, rituals, and taboos (Nso Ala) that structured yam cultivation, harvest, and consumption. Nso Ala forbade certain conduct and practices, such as stealing yams and cocoyams. Edward Morris Falk, remarking on the customs and practices of the Igbo in the Aba Division in 1920s, observed:

> Stealing from a farm or farm produce, livestock etc. is considered to be a very serious crime indeed, even though only a trifle of small value be stolen. The Chiefs never fail to complain that the punishment which they are allowed to inflict is utterly inadequate to act as a deterrent.[98]

Breaking these taboos called for an elaborate cleaning ritual to appease Ala. The domination of this ritual knowledge by the Nri Igbo, which reached its peak of power between 1200 and 1640 but declined between 1650 and 1900, gave the Nri a level of imperial control over the rest of Igboland.[99] According to Dike and Ekejiuba, the Nri rose to prominence because "they set themselves up as the chief representatives of powerful spiritual forces, especially the key agricultural deities of the Igbo."[100] In essence, Echeruo argues, "Nri was the immediate outgrowth of an agricultural revolution in the fertile valley of the Anambra River. That revolution, consecrated in the worship of the Earth (*Ala*) and of the Yam (*Ji; Fejioku*) was accelerated by the introduction of metal work, and the implements of agriculture which followed."[101] The precious artifacts discovered at Igbo-Ukwu, while attesting to the wealth and influence of the Nri Empire, are also evidence of the wealth generated by its agriculture-funded, long distance trade and exchange.[102]

Yam signified life, masculinity, cohesion, and the value of hard work among the Igbo people. Since yam was closely linked to male identity, an Igbo man's yam barn was his most important sacred space. Yam-based status was perhaps the most valued of commonly accepted male identity-linked pursuits, until the early twentieth century. Local political, social, and economic lives were woven around a man's success as a yam farmer in many parts of Igboland. "What is a farm without yams; what is a man without a yam barn?" asked an Mbaise elder.[103] Nze James Eboh of Alike Obowo, noted: "A man would be proud of all his assets, but he was nothing if he had no yams. This is how a *dimkpa* [successful man] was measured in the past. Things are not the same anymore."[104] In Ohafia, like many other parts of the Igbo country, part of a man's status or dignity derived from the performance of four yam-related rituals: (1) Ike Oba (Barn Raising), (2) Igwa Oba (Barn Sanctification) during which the Ifejuoku/Nfujuoku/Njuoku ji (Njoku Ji) shrine is constructed. Usually the chief celebrant places the biggest edible yam from his harvest on the Njoku Ji shrine; (3) Igwa Nnu (Consecration of the first 400 yams (*aka oba*)), and (4) Ime Okere Nkwa (a drumming celebration of achievement and the dignity of labour). To host the Ime Okere Nkwa festival, the farmer must have more than 2,800 yams (*aka oba ji asaa*). During the celebration, the chief celebrant places 2,800 yams at measured distances from his barn to his hamlet arena.[105] Such yam-based status conferred on men prestige at home and in the public arena, tied them together in associations of high achievers and differentiated them from less successful farmers.

Chinua Achebe's *Things Fall Apart* captures the Igbo attitude toward yams and the link between the crop and male identity. Okonkwo's work ethics and determination to build a barn through sharecropping exemplify this ethos.[106] The depiction of Unoka, Okonkwo's father, as a lazy, unsuccessful farmer whose perennially poor yam harvest was a source of ridicule in his village is reflected in the different statuses of the two men in their community. The link between environment and agricultural production is also reflected in this dialogue between Unoka and the priestess of Ala (the earth goddess):

> Every year before I put any crop in the earth, I sacrifice a cock to Ani, the owner of all land. It is the law of our fathers. I also kill a cock at the shrine of *Ifejioku*, the god of yams. I clear the bush and set fire to it when it is dry. I sow the yams when the first rain has fallen, and stake them when the young tendrils appear. I weed.[107]

The priestess's response reveals a deep connection between agriculture, Igbo cosmology, and agricultural practice and environment:

> You have offended neither the gods nor your fathers. And when a man is at peace with his gods and his ancestors, his harvest will be good or bad according to the strength of his arm. You, Unoka, are known in all the clan for the weakness of your machete and your hoe. When neighbors go out with their axe to cut down virgin

forests you sow your yams on exhausted farms that take no labor to clear. They cross seven rivers to make their farms; you stay at home and offer sacrifices to a reluctant soil. Go home and work like a man.[108]

Another fictional work, *The Seed Yams Have Been Eaten* by Phanuel Egejuru, further captures the importance of the yam among the Igbo and the transformations that have taken place since the Nigerian Civil War.[109] Egejuru wrote: "We young ones are happy now because we shall not go to the farm again as all seed yams have been eaten by the soldiers, both our soldiers and enemy soldiers. We now eat cassava and women go all the way to Umuagwo to collect cassava."[110] The dilemma faced by the protagonist Jiwundu [yam is life] in the story and his resolve to perpetuate his name by striving to build his life around yam production show the complexity of Igbo economic rationality as well as the centrality of yam in Igbo cultural life. The disappearance of the yam and the concomitant decline of the Ahiajoku ritual threatened communal cohesion and collective identities.[111]

Although these are fictional works, they depict the values attached to yam production among the Igbo, but most importantly, they provide an insight into how the Igbo speak about yams and the link with male identity. Both works also depict a society where notions of acceptable gender role affected the agricultural pattern – labour, crop, and identity. Achebe wrote in relation to Okonkwo's work ethic: "His mother and sisters worked hard enough ... but they grew women's crops like cocoyams [*ede*], beans and cassava. Yam, the king of crops was a man's crop."[112] Achebe was also presenting the Igbo work ethic in general. As the "male" crop, the Igbo granted the yam privilege over other crops and its production was closely integrated to male identity. Early colonial observers noted a distinctly gendered pattern in crops cultivated in Igboland. The district officer for Owerri District noted in 1928 that yams were the principal crop and "the remaining crops are all women's crops, grown, maintained and sold by them."[113] The celebration of the yam as a symbol of masculinity and the cultural and spiritual essence of Igbo manhood is obvious in such male names as Ezeji (a title meaning yam king) and Jiwundi. As an example of Igbo philosophizing about the importance of the yam, Jibundu means that the yam is life, and it builds a home, a community, and a town.[114]

That, to be sure, suggests a long antiquity of the cultivation and use of yams and the existence of deeply rooted cultural and ritual practices associated with yams.

While it is difficult to trace the origin of the gender ideology associated with yam production, some of the patriarchal ideas related to yam production may have become more elaborate following the introduction of cocoyams and cassava. The myths and rituals associated with yam cultivation, it seems, were part of the power dynamics that evolved in a predominantly decentralized society in which sociopolitical and economic power was very amorphous. The high social status derived from successful yam cultivation became an important means of affirming male authority, identity, and status.

The intensive nature of yam production required large amounts of labour and a high rate of soil depletion. Basden concluded in his study of Igboland in the 1920s that "From an agricultural point of view, the yam is a very extravagant vegetable to grow. Each tuber requires a full square yard of land, which in itself, is a big demand. For seven or eight months, regular attention must be given to its care, absorbing much time and labor." Basden noted that "If wages had to be paid, it is doubtful whether a yam farm would pay its way, let alone yield profit."[115] The concern raised early in the twentieth century became more pronounced by the end of the century, as farmers faced the problems of degraded land and dwindling availability of labour and cost. These factors have challenged the traditional cropping and dietary patterns, changing them in favor of cassava – the so-called women's crop.

However, when the Igbo talk about farming, they talk about yams, especially in defining male identity.[116] How did yams come to hold this prominent place in the agricultural and cultural life of the Igbo? The elevation of the yam to high status appears to have had its origin in the people's agricultural past. The yam was certainly one of the earliest crops to be domesticated by the Igbo.[117] Legends about the origin of the yam enable us to speculate not only about the prevailing ideology but also about the gendered nature of yam production in the Igbo habitat. One version of such a legend recorded by Basden states:

> In the olden times, there was nothing to eat, so Eze Nri considered what should be done to remedy the defect. He took the drastic

course of killing his eldest son, cutting the body in small pieces and burying them. His daughter shared a similar fate. Strange to say, five months later, yam tendrils were observed to be growing at the very places where the dismembered parts of the son's body had been planted. In similar fashion 'edde' [cocoyam] began to grow where the remains of the daughter had been buried. In the sixth month, the Eze Nri dug up fine large yams from his son's grave and 'edde' from the place where he had buried his daughter. He cooked both and found them sweet.[118]

While this legend may not explain the actual origin of yams, it reveals the prevailing ideology about yams and agriculture. In this legend, the yam represents masculinity, having grown out of the dismembered parts of the son, and the cocoyam represents femininity, for it grew out of the daughter's grave.

Yams were much more than a food crop. The development of a yam culture and settled life supported the emergence of the complex social and ritual systems associated with yam cultivation and the development of settlements and communal property.[119] Yam cultivation was the major incentive for the development of technology suited to the extensive clearance of the original rainforest and the more effective mastery of the forest. The rituals associated with the yam cult themselves also express important aspects of power relations in traditional society: namely, the domination of ritual functions by men and the social prestige attached to the status of Ezeji, a title only men could take. Yam production, perhaps, also triggered the era of property ownership, in which land was controlled by wealthy yam farmers, the emergence of the early formation of agrarian communes, and the distinction between the poor farmer and the rich one. Yams brought tremendous changes in the life of the Igbo, including population growth, the elaboration of archetypical social institutions, and the evolution of a cosmological system in which the Earth (Ala) became deified and occupied the central place as the ordainer and guardian of morality, the source of law and customs.[120]

Production relations, land tenure systems, and land use are shaped by this culturally defined ethos centred on yam production. Social status, gender ideology and identity, taboos, and rituals associated with yams and cocoyams form an essential part of the cultural ethos. The extended family (*umunna*) may have originally been organized to sustain yam production because yams required abundant labour and care from planting to harvest. This is probably true of most peasant agriculture. The labour required for yam cultivation led to the development of marriage alliances and the institution of polygamy. Marriage ensured the continued reproduction of the lineage and the availability of women's and children's labour. The consumption of yam followed the ties of kinship, comradeship, and friendship and its distribution generated new ties, trade, and social alliances.[121] By dominating yam production, men controlled the most important factors of production – land and the labour of wives, children, and younger lineage members.

Yet, women successfully exploited the flexibility inherent in the Igbo land use and tenure system to maximize their own production through intercropping a variety of other crops in the spaces provided by the yam mounds. This enabled women to perform important roles related to the overall diet and food security. They also performed important roles during the *unwu* period before the crops were harvested. Women dug up the first yams, known as *nwa agwagwa erie*, which helped people survive the lean periods, a task seen as demeaning for men.[122] These roles attest to the centrality of women in the food security arrangements.

PRE-COLONIAL COMMERCE

Until the mid-seventeenth century, the economy of most of Igboland was still largely a subsistence economy, based on complementary ecological niches suited for particular types of crop production, fishing, and craft manufacture. However, although pre-colonial Igbo farmers produced primarily for their own subsistence, they also produced for exchange and extra income. The commercial system served as an efficient means of exchange and distribution of agro-based goods among other consumables. Agricultural produce dominated the local market as Elizabeth Isichei noted of the traditional Igbo market: "Merchandise was plentiful and varied, but common to all markets were yams, cocoyams, meat, fish, salt, maize cobs, beans, leaves, banana, paw paws, groundnuts, chickens, goats, sheep and dogs, palm oil, palm kernels, pepper, kola nuts and so on."[123] The predominance of agricultural produce in the Igbo markets was an indication of the link between agriculture and commerce. This, of course, was the case in most pre-industrial societies where the need for food was of primary concern.

A high level of intensive internal exchange had developed in the region before the overseas trade in slaves.[124] Long-distance trade depended on what Ukwu I. Ukwu described as "specialist groups and individuals," and mostly relied on "covenants at the personal level, *Igbandu*, as the principal means of guaranteeing freedom of movement across the country and safety among strangers."[125] Both the Aro and the Nri-Awka, who dominated early long-distance trade in Igboland and beyond, relied on the existence of a relay of hosts and trade pacts that developed into a great network of ritually based commercial relations.[126] As in most such societies, the demand for high value goods was not entirely absent prior to the contact with Europeans. The trade in such goods, which included salt, beads, ivory, dyes, cloth, carving, charms, and ironmongery, Ukwu argues, was for the most part "irregular and scarcely created new markets within an area."[127] But the skills of the Awka people, whose prestige derived from the repute of their Agbala deity and their skills as diviners and doctors, were "useful in establishing markets for their wares," including luxury goods such as "ivory and coral beads from the Anambra and Igala country."[128] The Awka trading arrangement was integrated into agriculture as Awka people became Ahiakoku ritual specialists. The Awka trade diaspora did not establish permanent settlements, as the Aro did from the seventeenth century, yet the nature of their organization enhanced the exchange of goods and services throughout the Igbo country.

The Aro influence in the development of commercial relation in Igboland and beyond is legendary. The Aros' success was contingent on the establishment of well-coordinated trading networks based on settlements by members of the Aro diaspora and trade fairs.[129] The Aros' location in the escarpment between the central Igbo districts and the Cross River helped them to control several trade routes along the Awka-Orlu-Okigwi axis and as far away as Nike in northern Igboland. Their reach was felt throughout Afikpo, Uburu, Izza, and Izzi in the northeast. To the south, their commercial relations extended to Bende and Uzuakoli. Like the Nri-Awka before them, Aro traders distributed luxury item including horses and cattle "for ritual purposes."[130]

The Aro called upon their network of trade routes and settlements in the era of the slave trade, which they dominated from the eighteenth century.

Two separate but related economic systems existed in Igboland in the pre-colonial period: a trade in slaves and a trade in palm oil, which developed after the abolition of the Atlantic slave trade. The Atlantic slave trade provided the initial link with the external market and the impetus for greater European economic contact in later years. Until the mid-nineteenth century, slaves were the most important export from Igboland. David Eltis and David Richardson estimate that about one in seven Africans shipped to the New World during the whole era of the transatlantic slave trade originated from the Bight of Biafra.[131] This corresponds to the estimated 11.9 per cent of the total number for which data are available via the Du Bois CD-ROM database.[132] Trade in the region grew steadily from the 1730s, reaching a peak of 22,500 captives annually in the 1780s and expanded further to about 40,800 in the 1830s and 1840s.[133] Douglas Chambers suggests that, of the 11.6 million people estimated to have been shipped to the New World between 1470 and 1860, some 1.7 million were transported from the Bight of Biafra.[134] He estimates that 80 per cent of the people shipped from the Bight of Biafra were Igbo-speaking and reached the Americas in British ships.[135]

The Aro were the most important facilitators of the trade in the Bight of Biafra hinterland. The trade linked several parts of the Cross River trading system. Market routes branched in several directions within the Bight of Biafra hinterland from the major markets of Uburu and Bende. Nike or Aku in the north, Okigwi in the centre, and Awka in the centre-west were linked to these two major markets. Onitsha, Ossamari, and Aboh on the Niger, Akwete on the Imo River, and Itu on the Cross River were major routes linking the overland routes and waterways that moved goods.[136]

Important demographic and economic changes occurred as a result of the slave trade. The slave traffic resulted in concentrations of population in some coastal regions and clearance of the surrounding forest for agriculture.[137] Given the demand for local foodstuffs, local people understood the added importance of producing a greater surplus. Slaves were fed until they were sold in the Americas, in most cases on diets to which they were accustomed, on the odious journey known as the middle passage. The demand for provisions on slave ships increased the demand for local foodstuffs, including cassava, yams, and palm oil.[138] The Bight of Biafra became an important source of commercial foodstuffs as the external slave trade developed.

Yam production, in particular, rose in importance as an important source of food for those shipped from the Bight of Biafra into the overseas slave trade. There was considerable trade in yams in the seventeenth and early eighteenth century in the Bight of Biafra according to the accounts of John and James Barbot. European slavers were advised to carry no less than 100,000 yams for every 500 slaves on ships. James Barbot took only 50,000 and suffered because of a shortage of yams.[139] Although the seasonal nature, bulk, and difficulty of storage of yams constituted problems for slavers, the demand was substantial along the Bight. It is impossible to estimate accurately the quantities of yams bought in the Bight of Biafra. Jones estimated an annual export of more than a million yams based on Captain Adam's estimates of 20,000 slaves from the Rio Real and half the recommended quantity of 100,000 yams for a cargo of 500 slaves.[140] Between 1801 and 1867 when the external trade was winding down, about 217,781 slaves were exported from the Bight of Biafra, suggesting that a substantial quantity of yams was still being purchased in the Bight.[141]

Thus, local production in the Bight and its hinterland was in part structured around providing food items for the slave market. Slavers, on their part, were interested in the dietary habits of their African victims. Alexander Falconbridge, a British slave ship surgeon, recorded: "The diet of the negroes, while on board, consists chiefly of horse-beans, boiled to the consistence of a pulp; of boiled yams and rice, and sometimes of a small quantity of beef or pork." Different African ethnic groups were identified with particular kinds of food. "Yams are the favorites food of the Eboe, or Bight Negroes, and rice or corn, of those from the Gold and Windward Coasts; each preferring the produce of their native soil," Falconbridge noted.[142]

The early commercialization of food production created a diverse economy in which the Ngwa Igbo and Anang Ibibio became well-known yam producers who fed the Atlantic market. Indeed, the production cycle in the region influenced slavers' traffic in the region. Barbot made a connection between the availability of yams, the famine (*unwu*) period, when yams had been planted, and the ability of European slavers to achieve a quick turn around in the Bight of Biafra. "Yams are not fit to be taken out of the ground before July and August. Europeans account these two months the best season of the year, because of the continual rains which refresh the air and give natives an opportunity to apply themselves wholly to commerce up the land for getting slaves with expedition but more especially in August and September," Barbot wrote. "But in August and September and so on to March these eatables grow very dear among them insomuch that some ships have been forced to fall down to Amboses and Cameroons river in May and June to buy plantains.... To avoid this long delay it is much better for a ship bound to this place from Europe to stop at Cape Tres Pontas or at Anamabon on the Gold Coast to buy Indian wheat or corn there," he warned.[143]

Although Barbot was wrong about the time when yams were scarce (usually from March to June), the seasonal nature of yam production and African marketing strategies influenced the time Europeans felt it was best to trade – the dry season. Grazilhier, who traded in the Bight at the beginning of the eighteenth century, understood this, according to the information he gave to British slaver James Barbot:

> In the months of August and September a man may get in his complements of slaves much sooner than he can have the necessary quantity of yams to subsist them, but a ship leading slaves in January, February etc., when yams are very plentiful, the first thing to do is to take them in and then the slaves.[144]

These descriptions provide insights into the link between the developing Atlantic commerce and the rural agricultural economy of the regions on the West African coast. Slavers were forced to use local foodstuffs, thereby stimulating increased production of these food items in response to the new demands. We can assume that most of the yams bought at Bonny and New Calabar came from the forest region bordering the Imo River and from the marginal savannah of the Niger riverain area.[145]

The long-term impact of the abolition of the slave trade was also agrarian in nature. By the middle of the nineteenth century, the market facilitated effective exchange, which helped in the capitalization of aspects of the agricultural system.[146] Increased interest in tropical agricultural products followed the abolition of the Atlantic slave trade.[147] The growing industrial development of Europe created a need for palm oil and palm kernels. The Igbo responded to the transition from the slave trade to a commodity trade by introducing innovations that enabled them to increase export quantities. By the 1860s, the trade in palm oil, which began in the seventeenth and eighteenth centuries, had replaced the trade in slaves. The development of the trade in palm oil and later kernels helped in the further occupation of the rainforest and perhaps the further expansion of food crop production to support the demands created during the period of the slave trade. As early colonial reports suggest, the Igbo were expanding into the southeastern portions of the Guinean forest from the seventeenth century onward, a process that led to the cutting down of the forests and "replacing them by villages and hamlets surrounded by oil palm groves, farmlands and forest remnants."[148] By the beginning of the nineteenth century, a considerable quantity of palm oil was being exported to Liverpool and other European destinations.[149] The original development of the palm oil trade was therefore associated with the slave traffic but increased in importance after the slave trade's abolition.

By the mid-nineteenth century, some British officials and traders on the West African coasts strongly espoused the "liberating" tendency of so-called legitimate commerce within official colonial circles. When Captain Joseph Denman testified before the Select Committee on the West Coast of Africa in 1842, he assured its members: "When once the trade is interrupted at any place, people are not in the habit of sending traders up the country for slaves, and traders from the interior cease to bring slaves down to them there, and there is great difficulty felt in resuming 'the trade'." He argued that such disruption "in almost every instance" caused legitimate commerce to grow and helped to supply the "wants of the natives."[150] Benjamin Campbell, British consul at Lagos from 1853 to 1859, and Richard Hutchinson, consul for the Bight of Biafra from 1855 to 1860, both declared the liberating potential of "legitimate" trade.[151]

The export of palm oil to Liverpool from the Bight of Biafra in 1806 was 150 tons, and by 1829 it had reached over 8,000 tons.[152] The trade from the Bight of Biafra increased from 4,700 tons of palm oil in 1827 to about 13,945 tons in 1834.[153] By the 1830s, Britain was importing about 10,000 tons of palm oil a year.[154] Between 1855 and 1856, the entire African production of palm oil was about 40,000–42,000 tons. Out of this figure, 26,000 tons exported through the Bight of Biafra, although the amount that came from

An early method of commercial palm oil production. (Reproduced with the kind permission of the Bodleian Library, University of Oxford.)

the interior is not clear.[155] The output of palm oil from all the Niger markets between 1886 and 1888 was about 6,000 tons.[156] From the 1860s onward, when palm oil prices fell, and the 1880s, when the long process of establishing colonial hegemony began, the implication for the local agrarian economy of the dependence on palm produce became apparent, a legacy that was exacerbated with the imposition of colonial rule. Although the volume of the trade in palm oil was considerably low before 1900, Captain H. L. Gallwey, Her Majesty's vice-consul for the Oil Rivers Protectorate, had noted in the 1890s that trade on the Benin River was chiefly in palm oil with a substantial trade in kernels as well. This was the case for many other communities in the Niger Delta.[157] The limited quantity produced in the decades immediately before 1900 accounts for the reportedly high price demanded by native producers on the West African coast for palm oil.[158]

Palm kernels became an export item after 1860 as a result of the demand for soap making, margarine manufacture, and residue for cattle food.[159] Women and children performed the slow laborious process of cracking the nuts: "In 1862, 272 tons were imported into Great Britain."[160] Exports from

Old Calabar reached 1,000 tons at two shillings per bushel when the trade began at this port in 1869 and exports had doubled two years later at eight shillings per bushel. Exports had reached 10,000 tons by 1875 and 22,031 tons by 1885. Substantial quantities of kernels were being exported to other European countries, especially Germany, toward the end of the nineteenth century. By 1903, Nigerian kernel exports had reached 131,900 tons, mostly going to Germany.

This trade was buttressed by new waterways that were developed to move oil and kernels to the coast. The development of agriculture, particularly the palm oil trade, led to the large-scale development of the internal slave trade and an extensive use of unfree labour in parts of Igboland. The trade in commodities also engendered significant speculation and investment in land, with significant sociological consequences. The demands of the oil palm trade, Michael Echeruo says,

> ... created the adventurer class in Igboland, the dare-devil entrepreneur who knew how to accumulate capital. Banking, for example, which under Nri, was a glorified loan-scheme, guaranteed by the higher fees to be paid by new initiates, became a truly capitalist enterprise. It is not an accident that the resulting accumulation of cash capital was more pronounced among the Aro than the Nri who continued to commit their reserves into mansions and objects of art/prestige, while the Aro provided the cash credit on which a mercantilist economy could thrive.[161]

We can trace the domination of aspects of Igbo life (ritual and commercial) before, during, and after the slave trade by the Nri and Aro and the linking of Igboland into a sort of economic and ritual commonwealth, and we can also see how the internal economic crisis the Nri and Aro faced toward the end of the nineteenth century reflected on other parts of Igboland. When one examines the case of Nri, as Michael Echeruo argues, prosperity in earlier periods "led to urban and later to imperial interests – which in turn nurtured a parasitic tradition of rituals that in time dominated Nri life, and made agriculture itself secondary to the system of sinecures on which Nri opulence came to depend." In Echeruo's view, the rural agricultural virtues and ethos of the fertile valley "gave way to the aristocratic feudal lifestyle of a metropolitan capital and its satellite communities." The Nri tradition of ritualized culture had so "debased the economic function of the yam that production took a secondary place to celebration," while the very basis of the rise of the Nri – the internal sufficiency in food supply on which its dense population depended – was "seriously, even permanently, eroded, leaving the region "subject to the great hunger which would lead it to large-scale emigrations in the earlier part of the next century."[162]

The period was one of crisis for the Aro and other African traders who had to give up their long domination and control of the market. The production and marketing systems the Igbo and other traders in the Delta had enjoyed until toward the end of the nineteenth century underwent fundamental changes. Kenneth O. Dike and Felicia Ekejiuba have documented the extensive involvement of the Aro in the lives and fortunes of the Igbo and their neighbours, including the Efik, Ekoi, Igala, Ijo, Jukun, Idoma, and Tiv, a process that combined to transform eastern Nigeria socio-economically in the three centuries during which the Aro "rose to prominence."[163] Waibite Wariboko and Kenneth Dike, among others, have examined the attempt by African middlemen to hold their ground in the face of increased European encroachment.[164] African producers and traders frequently found themselves in unforeseen and unusual situations faced with the increasing interference in local trade by European traders, often encouraged by imperial officials. The volatility and unpredictability of the new trading and marketing relationships forced many Africans to respond in diverse ways. Two divergent socio-economic models competed for the upper hand in the African market: British ideas of free trade, which were attempting to break the old monopolies of African traders, and African trading systems that had largely relied on the structures of the Atlantic slave trade, which empowered some individuals and trading houses to act as major brokers of the palm oil trade.

Both the trade in slaves and the transition from the Atlantic slave trade to the palm oil trade had implications for gender. Anthony Hopkins called attention to changing gender relations that may have resulted from the high male slave export from West Africa.[165] How did traditional gender ideologies change in the wake of the transition to the palm oil trade? In a major contribution analyzing the gender dimensions of the commercial transition, Susan Martin argued that "the key to a man's success in commercial palm production lay in control of women's labor."[166] Martin maintained that despite women's contribution to the production of palm oil and yams, men claimed ownership of these products as heads of households. Martin suggests that by the end of the nineteenth century, the "division of labor acquired ideological significance, which strongly favoured men."[167] With specific reference to palm oil processing, Martin argued that "given the sexual division of labor and property rights, the opportunity clearly existed for Ngwa households and lineage heads to acquire wealth." This they did by "organizing yam and palm oil export production to serve the nineteenth century provisioning and palm oil export trade." On the other hand, "little pressure," she argues, was "put on their male labor resources by the palm oil industry in particular, since palms were harvested throughout the year and men's contribution to processing though vigorous, was brief, while women's work in palm oil extraction was both laborious and time consuming."[168]

Margaret Stone, like many other scholars, has shown that the imposition of capitalism on subsistence economies had fundamental implications for the sexual division of labour in farming and ideological construction of male and female status and power.[169] Most importantly, the European attitude towards African women helped magnify the differences between men's and women's access to resources and participation in commerce. Morgan notes that "European intervention brought economic changes which are reflected in new attitudes to land, cultivation and stock rearing, and these have produced further landscape modification."[170] But the reality of the palm oil trade's part in the transformation is more mundane than the picture painted by these scholars.

Admittedly, the impulse created by the new commerce tilted in favour of men, but there was also contestation for control between men and women. But scholars have often overestimated the ability of Igbo men to control the new trade in palm produce. Susan Martin, for example, ignores the fact that palm oil had little commercial value for either men or women prior to its commercialization at the end of the slave trade in the nineteenth century. And women remained important in both production and marketing from the late nineteenth century onward. They were not only entitled to some of the palm oil; they were entitled to the kernels, which became quite important in the export market. The evidence from the late nineteenth and early twentieth centuries confirms the important role played by women in the commercial economy of the transition era. Adolphe Burdo, a member of the Belgian Geographical Society who visited Igboland in the late nineteenth century, for example, described Onitsha women who "traverse the country to collect palm oil and ivory."[171] Basden reported that women traders dominated the trade in palm produce and imported goods into Onitsha from the turn of the twentieth century until the depression of 1929.[172]

Women traders were active in the distributive trade from the time of the opening up of the countryside to European goods. Raymond Gore Clough, who joined McNair, Henderson and Company as an assistant agent in 1919, described the tenacity of the African trading "mammies," most of whom were certainly Igbo, including Madams Unuka, Osika, and Omvaro. "My customers were not only the canoemen who brought in palm oil; there were also the African trading mammies who took large quantities of a wide range of goods," he wrote in his memoir.[173] The women in turn peddled these goods through their own shops or through subagents in the various markets. Clough describes these women as "ladies of standing among their own people and I could not help watching, with fascination, their dignified movements as they examined the brightly coloured earthenware bowls or basins, or cotton goods on display in the store." It seemed that there was no anxiety about extending credit to these women traders. "There was a strict code of honour between the trading mammies and the traders, and it was extremely rare that credit granted was abused," Clough recalled.[174]

Richard Henderson confirmed that women controlled the trade with European firms until the early twentieth century. Indeed, women resented male intrusion, when many men began to enter the trade.[175] Up to the beginning of the twentieth century, Nina Mba notes, women in the riverain Igbo areas who "controlled the markets, constituted the local and long-distance traders, were in a position to amass more wealth than the men."[176] Although men were often taunted "for doing women's work," some men could not resist the opportunities offered by the retail trade and supply of European goods.[177] F. Hives, district officer for Okigwe, reported in 1917 that "trading in oil, or kernels in the market, in small quantities is done chiefly by women."[178] Women's roles as producers and marketers of palm oil in the nineteenth and twentieth centuries challenge the perceived invisibility of women in this era. Like the members of many other societies in Africa, the Igbo were not hidebound by custom.

Nonetheless, the changes that occurred with the emergence of colonialism at the beginning of the twentieth century, in addition to the patriarchal ideology of European officials, had a direct impact on the nature of agrarian change and gender relations of production. Evidently, Igboland faced significant transformations in the economy and social relations, as trade in palm produce expanded and as the Igbo adopted new sources of status and power at the beginning of the twentieth century. The transformative power of colonialism and market-driven changes fuelled by a cash economy challenged the old structures and institutions of Igbo society, including existing relations of production and means of acquiring income. Like other parts of Africa, Igboland was dragged further into the capitalist world following the establishment of colonial rule in 1900.

CHAPTER TWO

PAX BRITANNICA AND THE DEVELOPMENT OF AGRICULTURE

On March 15, 1903, British forces occupied the Sokoto Caliphate. A *New York Times* report aptly gave the event this headline: "Big Territory Added to the British Empire."[1] Frederick Lugard, who was to become the first governor of colonial Nigeria, was heartily congratulated by Chamberlain and other officials of the British Government. In recognition of this feat, the *Pall Mall Gazette* wrote in an editorial: "We have to thank the bold initiative of Sir Frederick Lugard, a typical specimen of the sort of tool of empire building which is, perhaps, the one product of British industry which our rivals cannot imitate."[2] This important event, which reflected the sentiments of many in Britain on imperialism toward the end of the nineteenth century, was the climax of the gradual occupation of what was to become Britain's most important possession in western Africa and an important milestone in the expansion of the British Empire.[3]

The economic goals were urgent and the mandate clear. The Hon. W.G.A. Ormsby-Gore, parliamentary undersecretary of state for the colonies, on a visit to West Africa in 1926 reiterated the most important economic goal of the British Empire. "The wealth of West Africa lies primarily in its agricultural and forest products," he wrote, and "the economic progress of the Colonies depends on the development of their vast agricultural and sylvicultural resources."[4]

Although several parts of what became Nigeria had already been incorporated into the British Empire a decade earlier,[5] the Igbo of southeastern

Nigeria were only incorporated into the empire at the beginning of the twentieth century. The expansion of agriculture, the transformation of the production system, and the restructuring of the Igbo political system remained the most urgent tasks of the colonial administration for nearly two decades. The expansion of the palm oil industry was at the core of colonial economic policy for the Igbo region and the other areas lying in the palm oil belt. This had important consequences for both the politics and the economy of the Igbo.

Behind the thinking of British officials at the time of the conquest of Nigeria was the idea that the northern and the western parts of the country had achieved higher levels of political development and civilization before European contact. Achieving the economic objectives and control of the region required considerable restructuring of the political institutions of the Igbo. The men who carried out these tasks had a mindset of a dysfunctional political system and evinced little regard for existing institutions. Revenue Commissioner H. P. Palmer, on a visit to Southern Nigeria in 1913, remarked:

> The Eastern Provinces is [sic] some centuries behind the countries west of the Niger in natural development. Consequently, the social organization of its peoples is less easy for a European administration to deal with than the National organization of peoples like the Yorubas and Hausa. "Native" ideas, Native laws, and Native administrative machinery are so far remote from their European counterparts, that the destructive force of any European administration at all is proportionately greater than when applied to countries whose conceptions are more advanced and obvious.[6]

These words represent perhaps the most important perceptions as well as goals of the early colonial administrators – the notions of a backward peasantry and a "primitive" political system – notions that would determine colonial policy among the Igbo for the next half century. They also represent what would be sources of frustration for both the British and the Igbo people in the Anglo-Igbo relationship for most of the colonial period. The British sought to reorganize the indigenous polity of the Igbo as part of the economic mission.

REORGANIZING THE INDIGENOUS POLITY

Official British intervention in the area of what became Nigeria started in 1849 with the appointment of a consul for the Bight of Biafra to protect British commercial interests and impose anti-slave trade. Britain dominated much of the trade in the Bight of Biafra and had concluded treaties of "protection" with several chiefs in the Niger Delta, but British control remained very loose and limited to the major trading regions, known as the Oil Rivers, until 1885. The United Africa Company, an amalgamation of all the major British firms trading along the Niger Coast and in the Niger Delta, was also busy signing treaties with the chiefs along the banks of the Rivers Niger and Benue. Establishing effective British control in Nigeria became rather urgent after the enactment of the Berlin Act of 1885. The notification of the creation of the Oil Rivers Protectorate (1885–1893) was published in the *London Gazette* of June 5, 1885, and read in part as follows:

> Under and by virtue of certain treaties concluded between the month of July last and the present data and by other lawful means the territory [*sic*] on the West Coast of Africa hereinafter referred to as the Niger District were placed under the protection of Her Majesty the Queen from the date of the said treaties respectively.[7]

The proclamation and the appointment of a consul-general at Calabar and the posting of consuls and vice-consuls at district levels was a major step in the establishment of effective British administration in southern Nigeria.[8]

The nature of the commercial and political relations between the British and Africans was influenced by the activities of the Royal Niger Company from the 1880s to the first years of the twentieth century.[9] The expansion of the company's administrative authority over the Lower Niger and its adjoining territories, from New Calabar to Oguta in the Igbo country, led to a steady curtailing of African economic control relative to European traders toward the end of the nineteenth century. Conflict often occurred between African intermediaries and European traders as the latter, by virtue of the company's royal charter, increasingly laid claim to what Africans perceived as their market. The company's regulations, Waibinte Wariboko argues, were "designed specifically to frustrate the African middlemen competitors."[10] The attempt by New Calabar traders, for example, to evade these regulations led the Royal Niger Company to expel them from Oguta, Omoku, and Idu in 1887. In 1888, the European agents in Brass observed that the Brass chiefs had "become [greatly] impoverished by the loss of their Niger trade."[11] By 1895, Nembe-Brass's middleman position had become "a thing of the past," according to E. J. Alagoa.[12] These developments had a direct impact on the hinterland communities as the delta intermediaries increasingly relocated trading stations to positions closer to the sources of palm produce.[13]

From 1891 onward, London asserted its authority as both British officials and traders began pushing beyond the coast.[14] The pacification of the region, the British argued, was necessary to eliminate internal slavery and expand the production of palm oil. Much of the coast had indeed been brought under British control and the protectorate was renamed the Niger Coast Protectorate with its headquarters at Calabar and Sir Ralph Moore as its first consul-general in 1893. The new protectorate, which covered a wider area and extended further inland than the old one, was placed under the supervision of the Foreign Office in London. The protectorate recorded very significant achievements in the exploration of the hinterland and the expansion of trade, such that by 1896, Major A. G. Leonard, an official of the Niger Coast Protectorate, had travelled over one hundred miles into the hinterland.

The Niger Coast Protectorate became the Protectorate of Southern Nigeria on January 1, 1900. The new protectorate incorporated the territories formerly administered by the Royal Niger Company and the Niger Coast Protectorate. The consul-general became the high commissioner, while consuls and vice-consuls became district commissioners and assistant district commissioners, respectively. The headquarters of the new Protectorate of Southern Nigeria remained at Calabar until 1906 when it was transferred to Lagos. The creation of the Protectorate of Southern Nigeria marked the end of the consular period. In 1906, the Southern Protectorate absorbed the Colony and Protectorate of Lagos. The Southern Protectorate was formally amalgamated in 1914 with the Protectorate of Northern Nigeria. With the amalgamation, British political control of what became Nigeria was fully established with Frederick Lugard as governor-general.

Sir Ralph Moore's tenure as head of Her Britannic Majesty's Government in the Bights of Benin and Biafra was crucial in the history of the establishment of effective British administrative control in Southern Nigeria. The British attempt to establish political authority in the region was achieved through military expeditions. In 1900, the Southern Nigeria Regiment, West African Frontier Force, was created and soon launched "military patrols" into many parts of Southern Nigeria. Military expedition beyond the coast began with the Aro Expedition of 1901–1902.[15] In November 1901, the Southern Nigeria Regiment attacked the Aro and the so-called Long Juju Shrine or Ibini Ukpabia – the centre of the Aro slave trading oligarchy and its cult of human sacrifice. The British attack on the Aro was justified as an attempt to end internal slavery and open up the areas beyond the coast to free trade, thus blocking the Aro from the control of trade, including the slave trade, over which they had exercised control for more than two centuries.[16] Although the British were mistaken in believing that they had destroyed the oracle, upon which much of the Aro influence depended, the expedition marked the beginning of effective incorporation of the Igbo country into British colonial territory and a considerable decline of Aro influence in the region. Yet the expeditions against the Aro did not end resentment against British control.

The expansion into Igboland met significant local resistance. Don C. Ohadike's *The Ekumeku Movement: Western Igbo Resistance to the British Conquest of Nigeria, 1883–1914*, captures the spirited resistance of the Western Igbo (Anioma) people to British colonialism. His work demonstrates that some of the strongest African opposition to British rule in Nigeria came from small communities.[17] In 1896, a punitive expedition was sent against Obohia and Ohuru in the Opobo District. In the same year, Sir Ralph Moore sent an expedition to Akwete for the opening of friendly relations with its chiefs and people. The extension of the British colonial frontier in the Igbo country was not a coordinated and orderly advance, but a spasmodic struggle for control of Igbo societies. Repeated waves of resistance by the Ekumeku movement had to be put down forcefully in 1902, 1904, and 1909.[18] Other military expeditions included those to Orokpo (1901), Uzere (1903), Etua (1904), Ezionum (1905), Ahiara (1905), Ezza (1905), and Achara (1905).[19]

Generally, the British conquest was difficult and unwelcome, and pockets of isolated resistance continued throughout the period of British colonialism in the region. For over a decade, following the first incursion into the Igbo country, the British could not claim control of the region despite their often bold attempt to extend colonial authority far into the interior. Yet there were some parts of Igboland where people resigned themselves to the presence of the British in their lives rather than face a British punitive expedition and a no-win situation.

THE IGBO AND INDIRECT RULE

A major characteristic of British colonial government in Africa was its decentralized structure characterized by indirect rule or native administration.[20] Indirect rule was conceptualized as an indigenization of the colonial administration – a process that allowed Africans a certain degree of internal self-control. The system, which was elevated to a political ideology by Sir Frederick Lugard, was applied vigorously throughout Nigeria from the 1920s onward. Lugard, the first colonial governor of Nigeria and often regarded as the model British colonial administrator, emphasized:

> If continuity and decentralization are, as I have said, the first and most important conditions in maintaining an effective administration, co-operation is the key-note of success in its application – continuous co-operation between every link in the chain, from the head of the administration to its most junior member, co-operation between the Government and the commercial community, and, above all, between the provincial staff and the native rulers ... with as little interference as possible with native customs and modes of thought.[21]

Despite the ethnocentric and racist views expressed in *The Dual Mandate* and other writings, Lugard recognized the practical realities on the ground, including the shortage of European personnel and funds. Indeed, not all colonial subjects found colonial rule an encumbrance as the British did not readily dismiss African institutions upon conquest. A 1928 circular from the secretary, Northern Provinces, to all Residents in the area, which outlined the methods for training of junior officers in the implementation of indirect rule, is informative on the rationale for the policy. As outlined in the circular, indirect rule was adopted because Lugard realized that he "could not effectively administer the enormous area to [sic] the Northern Provinces with the utterly inadequate staff at his disposal, and that therefore he must enlist the assistance of the Native Chiefs."[22] His initial policy, the circular continued, "was carried on by his successor Sir Percy Girouard, ... and especially the present Lieutenant-Governor, Northern Provinces, to whom, among other important matters, was due the creation of native treasuries, the most essential corollary to indirect rule."[23] The pressure of economic development, according to the secretary to the Northern Provinces,

> ... makes it increasingly necessary to utilize the productive and energising capacity of all native institutions (which is very great) to the full, and to avoid arbitrary or empiric changes of method to which the people are not accustomed and which may produce discontent and dissatisfaction or ultimately inefficiency in the essential duty of keeping order.[24]

The British conquest of Northern Nigeria curtailed the power of the old aristocracy, but Britain did not embark on a project of dismantling the Islamic structures and social institutions of the caliphate. As part of the compromise, the emirs accepted British authority, abandoned the slave trade, and cooperated with British officials in creating a new administration. Consequently, some structures of the caliphate, including the legal system, particularly matters relating to marriage, property, inheritance, and divorce, were incorporated into the colonial bureaucracy.[25] Through this process, the British generated patriotism among some elements in colonial Northern Nigeria and reduced local resistance.[26]

The success of indirect rule in Northern Nigeria, officials argued, was "so striking and the development of the Native Administrations was so rapid that the Colonial Office have adopted it as the basis of administration in every tropical African dependency where it is still not too late to introduce it."[27] It became the model for other European powers. A year before the outbreak of the First World War, Dr. Solf, the German colonial minister, who had visited Northern Nigeria, ordered that the Northern Nigerian model of administration be adopted in German Cameroons. This was followed by a visit to Kano by Von Raben, the German resident at Dikwa, to study the system. General Lyautey, a French administrator in northern Africa, followed the Northern Nigerian model in developing what was seen as one of the successful colonial administrations in the French African colonies.[28] The success of indirect rule in Northern Nigeria and its extension, with remarkable results in parts of the Southern Provinces, including Oyo and Abeokuta, was reason enough to attempt to replicate it in the rest of the Southern Provinces.[29]

Officials were convinced of the efficiency of the model. One such official noted: "It is not likely, in view of the results achieved, and of the backing of the policy by the distinguished officers mentioned above, that there could be anything wrong with the theory of indirect rule."[30] Another official concurred: "The policy of Lord Lugard in the Northern Provinces, of governing through the Fulani Emirs appeared so successful from the outset that the British Government decided to adopt it throughout Nigeria."[31] Lugard had argued that: "Principles do not change, but their mode of application may and should vary with the customs, the traditions, and the prejudices of each unit." The Lugardian ideology held that:

> The task of the administrative officer is to clothe his principles
> in the garb of evolution, of revolution; to make it apparent alike to
> the educated native, the conservative Moslem, and the primitive
> pagan, in his own degree, that the policy of the Government is not
> antagonistic but progressive – sympathetic to his aspiration and
> the guardian of his natural rights.[32]

It was Lugard's view that the so-called primitive tribes should be allowed to develop self-government and institutions along the line of those of the Muslim states without compromising their own local independence.[33] In fact, a delegation of the Empire Parliamentary Association, after visiting Nigeria in 1927 and 1929, recognized the inherent problem in enforcing colonial rule but also agreed that indirect rule offered the best prospects.

> The great problem of co-operation between European and non-
> European races also arises in Nigeria. It is being tackled there with
> at least as great energy and freshness of outlook as anywhere in
> the British Empire. The guiding principle is that known as indirect

rule. This envisages the extension everywhere of executive African administration, building on the units of African government
evolved locally by traditions and custom, and on the foundations
which Africans themselves have in the past laid down.[34]

The need to expand indirect rule to the rest of Southern Nigeria was not in
doubt, but the problem was how and when to do it. The British had the good
fortune to encounter centralized political institutions in northern and western Nigeria that allowed the successful implementation of the policy of native
administration. Although there were those who assumed that the chieftaincy
system, including village war leaders, which made the northern Nigeria experiment very successful, "existed with modification throughout Africa,"[35]
this idea was proved wrong among the Igbo and other societies in southeastern Nigeria, where the British imposed the indirect rule system in a radically
different context.

The Igbo, whom European ethnographers described as belonging to
stateless societies, did not share identical social forms, as existed among the
inhabitants of the northern and southwestern areas. With no hereditary chieftaincy institutions, save for a few exceptional cases such as that of Onitsha,
the British created the warrant chief system. Native authorities were created
by the regrouping of minor, less organized, units into provinces (divided into
divisions) headed by Residents.[36]

But Alex J. Braham, a district agent of the Royal Niger Company (1898–
1902), noted in his memoir that "Chieftaincy amongst the Ibos is rather based
on commercial principle, than on any courage, or special fitness of the leadership of men." He observed that "Any freeborn man possessed of land merely
has to acquire sufficient wealth to make certain presents, and arrange a fetish
feast; to which he invites the existing chiefs."[37] Although Braham confused
the acquisition of an honorary status such as *ozo* with chieftaincy, a lack
of centralized political authority dominated by a single individual was the
norm. Ormsby-Gore observed:

> There are no outstanding rulers of any importance, and it is not
> certain whether this state of affairs among the Ibos or Ibibios is
> due to the decay of a higher type of organization or whether they
> had ever evolved any organized form of government. As things are

now, they live in villages, and no one village bears any relation to, or recognizes any affinity with, the people of the next village who may be only a few miles away.[38]

The task of the government, Ormsby-Gore suggested, was to cultivate among these people "a sense of common interest sufficient to enable a native administration to be set up, bearing, if possible, a definite relation to the structure of the tribal institutions."[39]

Some British officials who worked among the Igbo echoed similar sentiments. Edward M. Falk, District Officer for Aba, noted that the so-called pagan, unlike the Muslim, had

> ... no institutions of which he is proud and jealous; he is being moulded like wax by his European teachers, he is as imitative of them as a simian and will retain little of his pre-twentieth century self but the vices which are inherent in the negro character to which are being added those of his conquerors.[40]

These biases lay deep even in the ivory towers of imperial Britain. The universities of Oxford and Cambridge played their part in the preparation of the administrative cadre by offering Colonial Service courses. An important component of the colonial officers' training was the acquisition of African languages. Hausa was the language of choice and it remained the lingua franca, even in the Nigeria Regiment as late as the Second World War.

Perceiving the Igbo political and economic system as primitive and archaic, colonial officials argued for a different rate of assimilation into a modern society for them. In the officials' view, the lack of chieftaincy and centralized political institutions among the Igbo called for a more measured approach to the task of incorporating the Igbo into the colonial state. Palmer remarked:

> It is true that permanent Chiefs comparable to the Emirs of the North or the Alafin or Oba of Benin do not exist. This is however not to be expected among people still in the "clan" stage – where the (Chief – Eze in Ibo) like the Arab Sheikh or Tuareg Amanokel is little more than the elder brother of the clan, the experienced

senior who is consulted in important questions, but not necessarily obeyed.[41]

Yet Palmer observed rudimentary forms of chieftaincy institutions among the Igbo that would serve as the basis for more rapid progress in the establishment of colonial institutions in line with existing traditional political institutions. "That there are Eze (Chiefs) of this nature all over the Ibo country is beyond question," he wrote. He maintained that

> The septs which make up the clans are the so-called "towns" or "countries" – The various towns in a clan speak of each other as "brothers," and invariably acknowledge one of their members as elder brother i.e. head or senior town. It will usually be found that the head of that town was the acknowledged leader of the clan in war, and entitled to certain perquisites when the clan went hunting and was the arbitrator in land disputes.[42]

Some colonial officials admitted the difficulty of implementing the system, yet they were more concerned with applying the idealistic philosophy of the native administration project than dealing with the contradictions inherent in the Igbo political system. Indeed the confrontations that would occur between the British and the Igbo when the system of warrant chiefs was imposed from the first decade of the twentieth century onward did not arise from a cultural misunderstanding on the part of officials, but an avowed prosecution of an imperial policy that had been elevated to the status of ideology.

Palmer noted the cautious approach in the introduction of British rule among the Igbo from the early days of occupation:

> Those responsible for the administration of the country sought to avoid the wholesome introduction of British Law and European methods. A means was sought whereby "Native Customs" should be applied in the settlement of Native Disputes, so controlled, as it were, by the Government that it should be "pure" and not "corrupt." District Commissioners felt and still feel the practical necessity of their having power to use common sense and summary juris-diction in dealing with primitive people. Courts called "Native

A colonial assistant district officer hearing complaints (Reproduced with kind permission of the Bodleian Library, University of Oxford).

Courts" were therefore designed to be … "Provincial Courts," in which the District Officer would always sit, but would not be bound by the letter of the English law when sitting there.[43]

J.G.C. Allen, who arrived in Nigeria in 1926 and served as a district officer in Eastern Nigeria, recalled in his memoir, "From the Government's point of view this system had the advantage of being decidedly more efficient than government by a vague and inchoate council of elders."[44]

Yet the process of selecting warrant chiefs was arbitrary and bizarre in some cases. Often the British appointed willing participants, most of whom had made their wealth in the palm oil business as produce buyers. One official, A.F.B. Bridges, who joined the colonial service in 1921 and became district officer for Onitsha, recalled in his memoir:

THE LAND HAS CHANGED

It was, therefore the practice, when a Warrant Chief was to be appointed, for the D.O. to visit the town or quarter concerned and hold a public meeting. At this meeting all candidates would be put forward and required to weigh their claims before the public, the D.O. endeavoring to weigh the claims against one another and gauge both the validity of the claims and the degree of support for each candidate. If one of them had some obvious standing as a descendant of the founder, he was likely to be chosen but whatever the claims it was usually the majority vote that decided the issue, unless there was evidence that the meeting had been rigged. Other things being equal, the people tended more and more as time went on to vote for candidates who were literate and knew something of the outside world.[45]

The British approach to elevating the people in Eastern Nigeria in political and economic terms was often contradictory to the philosophy of indirect rule that they sought to implement. Those on the ground realized very early on the difficulties of applying indirect rule among the Igbo. Another official recalled the difficulties of the early days of colonial rule among the Igbo:

The interpreters were either content to tell the Europeans what they thought they wanted to know or to extend a helping and well-greased hand to ... ambitious individuals, and the difficulties of the Ibo language and the timidity of the people prevented the Administrative Officers until many years later from learning the truth. Therefore, at the outset of the British regime each village was called upon to nominate its "Chief" who was promptly presented with a cap and staff of office and a warrant formally appointing him as a representative of the Government.[46]

Other cases did not fit this model. In the case of Umuchieze, a village in Mbaise, Owerri Province, for example, oral sources say that the warrant chief was a twenty-two-year-old man, Philip Eluwa, who, out of curiosity or stupidity, refused to run away when other members of the village took to the bush on the arrival of the British in his village. For this act of "bravery," he was appointed the warrant chief of the village.[47] Under normal circumstances, Umudionu, the oldest lineage in the village, would have produced the warrant chief. But Chief Eluwa was chosen from Uhuala, the last in the hierarchy among the four lineages that make up the town. J.G.C. Allen, a former district officer in eastern Nigeria, recalled similar circumstances. In some cases some powerful men "persuaded their people to nominate them but in others the people, fearing a trap, selected the most insignificant individual in the village – in some cases even a slave – in the hope thereby of being freed from official interference with the village administration." Allen further noted: "The Government, believing that they had won the support of the traditional authorities exaggerated their status and granted them an increasing share in the day to day administration of the area."[48]

At the centre of the native administration system were the Provincial Courts, operated by European administrative officers, which tried serious offences or any other cases regarded as repugnant to Western moral concepts, even if they were accepted under local law and custom. The native courts, which dealt with offences against customary law or similar civil cases became part of the new administrative changes in Igboland. The warrant chiefs adjudicated cases at the native courts. The new authority given to the warrant chiefs and enhanced by the native court system led to the exercising of power unprecedented in pre-colonial times and challenged the core of the Igbo traditional political system – government by consensus – in which decisions were made by protracted debate and general agreement. The men who became warrant chiefs, one official wrote,

> Soon enjoyed power and authority to which they possessed no traditional right or title and most of them created an entirely artificial administrative structure in which the elders were pushed aside and authority was granted to a cadre of "Headmen" whose only qualifications [sic] for their position were a tough and ruthless personality and a capacity for blind obedience to the dictates of the "Warrant Chief."[49]

Chiefs sitting in court with a district officer (Reproduced with kind permission of the Bodleian Library, University of Oxford).

These headmen known as *idimala* (a corruption of headman) in some areas, became as arrogant as the warrant chiefs and a source of exploitation of villagers. In some areas, the "Warrant Chiefs established a form of personal dictatorship and the traditional administrative system started to disintegrate."[50] Their arrogance and intimidation are still remembered. Onyegbule Korieh remembers these men as gods. "They seized other people's land, bicycle, and took young women without paying bridewealth. Some took other people's wives or other properties that they fancied."[51]

There were practical reasons for adopting the warrant chief system and for the relatively high level of support given to the new administration by many of the newly appointed warrant chiefs. An official noted that the warrant chief system and the judicial system that was adopted "formed a cheap and expeditious form of justice."[52] African officials had some practical reasons for embracing the system, since it offered opportunities for the accumulation of wealth and the achievement of high status in societies where such desires fitted into the social dynamics of the Igbo. Part of the tax collected went to the native administration and for the payment of the chiefs. The political changes that occurred due to the introduction of native authorities and the creation of the warrant chief system created a class of wealthy individuals by the 1920s. The chiefs, who had become salaried officials of the colonial government, reinvested their wealth in the produce trade and invested in people by marrying several wives and having numerous children. They built mansions roofed with corrugated iron sheets. Court clerks whose "opportunities for talking bribes are great," as A. E. Cooks, district officer for Owerri District wrote in his memoir, also exhibited their newfound wealth by investing in wives and building large homes.[53]

Falk, who became the district officer for Aba in December 1920, observed the changes during the decades following European colonization:

> Rapidly increasing wealth is everywhere raising the standard of comfort with amazing rapidity. Chiefs who a few years ago could hardly boast of the possession of a decent garment to cover their nakedness now dress in the latest European style complete from sun helmet to patent leather boots and drive about in their own motors. The progress of the people apart from the villagers who still till the soil is corresponding.[54]

Palmer recalled that some form of strong individual authority in the villages was undoubtedly necessary in the early days of pacification, but the administration had not "divined that it had quite inadvertently and with the best intentions imposed on the hapless villagers a form of autocracy entirely foreign to their custom."[55] Many colonial officials, like Palmer, admitted that the native courts had, however, "developed in a manner quite contrary to the spirit in which they were designed."[56] Indeed, J.G.C. Allen noted that many of the warrant chiefs "led a very precarious existence the perils of which were not always offset by their illicit gains." Many of them, he noted, "died suddenly in suspicious circumstances and although their bodies were exhumed and examined it was never possible to discover sufficient evidence to support a charge of murder."[57]

The new administrative structure was very patriarchal, despite the visible role played by women in both the economy and politics. While some Igbo men participated in the local administration as interpreters, messengers, policemen, army recruits, and warrant chiefs, the British ignored women. Colonial anthropologist, Sylvia Leith-Ross, commented on the women and on the failure of the colonial administration to recognize their role in Igbo society:

> It is a pity that, from the beginning, some executive local powers had not been given to them. At that time, the men would not have resented it, as it would have meant little more than official recognition of powers the women already possessed. Now, too many vested interests are involved.[58]

This created significant problems for the administration and resentment among the Igbo. The 1929 Women's Revolt and other protests (discussed fully in chapter 4) were largely a reaction to the warrant chief system among other issues, such as taxation and the Great Depression.

Overall, some officials, notably Palmer, acknowledged the ideological consequences of native administration on societies such as those of the Igbo. In his view the attempt to preserve native custom "have so far resulted in steadily destroying it and among these relatively primitive peoples full Europeanized individualistic government is being introduced.... The Government machine is steadily grinding to powder all that is 'Native' and transforming the people into 'black Englishmen.'" Complimenting the state on effecting this change, he argued, included the "spread of Missionaries and Education."[59]

There were those who disagreed with Palmer's assessment. The commissioner of Calabar Province did not see any wisdom in challenging a policy formulated by officials highly regarded within colonial circles:

> It will be interesting to ascertain if when making such a sweeping condemnation of the Government's Policy, the writer considered who were responsible for the building up of the existing system of administration. The names of Sir Claude MacDonald, Sir Ralph Moor, and Sir Walter Egerton might have occurred to his mind. All these were men of mark and each one continued the policy on the lines of his predecessor.[60]

Reiterating the so-called backwardness of the Igbo, the commissioner remarked:

> It is obvious that the inhabitants of the Eastern provinces are of a much inferior mental caliber than those in the West of the Niger.... This fact was very quickly ascertained by former Administrators of the protectorate (old Southern Nigeria) and the system of native courts instituted for the simple reason that the object aimed at was to try and educate the native and to bring him on a higher scale also by the said system of Native Courts to make the Chiefs (so called) feel that they were Chiefs not only in name but as such were responsible for the welfare of their country. To this end not only successive Governors but all Political Officers have striven and striven hard. Mr. Palmer beyond his tour through the Provinces has I believe no experience of the type of native his report deals with.[61]

This reaction to Palmer's observations points to the extent to which policy was driven by ideology and why it witnessed little in the way of a shift until the 1929 Women's Revolt.

The new structures and institutions, nevertheless, became important instruments for the expansion of an economy based not only on the production and marketing of palm produce but also on the greater opportunity offered for marketing an extensive array of foreign commodities.

THE DEVELOPMENT OF AGRICULTURE

The potential wealth from the trade in palm produce was not lost on British imperial and trading interests. Imperial policies and the rhetoric of the era bears this out. The opening of the Igbo country or what Governor Sir Walter Egerton described as the "richest part of the country"[62] led High Commissioner Sir Ralph Moore to predict that effective occupation of the interior produce market would "result in largely increased prosperity both to the Administration, the commercial community, and to the natives themselves."[63] The quotations above reflect the attitude of European officials as they embarked on the mission to civilize and develop the region in the early part of the twentieth century.

Indeed, advocates of the abolition agenda of the mid-nineteenth century outlined the revolutionary potential of so-called legitimate commerce. To them, the trade in agricultural and other commodities had the potential to liberate African slaves, provide opportunities to all and sundry to participate in commerce, increase accumulation, and modernize African economies. While a few economic and political leaders dominated the slave trade, the trade in oil was not capital intensive. Wealth, therefore, could be gained from individual enterprise.[64] Although the new trade was influenced by the existing social, political, and economic structures, access to oil palms and labour remained the most critical factors in the new trade for many communities in the Bight of Biafra and its hinterland, where the transition to the palm oil economy necessitated structural changes to meet the labour required for oil production. New economic relations developed upon the ashes of the old trade in slaves, but palm produce was a driving force for the economic and political developments in this region from the mid-nineteenth century. The quantity of palm produce exports after the abolition and the expansion from the period of colonialism reveals major changes in the rural economy in response to new demands. But this does not in itself tell us much about how this expansion was achieved. The development of trade and commerce in the post-abolition era did not depend upon local initiative alone. It was achieved through the double imperial ethos of "civilization" and "commerce." This imperial ethos had fundamental implications for both the socio-economic and the political structures of the Igbo and other African communities.[65]

Colonial officials were obsessed with the agrarian question. To many officials, this was the main purpose of imperial acquisitions in Africa. Therefore, until the later development of minerals as an important part of the colonial economy, the African population, the European administrators, and the European traders were linked in the desire to expand African production of cash crops. Although the Igbo and other peoples of the Bight of Biafra hinterland were already exporting palm oil and kernels from the time of the abolition of the external slave trade, the incorporation of the region as a British colonial possession created a new market for palm produce and increased access to the world market for many small-scale producers.[66] As in many other societies in western Africa, relations between Europeans and Africans during the colonial period were rooted in the attempt to transform agriculture. From the beginning of the twentieth century through the end of colonial rule, officials and local farmers alike redirected local labour and resources to meet the demands of European markets for palm oil and kernels.[67]

The imposition of British rule in Nigeria was followed by the acquisition of land by the Forest Department, the precursor to the Department of Agriculture.[68] But the development of colonial agriculture can be traced to the work of Sir Alfred Maloney and the establishment of a botanical station at Lagos in 1887. The station started with a modest expenditure of £300 a year, and its superintendent was expected to grow different kinds of useful trees, plants, and herbs as well as a model kitchen garden.[69] In 1888, the Royal Niger Company began a plantation at Asaba, which was transferred to the government in 1901, in addition to a station at Nkissi. A garden was started in 1903 at Calabar under the administration of Sir Claude MacDonald. Advances made between 1903 and 1905 led to the formation of the Botanical, Agricultural, and Forestry Departments in the colonies of Lagos and Southern Nigeria under the headship of Mr. H. N. Thompson. By 1910, the departments were separated and the Agricultural Department became autonomous. The new department focusing on agriculture was firmly established from this time onward under the directorship of Mr. W. H. Johnson, who rapidly developed it into one of the most important departments in the colonial bureaucracy. After the amalgamation of Southern and Northern Nigeria, Mr. Johnson remained the director of agriculture in the Southern Provinces.[70] The agricultural structure began to assume new forms in a European colonial society obsessed with rural transformation and an African society forced to adapt "modernity" and a new consumer culture.

The advance into Igboland was prompted specifically by the desire to remove barriers to the palm oil trade. The colonial government took steps to encourage the production of palm oil and kernels among the Igbo and other communities lying within the palm oil belt of what became Nigeria. Most of southeastern Nigeria was particularly suited to the expansion of the palm industry. Wild groves dotted the landscape and little capital was required to produce palm products in minimum commercial quantities. But the transformation of the local agrarian society was the result of both local initiatives and the aggressive attempt by colonial officials to create a peasant society dependent on the production of agricultural produce. Fredrick Lugard, the first colonial governor of Nigeria, summed up British economic interests: "It is in order to foster the growth of the trade of this country [Britain], that our far-seeing statesmen and our commercial men advocate colonial expansion."[71] This expansion was essential because revenue from peasant production would support the local administration and provide the essential raw materials for British industries.

The expansion of colonial control and the development of roads and transport stimulated peasant production of palm oil and kernels. As early as 1902, colonial district officers were employing forced labour to construct roads, courthouses, and government rest houses. H. M. Douglas, for example, who joined the colonial service in 1894 and served in various capacities until 1920, personified what Felix Ekechi called the "uncrowned monarchs" and imperial autocrats. Douglas's role in planting British rule in Eastern Nigeria, including the use of forced labour to construct roads, is legendary. His philosophy of opening up the country through the construction of roads yielded significant results. By 1906, when he was transferred to Onitsha as district officer, Douglas had constructed over two hundred miles of motorable roads linking different parts of the Owerri district, earning him the nickname of *Beke ogbu ama*, the road-building Whiteman.[72]

The *Pax Britannica*, bringing new incentives and opportunities to expand domestic production for the market, was accompanied by increased production of palm oil from the Igbo region. In 1905, for example, Governor Walter Egerton, visiting Oguta, reported that there was an "Enormous quantity of oil in the store, ready for export on the rise of the river."[73] By 1914, exports from Owerri Province were valued at over one million pounds.[74] Opening up Igboland attracted European trading firms to Oguta, Owerri, Owerrinta, Ife, and other places. Coastal traders from Okrika, Bonny, and other areas also followed as the region opened up with the spread of roads.

Control over local production was achieved through political and administrative measures. Governor Hugh Clifford was clear about this when he wrote in 1920: "The administration was not to engage in commercial enterprises of any kind but would prepare and maintain the conditions – political, moral, and material – upon which the success or failure of such enterprises depended."[75] This policy was in response to the attempts of a Liverpool industrialist, William Lever, to develop oil palm plantations in Sierra Leone and Nigeria.[76] In a speech delivered at a dinner in honour of Sir Hugh Clifford, governor of Nigeria, at Liverpool on 9 July 1924, Lord Leverhulme, used the occasion to attempt to persuade the colonial authorities to give freehold rights to Europeans for the development of palm oil plantations in West Africa. The Leverhulme experiment in Sierra Leone had shown that plantations could not be successful without a carefully controlled labour force and secure land tenure. Leverhulme argued: "You cannot have prosperous business without some security of the capital invested in it." He charged: "We had not [sic] right for the palm-trees in Sierra Leone. We had no rights to collect the fruit ourselves or to force the Natives to collect it, with the result that there was such an irregular supply of fruit that when our 20,000 [pounds] capital was exhausted we packed up and went away." He reminded his audience of the difficulty facing European investors and of a widely held assumption about the African producers: "I don't know of better material anywhere for labour in the tropics than the Natives of West Africa, but they are not organizers."[77] In Leverhulme's view,

> Whatever merit the African native has, it has been proved by the opportunities he has had in the United States that he has not got organizing abilities. Now the organizing ability is the particular trait and character of the white man.… Do you think we shall be allowed to keep our present position of responsibility to the countries within the Empire if we do not make some organized effort to develop them on sound lines? That can only be done with capital, but capital will not flow without security. There is no rabbit so timid as your capitalist. The remedy for trade stagnation is to restore confidence. Leave our business men alone, but give them security in all that is reasonable and right. Our immense Empire will only be held if we administer it for the best interest of the people, on a wide, broad policy, without maudlin sentimentality or brutality.[78]

Indeed, Leverhulme stated with much confidence that "the African Native will be happier, produce the best, and live under the larger conditions of prosperity when his labour is directed and organized by his white brother who has all these million years' start of him."[79]

Critics like Governor Hugh Clifford were opposed to the creation of European owned and managed plantations for both political and economic reasons. They believed that local agricultural production had a "firmer root" in the hands of the local farmers since they would be self-supporting in terms of labour. Moreover, the prevailing land tenure system and the population density of the Igbo region, which was as high as 1,000 per square mile in some parts, would impede the plantation system. Lugard also recognized the peculiar population problem in Eastern Nigeria. Unlike Northern Nigeria, where the Native Ordinance of 1910 vested the freehold of the land in the governor, in the east the colonial government adopted a laissez-faire attitude to the land question. The Land and Native Rights Ordinances of 1910 and 1917 ensured that land for agricultural production remained in the hands of the local farmers.[80] The implications of the plantation model for the eastern part of Nigeria were clear in 1932 when the director of agriculture stated:

> Speaking generally and broadly, the people of the "palm belt" have already not enough land for food crops. If they start generally to plant palms on their farmlands, they will have to cut down the groves and use that land for farming. Such a change-over might not compare quite so badly with grove replacement as might be thought: for ordinary farm crops can be and are grown between the young palms on a "straight plantation" for two or three years, whereas the shade from the old palms in groves under direct replacement ... is too heavy.[81]

In addition, conditions in West Africa were different from those in the Belgian Congo and East Indies where successful experiments had paid dividends. Oil palm plantations dominated by European firms would have created a class of landless labourers and social problems for the colonial authority in West Africa.[82] The government was also wary of providing compulsory labour for private profit or protecting monopoly rights, as was done in the Congo. Plantation experiments in the former Gold Coast and in Sierra Leone had shown that European-owned oil palm plantations could be profitable without government assistance and subsidies.[83] Peasant production seemed the cheapest method for officials and gave local people the ability to pay their taxes in order to assist in maintaining self-sustaining colonies. Colonial officials were convinced of the built-in efficiency of peasant production, for local producers were meeting the needs of European industries.[84]

For most of the colonial era, local farmers undertook production and trade in palm products with simple techniques often described as primitive and crude.[85] Yet great strides were achieved in palm oil production, which remained the single most important export product for the Igbo and the country as a whole. Indeed palm oil production and export had shown a steady increase in Nigeria from 1901 to 1943. The production of kernels also doubled from the 1911 levels and in 1942, "exports were three times those of 1901–1905 period."[86] A delegation of the Empire Parliamentary Association, which visited Nigeria in 1927/28, described the oil palm as "the axis of the prosperity of the West Coast rainbelt."[87] In their view, there will be an obvious link between the "prosperity of the whole palm belt ... and the power of absorption of British goods."[88] In 1920, Nigeria produced about 52,771 tons of oil, valued at £1,655,914. About 153,354 tons of palm kernels, valued at £2,831,688 were produced in the same year.[89] Exports of palm oil and palm kernels increased to 246,638 tons and 127,111 tons respectively in 1928,[90] contributing 62 per cent of Nigeria's export earnings.[91] The value of exports had increased more than sevenfold and export volume multiplied fivefold by 1929.[92] Revenue from palm kernels increased from £3,189,000 in 1921 to £4,429,000 in 1930. Government revenue from palm oil also increased significantly, from £2,520,000 in 1921 to £3,375,000 in 1930.[93]

Amid these broad economic shifts, daily life in the countryside remained much the same. Considerable human energy was expended in extracting the thousands of tons exported annually until the development of mechanical crackers. The increase in export quantities during this time was the result of large-scale exploitation of wild palm groves that drew on the labour of men, women, and children.

Oil extraction was a labour-intensive, laborious exercise. The children removed the nuts from the husk, and the women directed the extraction of the oil. In the soft oil process, the nuts were boiled in water until they became tender. They then were pounded in a big mortar (*ikwe nkwu*), and the nuts were separated from the fibre. The resultant fibre was pressed by hand to

Table 2.1. Exports of palm oil and kernels; five-yearly averages.

YEARS	PALM OIL (TONS)	PALM KERNELS (TONS)
1901–05	56,740	125,432
1906–10	70,435	143,072
1911–15	76,099	170,301
1916–20	82,868	195,305
1921–25	99,003	216,204
1926–30	124,233	252,839
1931–35	123,667	285,131
1936–40	135,501	314,286
1941	127,777	378,120
1942	151,287	356,588
1943	135,268	331,292

Source: RH, Mss Afr. s. 823(4), R. J. Mackie Papers.

extract the oil.[94] In the hard oil process, the fresh nuts were pounded, and water was then poured over the pulp. The resulting surface oil was skimmed and boiled, and the oil was extracted.[95]

Producing kernels was even more tedious as each nut was cracked between two stones to extract the nut inside the hard outer layer. Women and children obtained kernels by cracking the nuts between stones.[96] I can relate this to my own personal experience. The drudgery of cracking palm kernels between two stones was one of the tasks I most hated as a child.

The division of labour involved the allocation of routine and light work to women and children and the assignment of more arduous, "dangerous," and heavy duties to men.[97] However, women's work was much more varied than men's work. Christiana Marizu remarked that "It required a lot of work and effort to produce even the smallest quantity of oil.... You start by fetching the wood and water even before production can begin."[98]

Some early European administrators remarked on the changing landscape as the Igbo encountered European rule and embraced the new consumer culture that came in its wake. A major stimulus for African production was the desire for European goods. Raymond Gore Clough, a European trader, had suggested that Africans should be induced to bring their oil to the traders by stimulating their interest in "imported goods and creating a desire ... for all the attractive articles displayed in the factory store."[99] By the 1920s, the African population was demanding a greater variety of goods, beyond salt, gin, tobacco, stockfish, and cloth, which had been the predominant articles of trade in the past.

Edward Morris Falk, who became the district officer for Aba Division in December 1920, noted the changing structure of the Igbo economy:

> The native produces food stuffs, palm oil and kernels. He is a consumer of a long list of European commodities such as clothing and textiles, liquor, imported food stuffs, ironmongery, crockery, kerosene oil, gunpowder, soap, matches, tobacco, camp equipment, cheap imitation jewelry, cycles, in fact anything which the natives sees the white man use from gramophones and sewing machines to tinned salmon or boot laces.[100]

The buoyancy of the palm produce trade in the post–First World War era created a demand for tools and equipment that aided practical skills and enhanced leisure. Raymond Gore Clough, trading on the Oil River, recalled in his memoir that African traders and produce agents were demanding sewing machines, phonographs, and bicycles for resale to an increasingly prosperous African community.[101] But to many Igbo people, these items were not regarded as luxury goods. Michael Echeruo has argued that "Typically, the Igbo did not expend their meager resources on these essentially luxury goods. However, their availability meant that some members of the community that could afford them became visible and triumphant successes. What was otherwise considered as wasteful frills became marks of real success to be emulated."[102] Yet only a very few people invested their money in such luxury goods as radio and phonographs. On the other hand, bicycles, the most common means of transport, were used for haulage and trade. A. E. Cooks, district officer for

Owerri, noted that the bicycle was an important factor in the economic life of the Igbo in the 1920s:

> Not only is it used to transport palm oil in four gallon petrol or kerosene tins and palm kernels in sacks from the bush markets to collecting centres on the main roads where the produce is transferred to lorries for export, but it is also used extensively in internal trade.[103]

An important trade developed out of the introduction of bicycles. Cooks observed that bicycle repairing "developed into a major industry and repair shacks are to be found dotted all over the division, especially at cross-roads, large markets, etc. The Owerri Ibo is a genius at improvisation and it is astonishing how long he manages to keep a bicycle on the road considering how grossly he overloads it."[104] The bicycle was increasingly also used for passenger traffic, including in the urban towns.

By the 1920s, most of southeastern Nigeria had been drawn into the expanding market for oil. The palm produce trade was the means through which many households met their basic need for food and other expenditures. Eugenia Otuonye recalled that she relied on income from palm oil and kernels to clothe and feed her children and pay their school fees. "Many people had no other means," she said.[105] Cooks confirmed this:

> In the eastern area of the division generally known as Nguru area, the people are too thick on the ground to live off it and are obliged to buy large quantities of foodstuffs, mainly cassava, from other areas. Fortunately, this area is rich in oil-palm and it is mainly from the sale of oil and kernels that the people obtain the cash to buy the food they cannot grow.[106]

Agricultural development stimulated changes in production relations and internal trade. The large increase in male participation in the produce trade led to the creation of new economic opportunities. The development of the commodity trade led to the rise of produce agents, mostly men, who also controlled the supply of European goods.[107] Women engaged in the new trade as producers and distributors and supplied the increasing urban population with foodstuffs. These changes associated with the production of palm oil and kernels were unquestionably the most dynamic features of rural economic life and accounted for the growth in per capita income throughout the colonial era.

Attempts at innovations in production methods were made from the late 1920s to encourage production and improve the quality of oil. Experimental farms were opened at Umudike in 1926 and Nkwere in 1930.[108] From 1927, the department of agriculture introduced a policy of encouraging the development of locally owned palm oil plantations. Twenty-seven acres of experimental palm plots were established in the Aba Division in 1927. There was a substantial rise in the number of locally owned plantations from 9 in 1928, covering 70 acres, to 88 in 1931, covering 377 acres.[109] The number of plantations rose to about 200 by 1932 in the Southern Provinces.[110] However, progress was slow as rural farmers viewed government-backed programs with suspicion. F. D. Carr, the colonial Resident for Owerri Province remarked that while "agricultural officers strove valiantly and set up model and experimental farms and though interest in them was considerable they were, generally, written off by the Ibo farmer as being excellent for the European but of little practical use for him."[111] The lack of enthusiasm for oil palm plantation was strongest in Owerri Province, a region with a very high population and very little land. Chief Enweremadu, a local administrative officer during the 1930s, remembers that there was fear among rural farmers that they could lose their land and oil palms to the government.[112] The director of the agricultural department confirmed the high level of apprehension among rural farmers when he wrote: "So strong has this sentiment proved that repeatedly individuals or families have decided to try planting palms, but have later given in to the strong adverse public opinion of their neighbors."[113]

There were, of course, other factors that explain the lack of enthusiasm for innovation among Igbo farmers in Owerri Province besides the fact that Carr describes the Igbo as "intensely jealous of his land and strongly and violently resented any form whatsoever of Government intervention."[114] The environmental and demographic factors in the province also explain this lack of enthusiasm among farmers. In fact, Carr acknowledged that progress in improved methods of agriculture was slow due mainly to the system of land tenure and consequent fragmentation. In the region, the need for improved

methods of farming was obvious, for the rapid increase in population necessitated more intense cultivation, which had seriously depleted the fertility of
the soil. Carr recalled that by the 1920s the average size of yams, the staple
diet, "had decreased by almost half and the less demanding coco-yam and
cassava [sic] were rapidly becoming more and more in evidence."[115]

Local producers were also encouraged to purchase hand presses for the
extraction of the oil.[116] The Duchacher press, manufactured by a Luxembourg
firm, proved to be most satisfactory in terms of cost, quality, and quantity
of oil extracted. About 80 of these presses, capable of extracting oil with a 5
per cent lower free fatty acid (FFA) content were in use by the early 1930s.[117]
However, there was no large-scale adoption of these presses by local producers, despite the improvement in quality and quantity of oil produced with
these presses, because the oil producer did not make sufficient extra profit to
warrant such an investment.

Attempts were also made to increase the quality of produce meant for the
European market. In 1928, the government introduced produce inspection
– enforced grading and inspection of palm oil and kernels by government
agents to ensure quality. The government considered exempting plantation
quality oil from export duties following the demands of the United Africa
Company, which had argued that it could not compete with cheaper products
from Malaya and Sumatra. The issue of export duty was problematic for the
government in Nigeria. Downing Street was of the view that such an exemp-
tion should extend to Africans who produced products of similar quality.[118]

Colonial efforts to restructure the local economy achieved success due to
the collaboration of local people. Most Igbo peasants embraced the opportunities provided by the colonial economy because they were drawn inexorably
into it. As many commentators have noted, colonial demands of taxes and
labour imposed added burdens upon peasant households, although crop sales
and wages also provided them with cash.[119] Still, the change in the character of
the economy was not achieved by any large-scale transition from subsistence
economy to large-scale appropriation of land devoted to palm oil production,
for two main reasons. First, the palm oil economy was already well developed
following the abolition of the Atlantic slave trade and the transition to commodity trade based on palm oil. Second, the history of combining subsistence
agriculture with market production within a physical environment littered

with palm groves had proved effective in meeting European demands. Only in a very few cases were significant amounts of land devoted to planting new oil palms.

Although many farmers did not devote whole plots of land to oil palms, they took risks when the prospects for gain justified this. A small number of businessmen who were in a position to acquire land and risk capital answered the call of the agricultural research departments to invest in commercial oil palm production. The introduction of motor transport contributed significantly to this expansion as African businessmen invested in the transport sector. By the 1920s, trucks were moving people and goods and bringing the whole region into the vortex of the world market. Rural farmers understood the importance of the palm oil industry and cash crops to their economy. They devoted more time to the care and protection of wild oil palms on their plots and often planted new ones on the boundaries (oke) that demarcated individually owned plots of land. However, struggles over ownership of palm trees and land, Onyegbule Korieh noted, "became more frequent."[120] Some rural farmers also often planted a few cocoa or coffee plants on compound plots as the demand for these products increased.

At the same time, rural farmers experienced major changes in kinship and community organization, especially in access to land. Under the traditional tenure system, pieces of communal land were divided among each family in the town, and heads of families would in turn allocate land to their own family members. British officials observed in the 1920s that considerable changes in the land tenure system had already taken place among communities in Owerri Province, resulting in frequent reports of land disputes in colonial reports. "Land cases and disputes are of annual occurrence, this year though numerous, no untoward event has occurred," the Resident for Owerri Province wrote in 1920. Mr. Ingles, writing from Okigwe Division, noted: "The Division has been quiet with the exception of a few minor disturbances – chiefly over land claims." Mr. Ferguson in Owerri Division, observed that the "usual land dispute, customary at this season caused a certain amount of excitement at the time, but died away when yams have [sic] been planted." The communities in Aba Division had also seen transformations in access to land by this period.[121]

The ethos of village life and the values attached to farming were also changing rapidly. By the second decade of colonial rule, many young men were shunning farming for the prestigious and attractive opportunities in the expanding colonial towns and in the public service. New trade patterns, migration, and new consumer habits were disrupting the old kinship structures that redistributed wealth, as individuals chose to leave the village to seek their fortunes elsewhere. For younger people in particular, migration enabled them to escape the control of the elders in the villages, but many named rural poverty as their reason for migrating to urban as well as rural areas in other parts of the country. Over time, women effectively began to take up the opportunities offered by the emerging urban areas by establishing small retail shops or establishing themselves as prostitutes. "A purely consuming class of Government employees, mission school teachers, motor drivers and other callings is only now arising," an official noted:[122]

> The greatest danger ahead is that the number of people who have taken to other means of making a living than farming is rapidly increasing. Experience teaches that the school boy who has learnt the rudiments of writing or those in the services of Europeans will not go back to the land to work on it though they may go back to their homes to trade or live on their wits as letter writers or money lenders. As a result, there is a rapid increase of consumers and of non producers. The people are flocking to the larger settlements where money is made easily by honest and dishonest means without physical effort. The Southeast is thus faced with a serious food problem, which may prove to be a difficult matter to solve in the future. Already the prices demanded for locally produced food stuffs are soaring upwards.[123]

Trade and social relations increased in the aftermath of the establishment of British colonial control. Significant trading links developed between the Igbo interior and coastal middlemen who established trading posts along the water highways in the interior of the Bight of Biafra. The emergence of the so-called "water sides" connected hinterland producing areas to the coastal towns of Okrika, Bonny, Calabar, and other trading posts. The booming oil markets attracted coastal merchants, and thus established not only economic relations but also long-lasting cultural connections. Coastal merchants and the burgeoning numbers of local traders in the interior helped to create a fluid and multicultural town identity in trading posts such as Oguta and Ife, along the Imo River. The story of Joshua Adinembo, who established himself as a produce merchant after the First World War at Ife in Mbaise, is typical of the way the new trade created new forms of economic and social relations. An Izhon (Ijo) oil produce merchant and trader from Okrika, by the time of his death in the 1940s, Adinembo was married to several women, many of whom were from Ife and Ezinihitte along the Imo River in Igbo-speaking territory.[124]

FOOD PRODUCTION

Colonial officials did not show much enthusiasm for the development of local food production. Reporting on the food situation in eastern Nigeria as early as 1912, the director of agriculture noted:

> In some districts and even divisions, we have not yet been able to get even one tiny plot of palms planted by any farmer. And it is in these areas, with their heavy populations of trees and palms that the matter is of most urgency. This is not merely because this is the "palm belt" *par excellence*, nor even merely because neither the people nor ourselves know of any other export crop which will succeed on these poor acid soils. A more important reason for the urgency of this work in these areas lies in the poor food crops that the poor soil yields and their inadequacy to feed such a heavy population. If the yield of the areas that are occupied by palms could be doubled, as they easily could by the substitution of plantations for wild trees and groves, then more land would be available for food crops or much more money would be obtained wherewith the people could purchase food imported from other parts of Nigeria or from abroad.[125]

In his memoir as a colonial administrator, which included service in Aba District, Edward M. Falk observed the decline in the production of food items and the lack of interest in rural life and agriculture on the part of young people. "So low is the output of the individual that the question of feeding the non producers of food stuffs is already becoming a serious question," Falk observed in the Aba District in 1920.[126] This grim assessment of the food crisis only worsened over time.

O. T. Faulkner, one-time director of the department, admitted in 1922 that the department was focusing most of its attention on export production without a corresponding interest in food production. Faulkner explained:

> In reviewing the appearance of work, on which the department has embarked, it may appear that it deals too largely with products, which are, or might be exported ... but it must be remembered that it is much easier for us to help the farmer to take advantage of the changed conditions resulting from opening to him of the world's market, than to teach him how to improve his method of growing his foods – methods which are the winnowing of centuries of experience, sifted to suit his ancestral conditions by generations of his forefathers.[127]

In a report written after his visit to West Africa in 1926, Ormsby-Gore clearly argued for a balanced agricultural policy that would include food production:

> The foundation of sound agricultural policy must necessarily begin with the production of food for the people. Food comes first, and economic crops for export should come second.... The development of economic crops at the expense of production of food for local consumption is most undesirable, and a plentiful supply of cheap food, both for the native and non-native inhabitants of the countries, is the first essential.[128]

Officials did not often follow Ormsby-Gore's advice.

Productivity in the area of local foodstuffs declined rapidly following colonial intervention. The Chief Secretary's Office acknowledged that the decrease in the production of local foods resulted in higher prices for these food items.[129] In its determination to maintain the momentum and expand the production of other cash crops, the government ignored the threat that the expansion of the oil palm export sector posed for rural farmers and food security.

The rising cost of food was also linked to climatic and structural conditions. Colonial reports from the first decade of the twentieth century show an erratic rainfall pattern that often affected the fortunes of local farmers. This was a recurring pattern in the years for which records are available. Diseases such as the influenza epidemic of 1918, officials reported, forced farmers to abandon their farms.[130] The First World War, the annual report of 1918 noted, combined with other factors to increase inflation and the cost of local foodstuffs in most alarming manners.[131] The report for 1919 is worth quoting in some length:

> The much reduced import of foodstuff, deficient rainfall and the inflation of the currency have been effective together in continuing the increase in the price of agricultural foodstuffs, and the heightened cost of these has been further operative in generally increasing the price of native foodstuffs not directly of agricultural origin, such as smoked fish and meat produced by the natives. The importance of the latter circumstance lies in the fact that the consequently reduced availability of such products for purchase by the poorer classes leads to a decrease in the important protein, rendered all the more difficult to make up owing to the fact that prices of the chief vegetable protein bearers, beanstuffs, have in turn been greatly enhanced.[132]

In spite of the developments in the rural economy and the emergence of a relatively viable peasant economy, poverty and impoverishment were evident in the colonial period. By the 1930s, for example, reports of men pawning themselves, their spouses, or their children were rampant as peasants struggled to produce for the market and to meet colonial demands for taxes.[133] Although pawning was practised in pre-colonial Igbo society, Linus Anabalam confirmed that there was an increase in pawning due to colonial demands for taxes.[134] Some pawns lacked access to labour to exploit the land, as John Illife has noted. Others were young, able-bodied men, who lacked access to land (or other resources) and were unable to sell their labour power at a price "sufficient to meet their minimum needs."[135] J. S. Harris has shown that lack of access to land forced many to become casual labourers in the 1940s.[136]

The demographic features of the region had economic implications. In 1921, barely two decades into the colonial period, there were nearly 4 million people living in Igboland.[137] C. K. Meek, who studied the Igbo in the 1930s, noted that the high population density found in some parts of the region was a major source of economic problems. Using the Nigerian census estimates of 1921, Meek noted that there was already a shortage of farmland among the Oratta (Uratta) and Isu groups in the Owerri area. He observed that the population might increase "beyond the limits of the productivity of the land, thereby giving rise to a non-agricultural floating population with nothing to give in exchange for the food it requires."[138] Margaret M. Green, who studied the Agbaja Igbo in the 1930s, remarked: "The pressure of population on land is so great that every square yard of it belongs to definite small land-owning groups which may consist three or four males."[139] In 1952, for example, Okigwe Division, with an area of 587 square miles and a total population of 422,706, had an average population of 720 persons per square mile. Orlu Division had an average of 873 persons per square mile. In the areas around Owerri, the density was no less than 500 to the square mile.[140] According to Udo, the most densely populated areas included Mbaise and Ikeduru in central Igboland, and northern Ngwaland where rural densities of over 1,000 persons per square mile were common. The Onitsha and Awka areas of Igboland were already facing food insecurity by this period. They depended on other regions to meet their subsistence needs.[141] Although the crops and the farming system were similar to those of other parts of Igboland, Udo notes that palm oil was very important in ensuring the survival of the dense population in the uplands of Awka and other parts of Igboland.[142]

Overall, the colonial government's agricultural development strategy was seen as the quickest way to introduce to Nigeria the benefits of "civilization" and modernization. This was the strategy of Hugh Clifford (governor-general of colonial Nigeria, 1919–25) for the economic development of the colony. Yet the Western ideologies of modernization were problematic for African agriculture because they neglected local production systems and often imposed a top-down approach to modern farming.[143] The perception that the local farmer was primitive and resistant to change stems from a narrow understanding of the socio-cultural context that influenced economic production. Rural farmers tried to balance their risks. But according to G. I. Jones, the "experts" made no allowance for the fact "that the Nigerian farmer and his family have to produce enough food to support a large group of people and cannot therefore afford to take risks."[144]

Yet the local agricultural economy was more sophisticated, developed, and adaptive to changing socio-economic conditions than colonial commentators thought. The growth in the palm produce trade was not just a reflection of the encouragement given by European traders and officials; African initiative was important in the expanding trade. Still, in this region, as in many parts of colonial Africa, the state acted, in the words of Bill Freund, as a "tribute-taker rather than an organizing agency for capital producers as in a developed capitalist society."[145] The changes that occurred were also shaped by the colonial ideology of the "male farmer." The colonial government created boundaries of economic and social difference based on gender. However, while colonial policies reinforced patriarchy, the new economy liberated women in some ways. Ironically, women subsidized the colonial state and the peasant household as agricultural labourers and peasant producers. Overall, women and men reaped a "mixed harvest" due to the ambivalent nature of colonial policy. Commercialization did not often translate into economic well-being. Prices and demands for palm produce fluctuated according to European economic and political conditions, creating dependence on the export trade and instability in the local agrarian economy. Colonial taxation and the high cost of living deprived peasants of any surplus.

The first two decades following the establishment of British colonialism witnessed irrevocable changes in the rural economic base. The response of the Igbo to the commercial incentives of the colonial period was influenced by the various contexts – local, external, and colonial – in which they found themselves. Primary production provided new means for achieving power and influence in the society. However, the concentration of resources in cash crop production increased the vulnerability of households to food insecurity. Farmers had neither control over the price of their produce nor the power to influence the price of imported foodstuffs.[146]

Yet the period of conquest, ironically, was also a period of considerable growth in agriculture. But the level of growth in the first two decades of the colonial period owed as much to the initiative of the local population as to the political and social restructuring that occurred in the first two colonial decades, including the demand for labour, the imposition of taxation, and the growth in Western education. The transformations that followed and the forms of the changes that occurred were, however, shaped by the capitalist interest of the colonial state as much as by the precapitalist structures of Igbo societies, including their kinship structures. But it was the Igbos' desire to acquire wealth, their desire for Western education, and their desire and appetite for Western culture that would have the most important impact on the economic and social structures of their society after a very short period of European contact.

Local farmers invested more labour and resources in cash cropping and trade, and palm oil and kernels became significant in reshaping the local economy as much as the relationship between colonial officials and their African subjects. Colonial policies contributed to further commercialization of agriculture and increased the number of small-scale traders in the produce and retail trades. But the emphasis on the production of cash crops disrupted the way the local population balanced production for exchange and production for subsistence as the palm oil trade dominated the economy in most parts of the Igbo country.

The process through which the goals of economic expansion and political transformation occurred was complex. Clearly, the hegemonic power of the colonial state helped create new modes of capital accumulation, but these could not have evolved without the participation, and sometimes the initiative, of the local people.[147] In southeastern Nigeria, where local people were already participating in the emerging capitalist economy of the late nineteenth century, by producing palm oil and later kernels for the European market, the colonial state became a powerful instrument for creating new opportunities for capital accumulation. Yet its unique features lay in the way local people were often able to manoeuvre to pursue other interests. The existing social, political, and economic structures that the Igbo population called upon influenced the changes that occurred in what was obviously a rapidly changing society.

Colonial agricultural policy was at best ambivalent, sometimes encouraging export crop production and at other times limiting it, as administrators struggled to balance competing African and European interests.[148] The state influenced what was grown, how it was grown, and how it was priced and marketed. This intervention was partly effected by "policing the countryside," in Walter Rodney's phrase.[149] At the same time, foreign trade was increasingly controlled by European traders, mainly British. Over time, indigenous entrepreneurs were turned into agents of the foreign merchants. The monopolization of the most lucrative export and import trade in agricultural products and the repatriation of profits by foreign traders stunted the transformation of peasant agriculture and the flow of capital from trade into agriculture. For the Igbo, the changes that followed the colonial agricultural transformation were neither progressive nor regressive; rather, they represented what Hal Barron calls a "hybrid" of change and continuity.[150] Yet the classification of men and women as farmers and farmers' wives, respectively, remained an important part of colonial agricultural policy, which I shall examine in the next chapter.

CHAPTER THREE

GENDER AND COLONIAL AGRICULTURAL POLICY

Colonial records provide an index from which the official view of the role of men and women in agriculture in the colonial period can be derived. From the beginning of the colonial agricultural experiment, officials had a fixed mindset about what to produce and who should produce it. Whether they emphasized cash crop production, soil conservation, or improved production methods, colonial officials had one thing in common: they believed improved agricultural production was important for the state and the local people. But colonial officials constantly imposed European gender ideas on Igbo society. As it did in the political arena, British rule transformed the context of traditional society by creating institutions and class differences based on European notions of gender roles.[1] By focusing on men as cash crop farmers, bureaucratic efforts to improve agriculture encouraged the men even in areas that men and women had previously complemented each other. While official policy did not reflect a blatant discrimination against women, colonial educational and extension schemes aimed at improving agricultural production excluded women. Yet up to the end of colonial rule, the expansion in production in both the palm oil economy and food production was as much the making of rural women as it was the product of men's expanding interest in the production and marketing of agricultural goods.

Igbo societies had a well-developed sense of gender distinction, but the idea of a very strict division of labour along gender lines did not exist. Gender roles were marked by a high level of fluidity in the roles women and men performed.[2] Indeed gender structures and ideology, as social categories, and their relationship with identities, as Igor Kopytoff observed, were based on what people do (roles) rather than any existential category.[3] Thus, a separate sexual world for men and women was not the norm. But the fluidity and complementarity that defined Igbo gender relations of production were at odds with the rigid colonial ideology regarding the role of men and women. Although women brought their own experiences and labour into most areas of production, colonial policies fostered the notion of "male farmers" and farmers' wives, who were not seen as independent contributors engaged in farming.[4] The most important area of bias centred on the exclusion of women from all agricultural extension and support services offered to local farmers.

WOMEN, AGRICULTURAL EDUCATION AND EXTENSION SERVICES

As early as 1888 the *Kew Bulletin* suggested that the superintendent of agriculture consider the apprenticeship of refugee (ex-slave) boys and the industrial education of sons of chiefs as part of the early colonial agricultural policy.[5] By 1913, the colonial department of agriculture in eastern Nigeria was offering agricultural instruction and practical demonstrations to school teachers and their senior pupils.[6] By 1918, seven male pupils were offered theoretical instruction in farming by the Department of Agriculture.[7] With the merger of the agriculture departments of the Northern and Southern Provinces in 1921, the parent department widened its horizon and intensified its activities in the dissemination of agricultural techniques, research results, training, and extension services. Even these early experiments focused on boys.

In 1922, the Department of Agriculture explained the direction of subsequent educational services and extension programs to assist the native farmer. The Director of Agriculture, O. T. Faulkner, proposed two courses (a lower and a higher) of agricultural education at Ibadan in Western Nigeria to cater to the interest of Nigerian farmers. The lower course was for boys with only primary school education and the upper course for boys with the Cambridge Junior Examination.[8] The aim of the course was to introduce new modes of production, limited mechanization, soil management techniques, and general agricultural processing improvements to these boys. No such course or program was devoted to women.

By the end of the 1920s, innovations were introduced in the palm oil industry. In the Awka and Nsukka divisions in 1929, the Department of Agriculture was selecting boys for training in palm oil extraction and nut-cracking machines operation.[9] In 1934, the Department of Agriculture began offering instruction to those who wished to derive a living from farming by targeting "farmers' sons." In a memorandum to superintendents of agriculture throughout the southern Provinces, H. G. Poynter, assistant director of agriculture, instructed that the government's agricultural education should consist of a course of three months' duration in practical work and demonstration supplemented by lectures "for boys who wish to farm" rather than train them as assistants in the department.[10] According to O. T. Faulkner, Director of Agriculture, thirteen boys attended the first course when the program took off in March 1934 at the Moor Plantation in Ibadan. The course consisted of practical and demonstration work on the field as well as instruction in crop science, elementary mathematics, and bookkeeping. Faulkner triumphantly noted that "those who attended the course returned to their father's farms, except one who is acting as secretary of a cooperative cocoa marketing society."[11] Limiting such programs to farmers' sons had unintended consequences and remains an excellent example of colonial perception about women's role in agriculture.

These measures did not have the desired effect because they failed to attract the boys as intended. A district officer remarked:

> I find that boys whose educational attainment would qualify them for this course are not at all anxious to take advantage of the offer. It seems that any who has acquired a slight knowledge of reading and writing consider farming, or for that matter any manual labor, beneath his dignity.[12]

The acting district officer for Onitsha Division confirmed the unwillingness of boys to take advantage of the program. When B. C. Stone wrote to his superior in 1934, he was frustrated by the lack of interest in farming for young boys: "I have not so far found any candidate for the instruction course for farmers' sons. The only boys who have presented themselves were unwilling to proceed with the idea when they found that there was no likelihood of the course leading to a subsequent Government appointment."[13] District Officer B. W. Walter of Udi wrote that there were no candidates from his division.[14]

Special courses were also offered for school teachers under a school farm program. The first agricultural course for teachers under the rural education scheme began at Ibadan in 1937. Although some theoretical courses in agriculture had been offered to teachers at the Moor Plantation earlier, the rural education course aimed to provide instruction in practical agriculture and in methods of managing school farms and the teaching of elementary science. Norman Herington, agricultural education officer "insisted that the new courses must cover at least a complete growing season, starting with the planting of crops in March and finishing with the harvest in October or November."[15] Since most of the teachers came from mission schools, Herington also suggested that the Missions or Voluntary Agencies should send "only trained teachers with some seniority" and that married teachers "should be allowed to bring their wives." He hoped that such a group would give the work some status in the schools.[16]

Herington had laid a solid foundation for a similar agricultural course for teachers in eastern Nigeria. A large area of land near the Agricultural Experimental Station at Umudike belonging to the Government College, Umuahia, had been chosen when he arrived at Umuahia on 23 December 1938. Thirty teachers, mostly drawn from mission schools, arrived for the course in March 1939. The group included the headmaster of a Roman Catholic school at Egbu near Owerri and a member of staff of St. Charles' Teacher Training College at Onitsha.[17] The syllabus for the course was flexible and largely based on "the seasonal work of the farm" and included instructions on planting and cultivation of crops soil science, seed germination, the science of air and water and plant growth. Herington wrote: "I was anxious that the educational value of the school farms should be recognized and appreciated and that this work should be linked with the school curriculum as whole and give a practical introduction to suitable elementary science."[18] The teachers were awarded Grade I Teachers Certificate on successful completion of the course.

At Achi in Onitsha Province, the agricultural department acquired ninety acres of land for a practical school farm that officials argued would train the "sons of genuine farmers in improved agricultural practice."[19] The trainees, who were supposed to be "sons of bona-fide farmers," were expected to "return to the family land and practice what has been taught."[20] Again, officials were too optimistic that these young men would return to farming rather than white-collar employment. Officials failed to consider the attitude of young men toward farming and the influence wage labour had on rural life.[21] For many young men, farming was becoming increasingly unattractive while migration and non-farm income offered opportunity for independence from elders who controlled rural labour and land.

In 1940, the registrar of cooperative societies, E.F.G. Haig, recommended that the Southern Nigerian government should give serious consideration to the establishment of cooperative farm settlements for boys. He wrote that "The plight of thousands of Nigerian lads who leave school at the end of the elementary stage is a miserable one." Haig emphasized: "In view of their partial education they believe, wrongly, no doubt, but with a deplorable strength of conviction that they are fit for clerical posts and unfit for manual labor. Whenever a clerical vacancy is advertised they send in their applications by the hundreds."[22] It was his view that some form of legislation was necessary in order to compel or force these boys to take "their traditional occupation."[23] To ensure that the target group would be effectively redirected to farming, Haig advised that the number of boys admitted into schools be limited through stiffened selection examinations. This would limit the number of candidates in schools and, conversely, increase the number of rural labourers for agriculture.[24]

By 1946, several schools had been established in which the teachers were expected to be instruments of rural transformation through school farms. By this time, The Niger Diocese at Onitsha had come to play a significant role in school farm inspections under Reverend Kenneth Prior of the United Church of Canada. Reverend Prior, who had been appointed as an Agricultural Missionary by the Church Missionary Society, played a valuable role in inspecting school farms belonging to the diocese. The school farms were not very successful in stemming rural drift especially for young boys. The persistent attitude was that "the schools were being used to get the boys back to farm work rather than give them the education needed for better paid work that most parents wanted." The curriculum of the rural school was criticized

Rural Education Course, Umuahia, 1946

Staff l.top.

A. W. Nwankwo

A. Mackenzie

G. N. Herington

W. E. Nwankolo

Rural Education Course for Teacher's School Farm Program, Umuahia, 1946. (Reproduced with the kind permission of the Bodleian Library, University of Oxford.)

for "being out of touch with the educational needs of village people" and was blamed for "the drift of school leavers into the towns in search of work" because the curriculum lacked an "agricultural bias."[25]

In a speech to the Conference of Christian Rural Workers held at Ibadan in 1955, Herington remarked: "The schools cannot be blamed for the drift of population brought about by changing social and economic conditions. In the badly over-populated rural areas of the Eastern Region many people are obliged to go elsewhere to find work and the village schools should assist in this development rather than try to discourage it." The attitudes toward farming had worsened by the 1950s. Herington accepted: "Boys cannot be forced to become farmers against their will and attempts to impose an agricultural bias would probably be resented by parents and teachers alike."[26] He suggested the course for school teachers should be called Rural Science rather than Agricultural Education and the centres at Ibadan and Umuahia became Rural Education Centres.[27]

Well into the 1950s, colonial officials were still very frustrated with what they termed the conservatism of Igbo farmers and the lack of interest towards farming by young boys. Michael Mann, community development officer for Okigwe Division, wrote:

> The conservatism of the Ibo farmer is well known, and indeed he does not differ from farmers all over the world. Many Ibo [sic] farming practices are not only inefficient, but are actually harmful and destructive to the soil. 'Ala' – the Okigwi Ibo word for land, is also used for 'our soil', 'our country' and it is their most precious possession, which is most jealously guarded, and over which suspicions and ill feelings are most readily aroused.

Mann was particularly frustrated with the number of Igbo young people migrating to the towns for non-existent jobs instead of settling to life as modern farmers in the rural areas. This was particularly so for people in Owerri Province, where education was seen as the route to a better life for the boys as well as their families.

> One of the major problems confronting us in the Eastern Provinces, and especially in Owerri Province, is that of the Standard VI schoolboy on leaving school. The family contributes in order to send the young boy to school, and often at great sacrifice, maintain him until he reaches Standard VI. Then the young boy, for a multitude of reasons – lack of funds, shortage of secondary education, competition in examinations – is thrown on to the labor market and his own resources. He as a result of his education, considers himself a cut above his uneducated brethren, and looks for employment for which his education has been training him.[28]

However, very few could find suitable employment based on their level of education, leading to high levels of what Mann described as "acute disappointments and disillusionment."[29]

Given their investment in their children's education, families expected some return for the "money expended, and are angry and disappointed the young boy is unable to secure employment." Despite the low prospect of jobs in the urban areas, officials described as alarming the drift of the youth towards the towns, where, away from the discipline of their village and advice of their elders, they became easy prey for the disgruntled politicians. These youths became a class of "unemployable and irresponsible fortune seekers."[30]

How to resolve this concern was a recurring problem for many officials. The obvious answer according Mann "is that the boy on leaving school should be absorbed into his village." To solve this problem, he proposed that: i) The village must be made more attractive by means of increased social services and social improvements; ii) the prestige of farming must be raised and methods brought up to date; iii) local industries be set up to employ the young girls and boys in their own villages.[31]

Mann set up an agricultural team, consisting of a member of the agricultural department and some labourers who would offer demonstrations, advice, and encouragement to "those farmers who will listen.... Time is not wasted in talking to those who are opposed to change, but a great effort is made to convert the young men and the more progressive farmers to such simple improvements as composting and feeding of livestock."[32] The team would select the most intelligent of willing farmers for a course at the Uturu Trade Centre. There they would farm under instruction on poor land with improved methods in order to show them that "improved methods will bring them more money as well as improving their land."[33] The trainees were expected to return to their farms in the villages and continue to have advice and regular visit from officials of the agricultural department. It was hoped that, by adopting improved farming methods, the progressive farmer would see the benefits and advantages of improved methods more easily than "his neighbor who carries on as usual."[34]

When the government of the Eastern Region introduced an agricultural extension services known as the "On-the-Job" training program, the scheme was initiated by the department of agriculture to enable the "men to improve their farming."[35] The evidence we have for the Onitsha Province shows that men came in groups for a month at a time to receive practical training in soil conservation, compost making, and the care of seedlings and of such permanent crops as citrus and oil palms.

There were a few farmers who attempted to modernize their production methods. One such person was seventeen-year-old Mark Nwadike of Ubahu

Okigwe whose father died when he was young and left him a lot of land, but little money. When Mark finished schooling at the Methodist School Ihube, reading up to standard VI, he returned to his village and tried to put into practice some of the things he had learnt on the school farm. Colonial Administration recorded that Mark travelled to Okigwe to look at the Government Agricultural plot and to consult the Agricultural Assistant. He started to put compost around some of his yams, and dug a compost pit near his house, into which the compound sweeping are placed each day.[36] Mark witnessed improvement in his farming as well as in his living condition. Such success stories were part of the campaign to encourage young men to go back to the land instead of "wandering off to the towns." Felix Dibia of Umuariam Obowo in Okigwe Division was another of the few who went back to agriculture in the village. Felix had been trained at the Government Agricultural Farm at Umudike. Despite "skeptical remarks and jeers of his neighbors," he was determined to put what he had learned into practice in his own farm. He had set up a small poultry with a Rhode Island Red cockerel and five pullets he had obtained from the Government and planned to expand into piggery. His feeding of the chicken, according to Michael Mann, "caused conversation amongst his neighbors who asked, 'Why feed a chicken, it is meant to feed us?'"[37] Such was the prevailing view of many people and their attitude to innovation.

Overall, the attempt to modernize agriculture focused attention on the men and the boys. The colonial government did not really draw the female folk into its modernizing program, despite the central role they played in agriculture. For women, the concept of the "private sphere" informed the colonial development program that targeted women specifically. Although British officials were operating in a different and changing historical context, women's work was seen as belonging largely to the domestic sphere.

To address the issue of women in the developmental process, women's training centres were opened to create opportunity and provide training for women. The Women's Training Centre, opened at Obowo in June 1950, provided a daily class where women were taught the making of fairly simple embroidered mats, tray cloths and other linen work. They paid a monthly subscription, which covered the cost of materials, and they sold their work for "their own profit." Man commented that this system has made these centres

Felix Dibia feeding corn to his chickens. (Reproduced with the kind permission of the Bodleian Library, University of Oxford.)

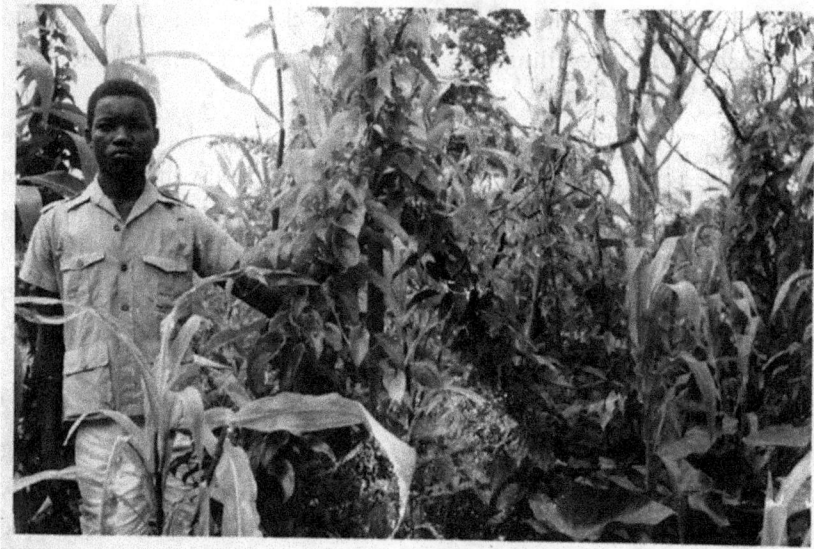

Mr. Mark Nwadike with his yams grown with composting. (Reproduced with the kind permission of the Bodleian Library, University of Oxford.)

extremely popular, as the "women are not considered by their husbands to be 'Wasting Time.'"[38]

Christian missions were also involved in the broader rural development agenda, but they had a clear notion of the role of women in a Christian home. Mission schools played a significant role in actualizing this ideology through the establishment of a rural development centre and by actively collaborating with the Government under the school farm program. The Anglican Diocese of the Niger, the Methodist Church in Eastern Nigeria, and the Church of Biafra (Presbyterian), for example, were aided with a grant from the Government and worked closely with agricultural and educational agencies to offer training to young men and women in the rural areas. The information and visitors' guide of the Rural Training Center at Asaba declared: "The aim of the center is to help young people become better farmers and home makers, that they may lead the way towards an improved and enriched Christian rural life."[39] Fourteen ex-Standard VI boys began a two-year course in farming in November 1951, followed by twenty other boys. On their part, each boy was allocated a half-acre to farm: 3/10 acre of yams in the flood area, and 2/10 acre rotation plot in the upland. After deducting cost of seed, labour, and ploughing by tractor, etc, all profits went to the student.

Colonial programs were male driven, regardless of obvious female participation in agriculture and the difference women farmers could have made to agricultural productivity. The first class of girls at the Rural Training Center at Asaba began with seven girls for a two-year course in domestic science and homemaking. The Girls' Cottages offered the practical training useful for homemaking:

> The girls learn how to be good homemakers by doing just that in their own cottages. They do their own marketing and food preparation, cooking on a smokeless Indian stove, built chiefly of mud. They eat together in family style, and have their own living room. They learn to sew, not only by hand, but on the Centre's sewing machines.... The girls operate their own kitchen near their cottages. They also have some work with poultry, and will later have their own flock of chickens.... Improved diets, a more hygienic mode of living and wiser motherhood are emphasized in the course.[40]

Yet in the rural context of the colonial society, officials clearly ignored women and concentrated efforts on men, who were more likely to leave the rural areas. This ideology was more often the result of ignorance or ideological bias than an objective assessment and a clear perspective of the local agricultural system. The improvement schemes targeted a group that was becoming increasingly interested in earning cash income outside agriculture and neglected the women, the major agricultural labour force.[41]

Behind the thinking of colonial officials was a model in which agriculture would continue to be central to the income of the overwhelming majority of the rural population and one in which it would be modernized to attract young men to continue to take farming as a profession. They, however, failed to recognize several factors that worked against such thinking and the potential of rural agriculture. A report by the secretary for community development to the Eastern Nigerian government noted the improbable conditions that worked against government policy in Igboland:

> The flight of the Standard VI boy from the land is the theme of especial lament by both educated Africans and Europeans, but it is difficult to see what there is to keep him. The social organization of the village does not encourage individual enterprise; many of the amenities, the existence of which his schooling has acquainted him with just do not extend to his village; and the main form of employment – agriculture – has such a low status and gives such a low return that the greater variety of employment offered in the towns and the social status of the white collar jobs are bound to exert a strong attraction.... Should the Standard VI boy stay in the village there is little enough to maintain his educational attainment, let alone stimulate him to improve it. The intellectual isolation of the educated man is enough to dissuade all but the really dedicated and the mediocre from taking up rural teaching.[42]

There were other reasons alluded to by a colonial official. Prospects for a better life and opportunities remained a major motivation behind the "drift from the land," he noted. "To the individual this might mean regular income, better health and educational provision, and the stimulus of the urban environment with its variety of cultural outlets. But deep down beneath this reasoning lies

Girls making pottery products at a Catholic Community Development Center. *Source:* PRO INF 10/241.

the knowledge that there is no future on the land for the majority. The third or fourth son knows that his chances of inheriting the father's farm are slight." He also alluded to the population density in parts of Igboland and the difficulties of surviving on farming in such areas where land hunger remained the greatest source of litigation in the Native Authority courts. In parts of Owerri Division, with an estimated population as high as 1,000 per square mile, the drift from the land "is sheer necessity," this official argued. In his view, "The productivity of the land cannot be significantly increased without capital, and this will not be available in sufficient amounts until a smaller farming population is able to win a reasonable surplus from its land and efforts."[43]

British policy was similar in other colonial possessions but some colonial officials recognized that this policy was wrong. J. B. Brown, an agricultural officer in Bamenda, Cameroon, tacitly acknowledged this when he wrote:

"Programs for future agricultural development ... ought to be aimed at women and at making the younger generation of males favourably disposed towards agriculture."[44] Despite women's indispensable role in the agricultural economy, colonial authorities continued with a policy that often misdirected services to improve agriculture throughout the country.

Changes to the dominant notion of the male farmer remained in place until the beginning of 1950. The increasing role that the African elite began to play in domestic politics from the early 1950s and the regional autonomy achieved in 1954 would eventually lead to some cosmetic changes in the gendered nature of state policy in agriculture. This period was a turning point in economic policy because the government of the Eastern Region introduced some changes to existing policy, particularly in the agricultural sector. The main emphasis beginning in 1952 was on extension work and agricultural shows to stimulate the farmers' interest in new farming techniques and "proper" land utilization.[45] This scheme continued to be offered to male farmers until 1956, when the Agricultural Department started a scheme of "non-residential" practical farm schools in two villages. The farmers, who included women, received training in the use of fertilizers and techniques for citrus farming and soil conservation. The new scheme, according to the government, was to require minimum intellectual effort and disruption of rural life, and "fitted into the normal social and economic life of the village."[46] This change in policy is a tacit acknowledgment that past policies had been misdirected and ignored local realities, but it also showed that officials still believed that local farmers were ignorant and incapable of adopting sophisticated innovation independently.

There was still no consistent effort or commitment to see women as important actors in agricultural development thought. For example, by 1960, the School of Agriculture at Umudike, Umuahia, which revolutionized agricultural training and extension work in the region, had a class of forty-one agricultural assistants and seven field overseers, which included only one female student, the first to be recruited since the school's inception in 1955.[47]

The gender bias in colonial support for local farmers was not limited to the extension services and improvement schemes. Colonial financial support schemes discriminated against women farmers. The co-operative section of the department of agriculture introduced a loan scheme in 1931 to improve family production and assist farmers in buying palm oil processing presses. By the 1940s, farmers were applying for loans as individuals or collectives from the Nigerian Local Development Board to develop palm oil and cocoa plantations. This pattern continued for most of the remaining part of the colonial period. In 1952, for example, the agricultural department in the Aro District supplied 1,250 palm seedlings to local farmers for the extension and establishment of plots.[48] All the private individuals supported in the scheme were males, a pattern seemingly replicated in other parts of the region. The available data on distribution of loans on 31 March 1958 indicates that 165 were given for general agriculture in the amount of £157,369. The share for agricultural industries, including palm oil mills, rice mills, corn mills, cassava graters, and copra plantation and factory, was 46 in the amount of £91,301.[49] What is remarkable about this is that all the borrowers listed from various divisions in Eastern Nigeria were men. Further, except for a few cases where farmers were given loans for mixed farming, all individual loans, which ranged from £250 to over £4,000, were given for the sole purpose of developing cash crops, mainly palm oil plantations.[50] These policies prevented innovations trickling down to women who carried out the bulk of farm work, and thus did not have the desired effect of transforming African agriculture.

GENDER AND INNOVATION

The first three decades of the colonial era were marked by continuity in agricultural processing technology in the palm oil industry. In this era, 90 per cent of palm oil and palm kernels were obtained from natural groves with minimal capital investment.[51] The production, marketing, and transport systems were described as still most "primitive" and wasteful, compared to the alternative method of mechanical extraction.[52] Emphasis, therefore, was placed on improved quality through produce inspection and the adoption of new processing methods and improvement schemes.[53] In 1915, the Colonial Secretary, Bonar Law, appointed a committee to study the trade in palm products and make a recommendation for "the promotion in the United Kingdom of the industries dependent thereon."[54] The committee reported in May 1916 and recommended improvement strategies that were eventually adopted.[55] By 1920, the government was ready to adopt measures to improve the quality of palm oil and reduce what was described as the wastage associated with the traditional hand extraction method. Toward the end of the 1930s, the demand for more efficient production methods, in general, and improved quality of oil, in particular, resulted in the adoption of new processing technology: palm oil presses and palm oil mills.

The colonial officials, as earlier indicated, began increasingly to advocate for the cultivation of oil palms on plantation lines.[56] So strong was the pressure on farmers to convert their farms to oil palm plantations that colonial officials relaxed an earlier policy that strongly resisted any attempts to influence indigenous production. By 1932, there were about 119 native-owned palm plots in Owerri region and about 36 in the Onitsha area.[57] In economic terms, the expansion of the palm oil industry, in particular, meant that Igboland was saddled with the burden of producing export goods to meet European demand, and this threatened the local agricultural economy. For most people in the region, the changing social and economic systems were affecting agricultural sustainability. As Dei argued, one can trace the harm to effective sustainable development in Africa to intensified appropriation of wealth from the rural peasants by the state.[58] The conditions fostered by commercialization eroded the agricultural base of Igbo societies.

The government's effort to encourage palm oil production received a boost with the introduction of palm oil presses. By February 1933, twenty-one Duchscher presses were in operation in Eastern Nigeria. The 1938 Annual Report of the Agricultural Department indicates that the number of hand presses increased from 58 in 1932 to 834 in 1938.[59] However, statistics are part of the story. Despite the high rate of increase in the installation of these presses between 1933 and 1938, hundreds of thousands of producers engaged in the oil industry still relied on traditional production methods. By 1938, Nigeria's position as the world's largest supplier of palm products had been surpassed by Indonesia and Malaysia, but some fundamental changes occurred in the way peasants produced oil and kernels.

As officials promoted new production regimes for local farmers, hand presses caused the re-distribution of traditional labour among men, women, and children, increasing men's participation in palm oil and kernels processing and marketing. Hand presses extracted about 65 per cent more oil than the traditional method, and they were reported to be in common use by the end of the Second World War.[60] A.F.B. Bridges compared the labour and quantity of oil extracted by the hand press and the traditional method in 1929. He recorded that it took 4 women about 9 hours to produce 331 lbs of oil, but 2 men and 2 boys about 2 hours to extract 491 lbs of oil from the same quantity of fruits.[61] Eno Usoro also compared the man-hours spent using the pre-war traditional method of preparing oil with the post–Second World War method using the hand press. Total production-hours spent by men increased from 600 to 1,050, while production-hours of women and children declined from 1,450 to 992 after the introduction of agricultural innovations.[62]

By the 1940s, the colonial government encouraged the new Nigeria Oil Produce Marketing Board and the Eastern Nigeria Regional Produce Development Board to supervise the installation of what became known as "pioneer oil mills." In 1946, the installation of palm oil mills began in the Igbo communities of Owerrinta and Azumini.[63] The pioneer oil mills were intended for large-scale production and processing of palm products. However, farmers continued to fear that "the building of mills meant that the Government was going to take over the palm trees."[64] Morgan is right to argue that changing circumstances may demand changes in agricultural technique, but all suggestions for improvement "must face the fact that Ibo farming practice is careful and based on long experience and that the environmental conditions offer limitations unknown to Europeans."[65] Nevertheless, the introduction of the pioneer oil palm scheme, marked by innovation in processing technology and methods, further affected the local agrarian economy in some parts of Igboland.

Morgan observed in the 1950s that despite the restrictions imposed by physical conditions and despite a conservative outlook, Igbo farmers were gradually changing their methods in response, "firstly to changing social and economic circumstances, and secondly to influence of Government Departments, particularly the Department of Agriculture."[66] The expansion of trade offered improved markets for palm products. Still, innovations were gradually entering the Igbo region. Export from Aba Division increased with fluctuation to 39,427 tons of oil and 21,523 tons of kernels in 1952. And by 1960, there were 3,236 hand presses, 153 powered nut crackers, 172 rice mills, and 30 cassava graters.[67] Obviously adequate attention was not given to the food sector.

The Agricultural Department could not persuade women to use hand presses even though they extracted about 20 per cent more oil than the manual method.[68] The installation of oil mills moved the production process out of the home and eliminated women's access to produce. The introduction of palm nut cracking machines also slowly challenged women and children's role in the production of kernels, a process that threatened women's income from the sales of kernels. The threat to women's participation in palm oil and kernel production was resisted in parts of Eastern Nigeria. Susan Martin documented how Ngwa women in central Igboland mounted spirited protests.[69] Women resisted the attempts by foreign men (European firms) to buy the uncracked nuts.[70] Women carried similar oil protests in other parts of the province. In 1951, for example, women attacked men as they returned from the market and seized the palm fruits they bought. Women's refusal to deal with the pioneer oil mills gave men an advantage that women protested. The District Officer for Opobo reported that the women refuse to deal with the Pioneer Oil Mills, as they were "unable to compete with the men." He added that women were losing their traditional means of income and their only prospect seems to be "ever-increasing dependence on their men-folk." This prospect, he argued was "not attractive to these vigorously independent women."[71] In Abak, Midim/Nung Okot in the Opobo Division, women protested the purchase of palm fruits by men from the local market.[72]

The protests against the installation of oil mills were an indication of the important role of palm kernels as a source of independent income for women. Moreover, the innovation in production methods was not uniform throughout Igboland. The oil mills and presses were concentrated in the Ngwa and Azumini areas. Yet the revolutionary impact of the new technology has been exaggerated. The new technology, it appears, was not the most significant element of change because mechanization did little to diminish women's control over production in many parts of Igboland, where they continued to use traditional methods of production. M. M. Green, a British anthropologist who studied the Agbaja people in the 1930s, recorded that there was only one hand press in the village in which she worked.[73] In Mbaise area, hand presses were not common until recent times according to several informants.[74]

Indeed, colonial attempts to shape the direction of agriculture and the exclusion of women ignored the importance of female labour. Yet, previous studies of the dynamics of commercialization of agriculture and technological innovation have sometimes fostered confusion over how these influenced gender relations. Susan Martin has pointed to the profound changes in gender relations of production following the expansion of the palm oil industry among the Ngwa of Eastern Nigeria from the late nineteenth century. As the commercialization of agriculture became a critical source of income and new forms of identity, the pre-commercialization division of labour, Martin argues, "acquired ideological significance," which strongly favoured men.[75] Margaret Stone has also drawn attention to what she refers to as the "ideological constructs of male and female status and power" and the transformation of sexual division of labour as rural societies were drawn into the capitalist world.[76] Basil Ukaegbu suggested that the production of palm oil was women's prerogative and was essentially used for household consumption until the trade with Europe developed.[77] The export potential of palm oil made it an attractive economic opportunity for men who could acquire goods such as guns and spirits with income earned from the palm oil trade.[78]

These works have drawn attention to the use of gender as a framework for understanding social transformation, in particular why allocation of resources was transformed in ways that often favoured men. This is probably true for most parts of Africa, although the real impact at the household level may have been exaggerated. These models emphasize social structures (culture) as the engine of history, when in fact social structures were products of historical processes of social change. Sara Berry's study of how cocoa production for export stimulated the development of capitalist social structures in rural Yorubaland, including the evolution of private property rights in land, is a good example of this market driven transformation of social structures.[79]

The expansion of export production led to fundamental changes in the local economy, most of which affected gender relations. The commercialization of agriculture in Igbo societies introduced new elements, particularly in the mode of production and gender relations, but like many other societies in colonial Africa, the Igbo were adjusting both their economic orientation and social relations, including gender, to commercialization and the demands of the colonial state. The changing economy was a "mobile" one and the opportunity provided by the palm oil trade allowed both men and women to advance economic opportunities – for women through their participation in the production process and role in marketing. Fewer women become large-scale brokers, but they were central to small-scale marketing, which provided women significant economic opportunity and livelihood. Indeed, women continued to play important roles as producers and marketers of palm produce at the local level. Reporting on the Ekwerazu and Ahiara Clans in the Owerri Division in the 1930s, G. I. Stockley, Assistant District Officer for Owerri Division noted that: "The only traders from outside who visit the area in any number are the Isu middlemen who buy palm oil in the local markets principally from the women and carry it to the European firms."[80]

Indeed, the control of palm oil was being renegotiated as the Igbo responded to new challenges from the mid-nineteenth century. The commercial importance of palm oil was very limited prior to this period. It largely fell under the purview of household subsistence need. In fact, its commercial importance did not increase in some parts of Igboland until the colonial period. An oral interview collected in Ogbe Mbaise in 1972 by A. M. Iheaturu is insightful. Eighty-year-old Andrew Anyanwu recalled, "People started of late to boil palm nuts and to sell oil and kernels. In the olden days, our people only made *eketeke* from fresh palm nuts, which they used for cooking. Palm kernel was thrown away, nobody bought them in the markets. Before we started selling oil and kernels, it was *ohu* [slaves], that people bought and sold to Nkwerre and Aro people."[81] But the commercialization of the oil palm industry and the income it offered was attractive to both men and women. Both men and women, it appears, re-negotiated their production relationships in the face of new challenges and opportunities. Colonial demands in taxes and rates, often increased the burden of rural households, particularly those of the men, and they explain the interest men developed in the trade. Moreover, the incorporation of households into a capitalist market economy required fundamental adjustments in resource allocation. The nature of the transformation in the palm oil industry was also shaped by the fact that men controlled the most important factors of production: land and household labour. In spite of a gendered colonial policy, however, women continued to play important roles in both the formal and domestic economy.

PERSISTENCE OF WOMEN IN PRODUCTION

Women participated in palm oil industry despite colonial neglect. Although men were increasingly drawn into the new economy, and disproportionately favoured by it, the trade in palm oil was not entirely dominated by men. Women played important roles as producers and marketers of oil. Early twentieth century European reports noted that women usually brought palm oil to the markets.[82] Some women, such as Omu Okwei of Ossomari and Ruth Onumonu Uzoaru of Oguta, acted as produce agents, buying from local producers and re-selling to European factories from the late 1920s.[83] Besides acting as buying agents for produce in the palm oil industry, women performed the task of "bulking and of breaking bulk in produce-buying, both activities facilitating exchange at quantity and cost levels appropriate to the scale of production and buying habits of the customers."[84] Women almost dominated cash crop trade at the local level, and they often combined their activities in the palm oil trade with the sale of cooked food, as the expanding commercial sector offered new opportunities for capital accumulation.[85]

Oral accounts from the Mbaise region confirm the significant role of women in the local palm oil trade and their reduced reliance on farming in the colonial period. According to one informant, women frequented the buying stations at Umuahia, Ife, and Udo beaches in Mbaise, from where palm produce was transported by river to the coast.[86] Linus Anabalam, who was a small produce buyer in Mbaise in the 1940s, recalls buying mostly from women.[87] An informant stated that women were in control of production and marketing at the local level. She argued that the expansion of the palm oil market actually enabled women to obtain independent incomes.[88]

Given bicycles and motor lorries, however, men were able to cover longer distances to market produce that women had previously been able to sell. The ability to move oil and kernels in bulk and men's domination of the system of haulage gave them a substantial advantage over women produce buyers.[89] Women only joined the long distance trade in later years, when motor lorry became the major means of transportation to distant markets.[90]

Yet, both men and women sought advancement in the new economy by dedicating more time to the production and marketing of palm products.[91]

Women particularly cashed in on the increased demands for palm kernels, despite the laborious nature of its production.[92] Therefore, primary production drew men and women away from subsistence production under which subsistence increasingly depended on the market. In historical perspective, changes in the colonial economy diminished women's levels of control of commercial production and significantly influenced the subsistence economy as well. Overall, commercialization in the Igbo region created an environment for men and women to diversify income generation strategies in the area of trade and production while significantly reducing farming activities over time. However, as the importance of cash crops for export grew, men and women became vulnerable because of their dependency on the export market and their reduced reliance on the traditional system of household subsistence. This not only diminished interest in other aspects of agriculture but also increased household food insecurity. The struggle over spheres of influence in the new economy increased tension between women and men, as women often threatened the men, and at times, called on the colonial authorities to protect their interest in the export or domestic economy.[93]

There was a dramatic response by both men and women to cash incentives offered by the produce sector. Apart from the cash incentives following the increased export of palm products in particular, the limited resources of the early colonial administration, as Martin Klein argued in another context, forced colonial officials to systematically extract wealth from colonized peoples.[94] The need to meet increased financial obligations (direct taxes and rates to the colonial administration) and the export potential of palm oil encouraged men to participate more in its production and marketing. Changes occurred in Igbo gender relations of production when the export of oil and kernels became the source of cash income and an important means to purchase the European goods that increasingly became part of the new consumer culture. The new cash nexus that emerged with the oil trade dissolved the old economic order based on subsistence agriculture, while rising consumption needs of the new society and colonial demands for taxes and rates forced upon the Igbo peasantry a burden their subsistence economy could not bear.

Some scholars have argued that the export-crop innovations took place and flourished without or perhaps in spite of the advice of European experts. Others stress the important contributions of research and innovations brought by the colonial regime. The truth, Hart suggests, is "probably somewhere in between."[95] The colonial administration included some basic training in its agricultural development program, but agricultural extension schemes represented the most ambitious effort by the government to enhance agricultural production. Although ignorant of the complementary nature of production and resource allocation at the household level, these improvements nevertheless offered a chance for higher productivity and improved quality.

In the end, the expansion of the agricultural sector, particularly the oil palm industry, was a result of colonial intervention as much as the outcome of the actions of African households. Although women constituted the main subsistence producers and an indispensable part of commercial production, they were generally denied extension services, agricultural loans, and agricultural training provided by the agricultural department. Women were an invisible factor in rural agriculture. However, women's labour played an important part in sustaining peasant farming, despite the constraints imposed by colonial rule and the ideology of the "male farmer." Women continued to participate in the production and marketing of cash crops and adopted new strategies to meet the demands of the changing agricultural economy. Contrary to the findings of some scholars, the commercialization of palm oil production did not radically alter the production process and women's participation in the economy at the household level. While men increasingly dominated the "middleman" position in the trading system, women continued to control the marketing of palm oil and palm kernels at the household level. But production at the household level continued in the same manner except for the isolated installation of oil mills and other innovations in parts of Igboland.

But, the new economy also liberated women in some ways. Women engaged in the new trade as producers, while some became distributors. Many became suppliers of foodstuff to the increasing urban population and became involved in the wider colonial and international economies. However, the government's agricultural and development policy, the changes in the gendered division of labour, and the control over and exploitation of the local agricultural resource base did not often stimulate peasant interest. Since the new economic structure was predicated on the patriarchal ideology of the "male farmer," the neglect of women farmers that intensified with colonial exploitation of peasant agriculture is important in explaining the political economy of colonialism, the impact on men and women's autonomy, and the responses of women in particular to their exclusion from official agricultural programs and the transformation of their society. Ironically, women, as agricultural labourers and peasant producers subsidized the colonial state and peasant household.

European commentators observed the persistence of women in food production. Writing in 1955 about the Ngwa and Ikwere Igbo, Morgan delineated a clearly gendered pattern of crop production and control:

> Generally, amongst the Ngwa and Ikwerri the men plant and tend those crops and trees needing most attention, leaving the remaining food production to women. Thus men's crops include yams, pineapples, oil and raffia palms, coconut palms, plantains and bananas, and oranges, African pear, kola, oil bean and 'vegetable leaf' trees. Women's crops are cassava, maize, cocoyams, beans, groundnuts, pumpkins, calabash, melons, okra, chillies and peppers.[96]

Women in some parts of Igboland devoted considerable time to cassava production, which became quite lucrative in the early 1940s. The demands for processed cassava flour [gari] in the urban areas increased women's interest in cassava production.[97] As elaborated in chapter five, the food crisis that occurred during the Second World War also increased the importance of gari as a substitute to imported food items such as rice. In 1949, for example, about 5,530 tons of gari was railed to the North, and by 1952, it had increased to 22,170 tons.[98] Farmers concentrated on food production in this period because of the lucrative market for food items. As the expansion of the economy, based on cash crops, forced peasants – male and female – into trading and other activities, they became increasingly less reliant on farming. In addition to the thousands of people whose livelihoods were directly affected by participating in the production of oil and kernels, many more became involved in small-scale trade. They would buy a bag of kernels or a few tins of oil at a time for resale to middlemen, for whom large-scale bulking of these goods became a full time occupation.

The expanding bureaucracy and the emergence of urban towns and cities inadvertently created the opportunities that arose from European intervention. As elsewhere in Africa, the urban population in Nigerian cities grew at an astronomical rate of more than 15 per cent annually in the first decades of the colonial period. By the 1920s, the population of cities such as Onitsha, Enugu, and Port Harcourt in eastern Nigeria had increased tremendously. Enugu and Onitsha grew from a combined population of a little over 1.1 million in 1931 to a population of about 1.77 million in 1952.[99] The urban population created a market for food produced in rural areas. Thus, the production of foodstuffs to feed the urban population increased despite the lack of colonial support for the sector. In this sector, women held advantage.

Overall, colonial officials navigated an uneven terrain in their attempt to balance competing gender interests, colonial policy, and the maintenance of a stable political economy. The development model pursued by the Department of Agriculture and other colonial officials aimed to transform the local economy and to meet European demands for palm produce. However, it also broke down "traditional" production relations, but balanced competing interests effectively enough to ensure continued production. At the same time, the development ideology of the colonial agricultural department, particularly the neglect of women farmers, combined with colonial extraction measures, limited the progress that could be made in local agriculture.

CHAPTER FOUR

PEASANTS, DEPRESSION, AND RURAL REVOLTS

I never saw women demonstrating in that manner before. I have seen them play many a time but this was obviously entirely different and there was no doubt whatever that they were out for trouble. – *Henry Alexander Miller, Aba Commission of Inquiry, 1930*

I wish to tell you what made the women move about and remain here for about five days. We do not want women to pay tax and we want the tax on men to be abolished. – *129th witness, Oguta, Aba Commission of Inquiry Notes of Evidence, 1930, 278*

The market is our strength. When the market is spoiled, we are useless. – *Witness, Aba Commission of Inquiry Notes of Evidence*

James Scott's *Weapon of the Weak* outlines why open peasant revolts have been rare and "everyday forms of resistance make no headlines."[1] According to Scott, peasant actions, where they occur, are often limited in scope and lack a collective consciousness and well-outlined plans for action. They are also characterized by informal structure and a lack of direct confrontation with authority.[2] Yet, whatever forms of consciousness have developed are rooted in what Scott described in his earlier study in Burma and Vietnam as the "moral economy of the peasant."[3] This moral economy, and the system of values that it contains, explains why peasant revolts are irrevocably linked to issues of peasant subsistence and survival. African peasants have shared values similar to those of their counterparts in other parts of the developing world, even though Goran Hyden claims that the poor sectors of the population in Africa "are much less aware of their exploitation than their counterparts in Asia and Latin America."[4] But have African peasants in truth been less conscious of their exploitation than their Asian and Latin American counterparts?

This chapter explores the economic roots and the consequences of these revolts and protests. It traces the responses of the local population to the declining price of palm products and the introduction of direct taxation in 1928 and shows how these events became part of local political discourse. In doing so, the chapter places the economic and social conditions of Igboland in this period within the context of the political restructuring of the indirect rule system and the worldwide depression heralded by the slump in 1929.

Indeed, agrarian concerns remained the major source of rural protest in colonial Africa. Even recent competing claims to the state in contemporary Africa have been linked to rural consciousness. Contrary to the urban-centred model of war proposed by many scholars, Paul Richards has argued that the roots of the recent Liberia and Sierra Leone conflicts, for example, were agrarian and that these conflicts arose from rural poverty.[5] The Igbo provide an important example of rural people in colonial Africa protesting when their subsistence came under threat from colonial from colonial policies.

Rural Igbo men and women were certainly aware of the impact of state regulations, controlled prices, and market forces on their income. They did something about it by protesting against colonial policies in a variety of ways. From the 1920s onward, and for good reasons, Igbo farmers and petty traders frequently linked their declining fortunes in the palm oil trade to the policies of European officials and the activities of foreign trading firms. Subtle acts of resistance such as adulteration of produce and refusal to pay taxes were employed by rural men and women in Igboland to address their concerns.[6]

But Igbo farmers and traders also adopted strategies that included direct confrontations with colonial authorities. These confrontations, which were mostly organized by women, became very frequent from the mid-1920s onward. The articulation and framing of these protests show that Igbo women were not passive recipients of change but articulated their interests and acted in the interest of their class. Because the actions taken by Igbo women had long-term political consequences, and because those actions were designed by them to protect their own interests, the women who articulated these movements among the Igbo would emerge as "peasant intellectuals," in Steven Feierman's appropriate classification of such groups in his study of Tanzania.[7]

In Nigeria, the rural population did not accept the impacts of colonial policies and the depression on their lives with passive resignation. The major peasant revolts that occurred in Igbo society during the colonial period were directly linked to the oil palm industry. The local way of life, which had been tied to the oil trade, became very precarious due to the depression in the global demand for oil toward the end of the 1920s and the periodic decline in prices that occurred afterward. The centrality of the palm oil trade to the income of many rural households and the consciousness among rural peasants that the colonial state, its agencies, and the European trading firms had means of exerting hegemonic control over their lives and the rural economy led to discontent among rural people.[8]

One important example of the rural protest that occurred was the Women's War that broke out in Owerri and Calabar provinces in 1929. Although not all of Igboland participated in the 1929 revolt, the economic and sociopolitical conditions that gave rise to it were not limited to the areas that participated in the revolt. The 1929 revolt was the most violent, but it was neither the first nor the last protest directed against the colonial masters that was rooted in the rural agrarian economy.

THE 1929 WOMEN'S REVOLT

On 23 November 1929, a remarkable incident occurred at Oloko, a rural community in Bende Division in colonial eastern Nigeria. Nwanyereuwa, a rural peasant woman, challenged the foundation of British authority in Nigeria by her simple act of refusing to be counted for the purpose of an impending colonial tax. The revolt, styled the "Aba Riot" by the British and the "Women's War" (*Ogu Umunwanyi*) by women, quickly spread to other parts of the Owerri and Calabar provinces, turning into an all-out revolt against all aspects of the political establishment and the European trading companies. Thousands of ordinary women took matters into their own hands and stormed colonial administrative centres and later, the important commercial city of Aba, which housed several European trading companies. They took on the warrant chiefs appointed by the government, demanding their caps (the most important symbol of their authority).

The impetus for the 1929 Women's Revolt stems from the perceived deficiency of the enumeration exercise in April 1927, upon which the 1928 tax was based. The 1927 assessments had been based on what an Assistant District Officer Captain J. Cook of Bende District described as "incomplete and probably inaccurate" information.[9] In September 1929, Captain Cooks was sent to take over the Bende division temporarily from the district officer, Mr. Weir, until the return of Captain Hill from leave in November. Upon taking over, Cook found the original nominal rolls for taxation inadequate because they did not include details of the number of wives, children, and livestock in each household. Cook set about to revise the nominal roll, announcing the intent to a few chiefs in Oloko native Court; the counting began on or about 14 October 1929. Although this data was not required for accurate assessment of the tax rate on the men, colonial officials saw this information as necessary for the annual statistics and for accurately "gauging the wealth of the individual and community."[10] Consequently, when in 1929 District Officer Cook proceeded to elicit nominal rolls that would gather such information, instructions were given to the ezeala (traditional leaders) in some cases and to members of the native court tribunals to provide suitable individuals for the counting.

About five towns had been reassessed without incident but then came the incident at Oloko. But the incident that led to the revolt began when Nwanyeruwa, the wife of an Oloko man, was approached by a local schoolteacher Mark Emeruwa, who had come to assess her possessions for the purpose of estimating income tax. A quarrel broke out between Nwayereuwa and Mark, who had been appointed by Chief Okugo, the warrant chief of Oloko, to carry out the enumeration exercise. After this incident, the women of Oloko went to Okugo to demand explanations on why women were being assessed for taxation. When Okugo did not give a satisfactory explanation, the women of Oloko and neighbouring towns assembled at Okugo's house on Sunday, 24 November 1929, and employed women's traditional protest

A sketch of the areas affected by the Women's Revolt in Okigwe and Bende Divisions (PRO, CO 583/22).

strategy described as "sitting on a man."[11] This strategy involved gathering in his house, calling out insults, and sometimes exposing their nakedness to humiliate him. Tempers flared when Okugo ordered his servants to drive the women away. The ensuing protest drew thousands of women from different parts of Owerri Province. The women's protest would come to constitute the most significant challenge to British authority in Nigeria. The protest spread to most of the Owerri and Calabar provinces, stretching from Okigwe in the north to Andoni close to the Kwa Ibo River (in the south), and from Owerri on the west to Umon and Itu on the Cross River (east). The revolt left death and destruction in its wake, including the killing of fifty-three women, the destruction of a large amount of public and private property, and the looting of European-owned companies at Imo River, Aba, Mbawsi, and Amata.[12]

Scholars of Igbo Studies and other commentators have written more on the 1929 Women's Revolt than on any other single event in the history of colonial Nigeria. The early writings on the revolt centred on two important elements. The first is the general conditions prevailing in this part of the colony, relative to other parts of colonial Nigeria at the time. The second is the role, place, and condition of women in the colonial context, including the exclusion of women from colonial institutions. Sylvia Leith-Ross's seminal work, *African Women* (1939), addressed the above issues.[13] Margery Perham, Harry Gailey, and Adiele Afigbo saw the revolt as an early expression of African nationalism. U. C. Onwuteaka, like Afigbo, linked the women's revolt to the implementation of indirect rule in Eastern Nigeria, which they argued was foreign to existing political structures.[14] As I will discuss later in this chapter, as far as rural people were concerned, the impact of the political conditions in Igboland in 1929 was directly related to the economic conditions that gave rise to the revolt.

Feminist scholars have also found the revolt fertile ground for a gendered analysis of the colonial encounter and the visibility of the women of eastern Nigeria. Most have drawn on the revolt to assert women's autonomy and independence in pre-colonial times, the threat that colonialism posed to that autonomy, and the exhibition of female agency, as much as women's resistance to colonialism.[15] One of the early feminist commentators on the 1929 revolt, Judith Van Allen, portrayed the revolt primarily as a political protest in which women employed a feminist method of protest, "sitting on a man," to regain their pre-colonial political roles.[16] Nkiru Nzegwu suggests that the revolt was an attempt by the women to prevent the erosion of their rights and an expression of a female consciousness and solidarity.[17] Women's political consciousness was directed toward the restoration of equitable gender relations, which had been disrupted by the colonial patriarchal social and political order. [18]

Scholars have also stressed the exceptionalism of Igbo women in colonial Nigeria. The women who struck back at the colonial authority in 1929, when they feared that their livelihood and lifestyle were in jeopardy, were not led by well-organized political leaders. Their actions were not framed around any major ideology, but their peasant roots informed their consciousness. Indeed, they were ordinary women who led routine lives as peasants, wives, and mothers. Their exceptionalism, in comparison to women elsewhere in colonial Nigeria, can be found in the structures of Igbo society and the significant social and economic authority that women enjoyed in an area characterized by a high degree of complementarity of male and female roles. Even though the colonial institution did not include women, they were not completely disempowered. Their action in 1929 was perplexing to colonial officials and defied their expectations of women's behaviour. So, in the sixty years of contact between the Igbo and imperial Britain, the 1929 women's rebellion stands as the most notable of the many revolts and confrontations that characterized Anglo-Igbo relations and became important in redefining colonial policy in Nigeria.

While this incident has received considerable attention, we can draw conclusions that have hitherto been neglected, including its roots in the agrarian economy, by re-examining two important questions that are central to a re-evaluation of the incident. What were the causes of the revolt? And to what extent were they rooted in the crisis in the agrarian economy of the Igbo in the 1920s? The following sections will explore these questions, by emphasizing the rebellion's agrarian roots and by linking other political and social grievances raised by the agitators to the economic condition in the rural areas in this period.

THE AGRARIAN ECONOMY AND RURAL PROTEST

Few minor protests related to socio-economic conditions had occurred in Owerri Province prior to the 1929 Women's Revolt. But the declining economy in mid-1920s provided fertile ground for the protests that frequently occurred among the Igbo in this period. One was the 1925 women's "Dance Movement," which called for Europeans to leave. This movement came to be known as the *nwaobia la* (literally, strangers leave) protest. The protest started in Atta, in Okigwe Division of Owerri Province, as a result of a message said to have been received from God.[19] The message included forbidding men from growing cassava, regarded as the women's prerogative. Parts of the demand included banning the use of European coins, fixing prices of foodstuffs in the markets, and regulating the cloth that women and girls wore.[20]

The protest was anti-European and anti-Christian but its economic root was evident.[21] The "dancing women" were aggrieved about the high cost of staples in the market. This movement soon affected most parts of Igboland. The women complained about the moral laxity that came in the wake of colonialism and Christianity. Undoubtedly, women regarded colonial administrators, the missionaries, and the European traders as one entity – foreigners whose intervention was responsible for economic and social upheaval. They demanded that there be "no more Government and no more Native Courts" and that there should be a return to "old customs."[22]

Officials like the senior Resident for Onitsha Province were clearly concerned about what the Resident described as bands of women "preaching their own ideas of desirable reforms."[23] While officials interpreted the women's movement as a disturbance of the peace and order, they failed to comprehend the level of discontent among the African population as a result of the low price for palm products and the general economic distress in the rural areas, which was also caused by the incessant price increases in basic staples. Despite being framed around traditional values and a rejection of what was seen as European, the 1925 revolt was rooted in the peasant economy.[24]

Sporadic protests continued for the following two years, mostly related to the produce trade, particularly the introduction of produce inspection and a new system of buying produce by weight. In 1926, women protested against the low prices of palm oil and kernels as well as the new method of buying produce by weight, which replaced the old method of buying by measure. Most upsetting to rural farmers was the steep decline in revenue. In March 1927, palm oil sold for £20 per ton in parts of Owerri Province. By December, it had declined to about £18 per ton, although kernels sold for between £13 and £14 per ton in the same period.[25] In Aba District, the price had gone from 7 shillings to 5 shillings. Attempts by the buying agents and the international trading companies to control prices forced producers to demand price stabilization. Groups of women petitioned colonial officials and European firms asking for "fair prices" for palm produce. The demands made by Obowo women in a petition to the district officer of Okigwe District included fixing the price of one tin of palm oil at 10 shillings, and fixing the price of a bushel of palm kernels at 7 shillings. They threatened that "no products [will] be sold if these proposals are not granted."[26]

In these lean periods, the colonial authorities introduced new measures to stabilize and sustain the local economy – measures that rural people perceived as a threat to their own survival. The year 1928 also saw the introduction of produce inspection to improve the quality of produce. As the effect of produce inspections was increasingly felt among local producers, the discontent that grew, especially among women, culminated in petitions to colonial officials. Women complained about the inspection, which they regarded as undue interference with their trade.[27] In Okigwe Division, women complained about the interference of produce inspectors in local trade.[28] Some women suggested strikes and curtailing production of oil and kernels to force concessions and increase prices.[29] Certainly, withholding the supply of palm produce on a large scale could have had an impact on prices, since Eastern Nigeria supplied the bulk of the total world output. However, it was unclear how these protest measures could be implemented under a production system dominated by thousands of independent small-scale producers. The geographical spread of producing areas and the lack of rural organizational structure that could represent producers certainly precluded any outright cut in the supply of palm produce.

With their dwindling income, the region's farmers severely felt the impact of the depression years. The situation was probably worse for women because of their role in providing the bulk of household food needs. In subtle and less subtle ways, women in Owerri Province responded to the growing insecurity in the local economy. Caught by the fall in prices and insecure incomes, "women adulterated palm oil and cut back production and mixed palm oil with water to increase the volume," Eliazer Ihediwa remembers.[30] Producers partially cracked kernels, and mixed cracked kernels with uncracked ones, to increase the weight. Rural men and women grumbled against government, believing that officials and the European trading companies were responsible for the fall in the price of palm oil. This conviction and the impact of the depression on the local economy ultimately influenced the timing of the 1929 Women's Revolt.

THE DEPRESSION AND AGRARIAN ROOTS OF RURAL PROTEST

The world economy was in a depression toward the end of 1929 because the economic crisis that began with the crash of the American stock market in 1929 had dire effects all over the world. Obviously, as the most severe peacetime economic crisis of the twentieth century in colonial Africa, it hit the majority of rural Africans with great force.[31] But the depression began much earlier for many in rural Africa. Among the Igbo, rural producers whose livelihood was dependent on the sale of palm produce and who had not enjoyed relatively good prices after the First World War, found themselves struggling to sell their oil and kernels even at very low prices. Although the depression in 1929 was by no means the first time farmers had experienced economic hardship and low prices for agricultural produce, it was clearly the worst period and the effects were more biting than at other bad times.

Even European traders were not spared the changing economic fortunes brought about by the depression. They had been directed by their parent companies in Europe to reduce their purchases of oil palm due to rising unemployment and an economic downturn in Europe. It was without much enthusiasm that the directive was implemented by European traders. Many had come to develop personal relationships with African traders who had dealt with them over time. One such trader who witnessed the social and political changes taking place and who sympathized profoundly with local traders severely affected by the slump and the decline in the demand for oil palm was Raymond Gore Clough. Clough, who had joined the Niger Company a year before the outbreak of the Women's Revolt, wrote in his memoir that "things reached their depth" in the last quarter of 1929. "Merchandise ordered long before," he remembered, "continued on its unhurried way to the beaches of Olomo, at the same time as the traders were taking less and less palm oil."[32] The whole process was "bewildering to the Africans who saw the factories bulging with goods which the Whiteman had introduced to them, and which had become a need, and sometimes a craving."[33]

Company agents in Eastern Nigeria were left to manage the impending crisis. Clough noted the "sullen looks on the faces of the hitherto friendly canoemen as they went from factory to factory demanding better *Good For's* for their casks of oil."[34] Madam Umunna and Madam Osika, who had been regular customers at Megwana trading post at Olomo, expressed the sentiments of many local traders when they spoke to Alexander MacKay, the district agent of the Niger Company at Megwana beach:

> Mackay, sah, we come to tell you they be plenty palaver for bush
> – the people no savvy why *Megwana*, *Sunflower*, and the other factories say 'no' for the palm oil when they done bring am when for long time you and all the Agents give plenty *Good For* [,] for palm oil. The people want salt, cloth and gin, but the factories no fit to give am now, though the people know you all get plenty for store
> – plenty, plenty![35]

The sale of produce was governed by the demand from the commercial companies. Although European agents did their best to explain the worldwide depression in simple terms, local traders did not follow such reasoning or comprehend how events outside their local environment could alter their fortunes drastically within a very short period. Many local traders believed that the slump was a manoeuvre by the traders and was linked to the tax imposed by the government. With the small size of the police presence, many European factories and trading stations were at the mercy of the women. Trading factories were closed as agents waited to see a resolution.

In Nigeria, as in many other colonial territories, the kinds of social programs that ameliorated the effects of the depression in Europe and the United States were lacking.[36] The Igbo region, which depended on a single product (palm oil), was even more vulnerable when the depression set in as it was subjected to severe price fluctuation and shortages.[37] Basden wrote about the precarious nature of the palm oil trade and local perceptions of the change in the economy during the depression: "The fluctuation of price for this raw product is a very disturbing element. The untaught native does not understand the vagaries of world markets and, when there is a slump, he is puzzled, not to say disgruntled, when he cannot sell his oil, or can only dispose of it at a low price."[38] The depression left producers with "less money to spend, and that means that all prices depreciate proportionately," Basden observed. At the same time, "The cost of food has more than doubled" since the beginning of the twentieth century, while "there is a much wider variety of imported foodstuffs on sale."[39] Households faced rising inflation. Consequently, the difficulty of meeting household food requirements increased women's tasks in both agriculture and trade. The inability of many men to meet colonial tax obligations also increased the burden on women, who often paid taxes for their indigent spouses or sons.[40] These problems were compounded by the significant fall in produce prices. In fact, from 1929 onward, the value of palm produce trade dropped progressively, declining by over 70 per cent by 1935. This led to a substantial decrease in peasant incomes and government revenue.[41] Thus, the condition was ripe for the revolt that would occur towards the end of 1929. We must then look for the causes of the revolt in 1929 in the economic crisis originating from the severe economic depression of the late 1920s, characterized by falling prices for export goods, especially palm oil and kernels. The tax incident only lit the fire on an already tense situation.

Raymond Clough wrote in his memoir that "By some queer twist of reasoning," African women "associated the tax and the sudden recession in the trade with each other."[42] But their grievance was not an imaginary one. In Umuahia and surrounding districts, the price of mixed oil fell from 6 shillings and 11 pence in January 1929 to 5 shillings and 5 pence in December 1929. The price of edible oil fell from 7 shillings and 4 pence to 5 shillings and 11 pence in the same period. The palm kernel price fell from 5 shillings and 10 pence to 4 shillings and 5 pence in the same period.[43] Women's responses and testimonies further substantiate the agrarian and economic root of the women's protest. On 4 December 1929, for example, women gathered at Umuahia to discuss the low prices of produce. By this time, the price of a four-gallon tin of palm oil in Umuahia District had fallen from six shillings and eight pence to five shillings.

At the Aba Commission of Inquiry on the women's revolt, women seized the opportunity whenever they were asked to state their grievances to put forward the low price of produce as one of them.[44] One women's leader, Nwanwanyi, during a meeting with company agents at Umuahia said: "We wish to discuss the price of produce. We have no desire or intention of making any trouble but we have fixed a certain price for palm oil and kernels and if we get that we will bring them in. We want 10 shillings a [4 gallon] tin for oil and 9 shillings a bushel for kernels."[45] This paints a broad picture of the economic dilemma that rural households faced during the 1920s and 1930s as well as their actions to ameliorate their difficulties.

The effects of the slump were felt in some areas more than others. The Mbaise, Etiti, and Obowo areas of Owerri Province suffered the double effects of over-population and poor soil quality.[46] Intelligence reports from these areas confirmed the widespread dependency on palm oil exports by households and the general level of insecurity that came with the Great Depression.[47] The assistant district officer for Owerri Division reported that the principal products sold in this area in the 1930s were farm produce, palm produce, native baskets and clay pots, cloths, and other articles bought from European stores. The Ekwerazu and Ahiara clans were so poor in resources that they could not support themselves on foodstuffs produced locally. Yams and cassava were brought from Oratta and Ngwa areas to the south and west to supplement what they produced, noted colonial officials. The only commodities the people of Ekwerazu and Ahiara could offer to sell to the outside world were palm oil and kernels.[48] The threat to the palm oil economy, their sole means of livelihood, hit them hardest and threatened their very existence. N.A.P.G. MacKenzie, a British assistant district officer, noted the vulnerability of the Obowo clan in Okigwe Division. MacKenzie linked the inability of young men to marry and settle on the land to the poverty in the area. According to him, only half the women of marriageable age had husbands, and only half the men had wives.[49] MacKenzie's emphasis on marriage, which he saw as critical to the stability of local societies, suggests that the labour of women was very important in household production and economic stability. Women from such parts of Igboland would be active in the 1929 protests for obvious economic and social reasons.

To fully understand the timing of the revolt, however, we must situate it in the context of the political decision made to introduce direct taxation in 1928 and the impact of the taxation on household income.

THE RURAL ECONOMY, TAXATION, AND RURAL UPRISING

Taxes were used everywhere in colonial Africa to force peasants to produce more for the market or sell their labour.[50] Far from being the sign of humiliating servitude, taxation was seen by colonialists rather as proof that the African was "beginning to rise on the ladder of humanity … [and had] entered upon the path of civilization."[51] In Nigeria, the 1917 Revenue Ordinance, which applied originally to the Northern Provinces, was first extended to parts of Southern Nigeria, including the old provinces of Abeokuta, Oyo, and parts of Benin (including Asaba Division), Ondo (1919 and 1920). By 1927, it was extended to the rest of the Southern Provinces.[52] There was no income tax in eastern Nigeria until 1928. Since the cost of government was underwritten by export taxes, the need to finance government expenditure, including expenditure on public works, was an important reason to extend taxation to the Igbo country. For colonial officials, the answer lay in the introduction of taxes to be paid in European currency.

Taxation was seen as a corollary to the abolition of the slave trade, the civilizing mission of colonialism, and the uplifting of the dignity of the African population.[53] Frederick Lugard, first governor general of colonial Nigeria, and a main architect of British policy in tropical Africa, had also argued that direct taxation was of moral benefit to the people and would stimulate industry and production, promote the circulation of currency, and expand trade.[54] Another benefit of direct taxation, according to Lugard, was its "great importance as an acknowledgement of British Suzerainty."[55] In Lugard's judgment, contact between Africans and colonialists during the assessment and collection of taxes would bring African "tribes into touch with civilizing influences, and [would promote] confidence and appreciation of the aims of Government."[56] The debate around taxation was viewed as a moral crusade and part of the "mission to civilize," that would create and enforce native authority, lead to the evolution of "tribal" societies, and end internal slavery, which was still prevalent in this period.[57]

When Sir Graeme Thomson became governor of Nigeria in 1925, it was apparent to him that some reorganization was necessary in order to introduce direct taxation to the Colony of Lagos and the five provinces in the south where direct taxation had not been introduced. W.G.A. Ormsby-Gore, parliamentary under-secretary for the colonies, had emphasized the political and financial importance of taxation during his visit to West Africa in 1926:

> It is important to remember that the acknowledgement of authority and the rendering of some kind of tax or gift to that authority are inseparable conceptions in the native mind. No African recognizes the authority of the chief to whom he does not pay something in cash, or kind, or in service, and any refusal to pay tax amounts to an actual or potential refusal to obey authority.[58]

The justifications for taxation were contentious from the beginning for several reasons and did not sit well with many Africans who saw it as unwarranted interference in their lives. The local response was generally negative, although not coordinated between different parts of the protectorate. Considerable opposition came from local chiefs and leaders. The government was aware of the potential for conflict. The Legislative Council in Nigeria was asked to increase the police force by 500 men to curb any disturbances if the situation warranted such an action.[59] However, in Owerri Province, colonial officials reported that "the general attitude throughout the whole Province is most satisfactory."[60]

Other district officials were less optimistic. "Considerable difficulty is anticipated," the Resident of Onitsha Province wrote in May 1928.[61] "There is no active opposition, but the measure is very unpopular; escorts may be required for District Officers in the bush at first," wrote colonial officials from Calabar Province.[62] Deputy Governor Baddeley noted that there was considerable "agitation against the tax" from Awka.[63] However, considerable progress was made in the enumeration exercise in Awka until November 1927, when itinerate Awka blacksmiths began to return to the area. Agitation about the tax began and oaths were sworn in many areas of Awka to "refuse payment of the tax" and "boycott the Awka Native Court." The agitation at Awka led to reductions in the number of cases brought to the native courts. The district officer for Awka had hoped that chiefs would help in quashing the movements. W. Buchanan Smith, Resident for Onitsha Province reported: "the prospects of obtaining payment from the rest of the Division depend entirely on the payment first by Awka itself; the question of 'breaking the oath' is in this instance more important than usual."[64] Awka agitators began collecting subscriptions from their people to petition the English king. Buchanan has argued that delays in implementing the Native Revenue Amendment Ordinance of 1927 were creating an opportunity for the agitators to line their pockets with the money collected from what Buchanan characterized as "poor deluded people."[65]

It had become sufficiently clear to officials how Africans would respond to the imposition of taxation. W. Buchanan Smith, Resident in Onitsha Province, noted in 1928 that, despite many meetings held by the district officer in Awgu Division to explain his position, and although the temper of the male population was markedly good humoured, the women were less so and made definite attempts to break up some meetings by indulging in unceasing song.[66] The general attitude of the people from Obubra, Ikom, Ogoja, and Afikpo "seems to be one of acquiescence," although it was unclear what their attitude would be when the "actual demand for payment is made," the Resident of Ogoja Province wrote.[67] In early 1928, the district officer in Abakiliki Division found it difficult to obtain an accurate count of the adult male population, particularly among some of the Ezzi clans, as people evaded the enumerators. Awareness of the impending taxation was created by people moving between districts as traders or employees of the colonial government. The Ezzi chiefs, in their petition against taxation, seem to have been informed of the impending taxation by persons from Aba and Onitsha districts connected with road construction in the area.[68] Officials remained pessimistic and "prepared for considerable difficulty in the collection of the tax."[69] One tactic adopted by the Ezzi to avoid enumeration was to go to the farms, thus avoiding direct conflict with officials, forcing officials to prosecute a number of people before some figures could be obtained.

Perhaps the most frequent means of protest was that of individuals who petitioned colonial officials. Individuals made a great variety of requests and complaints to district officers or Residents, including petitions for exception from taxation or reduction of income tax based on their economic conditions. Native court chiefs in Abakiliki District petitioned the Secretary of State for the colonies in January 1929. In their petition, they noted that people under their jurisdiction were opposed to a poll tax, arguing that "many male adults in Abakiliki District do not at one time possess 7 shillings," obviously a considerable amount for many rural people.[70] Thus, many local people sought ways to evade taxes altogether or reduce their tax burden. A. E. Cooks, who served for fifteen years as an administrative officer in colonial Nigeria, recalls in his memoirs: "Tax evasion is a popular pastime, in fact it is hardly an exaggeration to say that each of the three thousand or so family heads in the Division ... is continuously engaged in a grim struggle to effect, by hook or by crook, a reduction in the tax assessment of his family."[71]

Almost all opposed the introduction of direct taxation. For the newly appointed warrant chiefs, the collection of taxes constituted one of their most important functions and symbolized their power over the villages where they reigned. The collaboration of the new powerful indigenous chiefs stemmed the tide of protest in some areas. Chief Onyeama of Enugu was one of those that stood on the side of the administration. As a willing collaborator, Onyeama was responsible for allegedly throwing out some "anti-tax agitators" from Awka.[72] Onyeama, like many other warrant chiefs, benefited directly from taxation. These chiefs' salaries and allowances were directly dependent on the amount of tax collected in their areas.

Following the enactment of the Native Revenue Amendment Ordinance, direct taxation was introduced in April 1928 based on 2.5 per cent of personal income, after the careful propaganda of the preceding year.[73] A special police unit, which then numbered 417, was used to maintain the peace. An additional 250 were to be recruited for the 1929 and 1930 tax period.[74] But opposition continued because the tax burden caused a significant amount of stress for the local population and forced many to devise ways of avoiding intimidation by local tax collectors and officials.

Additionally, the imposition of taxation coincided with the slump in palm produce prices, which was the principal source of income used for the assessment of incomes in 1927. The enumeration exercise in Oloko Umuahia, which included the counting of women and livestock, raised suspicions that women would be taxed as the men.[75] Women expressed the view that the tax on men was already a big burden on the household as some men had been forced to pawn themselves or their children in order to pay taxes.[76] In Ngwa region, taxation forced communities to collect palm fruit communally, for a period of three months in the year, in order "to give everyone a means of paying tax."[77] Indeed, the district officer for Owerri had noted in 1928: "The amount of petty trade done by women does not appear to warrant separate assessment as it consists largely of sale or barter of farm or palm produce."[78] However, the general belief that women were about to be taxed raised a general level of apprehension among them.[79]

Since the historiography is deficient in its treatment of the economic roots of the revolt in relation to the local economy when taxation was introduced, let us turn to the voices of rural women and men as recorded in the "Notes of Evidence" taken by the commission of inquiry set up to examine the causes of the revolt. Their voices allow us to understand the range of their emotions, their motives, and what they believed were their responsibilities to their families and communities and the responsibilities of the colonial government to them. Ikodia, one of the women of Oloko who participated in the revolt, summarized the feeling of women: "We heard that women were being counted by their chiefs. Women became annoyed at this ... as they did not wish to accept it.... We, women, therefore held a large meeting at which we decided to wait until we heard definitely from one person that women were to be taxed, in which case we would make trouble, as we did not mind to be killed for doing so."[80] Nwakaji, of Ekweli in Oloko, asked: "How could women who have no means themselves to buy food or clothing pay tax?" And Uligbo of Awon Uku, Oloko, asked how women who depended on their husbands could pay tax: "we cannot buy food or clothe ourselves: how shall we get money to pay tax?"[81] Nwanyiafo Obasi, whose mother participated in the revolts in Mbaise, confirmed that women were infuriated by the prospect of new taxation. According to her, "in Igbo tradition women were not required to make cash contributions to community development, but the white people were trying to introduce a new rule and women rejected it."[82] Enyidia, another leader of the women's movement from Oloko, lamented: "What have we women done to warrant our being taxed? We women are like trees, which bear fruit. You should tell us the reason why we who bear seeds should be counted."[83]

Women did not expect to be taxed on account of their femininity and reproductive roles. Adiele Afigbo noted that the reference to women as fruit-bearing trees or reproducers of humans and the perception that they would be taxed raised a "very strong moral and psychological dilemma" that "lies at the root of certain aspects of indigenous social and ethical philosophy."[84] Thus, just as "one cannot, in the interest of human beings deal lightly with the survival of fruit-bearing trees, one could not play with the fate of women."[85] The contrast between local ideas of taxation of women and British ideas of personal income tax also touched on certain ethical aspects of Igbo society.

Indeed the Native Revenue Amendment Ordinance was introduced with a comparatively small knowledge base on the social organizations of the southeastern provinces.[86] In Owerri Province, for example, the extended family or village group formed the responsible units, and the emphasis on "individual responsibility" challenged communal responsibility and the sort of communal humane living that defined Igbo social relations. S. M. Jacob, former government statistician, wrote that English law as applied to Nigeria did not admit "communal responsibility for tax payment."[87] The Aba Commission confirmed that:

> It was the expressed intention of the Government in extending direct taxation to these Provinces to fix and make definitely known the liability of the individual and the doctrine that the individual's default is to be made good by the community collectively seems to us a misinterpretation of the declared policy.[88]

The method of assessment, *West Africa* noted, was a cause for uneasiness. The newspaper noted that some members of the local population had become suspicious due to interference with their land tenure by the Department of Agriculture in order to "establish small palmeries." When the enumeration exercise of 1926 was carried out, the Resident of Owerri was not open about why the enumeration was being carried out; the people felt deceived when taxation was later introduced. There was complete loss of popular faith, both in the administration and in those chiefs who sought to carry out its orders. This popular mistrust reached what *West Africa* termed a "dangerous level" three years later when the acting district officer for Bende, on his own initiative tried to update the tax system.[89]

The timing of the revolt, therefore, has to be linked to the frustration of the local population, who associated the introduction of taxation to the economic conditions of the time, particularly the fall in the price of palm produce.[90] The low price of palm produce and the high price for imported commodities, now out of the reach of many households, offer some perspectives on what fuelled women's anger in 1929, the reasons for their hostility to the colonial state, to European firms, and to the firms' representatives.[91] Because the colonial administration failed to recognize the interdependent nature of the domestic and formal economy, the officials had the false impression that men were better off than women. But the women's response suggested otherwise. Hence, the women's demand extended to the abolition of taxation on males, an increase in prices of produce (palm oil and kernels), and a decrease in the prices of imported goods.[92]

There were also other factors responsible for the revolt. Beyond the economic factors was the deep distrust for the new political institutions introduced under colonialism, especially the indirect rule system imposed on the Igbo.

POLITICAL AND SOCIAL ROOTS OF THE REVOLT

The political and class divisions created by the imposition of colonial rule had deepened by the late 1920s. The Igbo were antagonistic to the warrant chief system and the political and economic privileges the warrant chiefs enjoyed under the new dispensation. The emerging political elite with economic and political interests to protect clearly sided the British as indicated earlier. In response, local people openly expressed detestation for the warrant chiefs, who acted in a manner contrary to local political ethic. Besides, their methods of dispensing justice under the new administrative system drained the people's resources. The testimony of many people at the Aba Commission of Inquiry reflected this antagonistic relationship between the chiefs and local people. Nwanyeruwa of Oloko told the commission: "Okugo became a rich man because of the money he got from us. If he had not got money from us, he would not have been able to provide for himself."[93] In addition, some warrant chiefs were noted for their ability to exploit their subjects with impunity. Okugo, for example, was said to have imposed levies on the entire community in the pretext that he had the mandate of the district officer. Nwanyeruwa narrated such incidents to the commission. On one occasion, Okugo had called both men and women together and told them that the district officer had ordered that money should be collected for him to build a house. The villagers contributed 20 pounds towards this project. On another occasion, Okugo told villagers that the district officer "had been worrying him for a young wife and that both men and women should collect money to pay the dowry of a young wife for the District Officer. We collected the sum of 20 pounds sterling and gave it to him.... We are sure these women were not given to the District Officer."[94] Although most villagers were aware that European officials were not responsible for these levies, many could not speak out for fear of reprisals. These practices were common among warrant chiefs and members of the native courts.

In the memories of those who lived through the era, native court members were worse than many European officials. They were often bribed by litigants and many grew rich and powerful in the process. The warrant chief of my own town, Philip Eluwa, was illiterate, yet he, like many others, sat as a judge of civil and criminal cases. The warrant chiefs learned on the job, Eze Enyeribe Onuoha remembered, and soon became "experts" in "handling cases."[95] "I wish to say something about Chiefs," Ahudi, a female witness from Nsidimo, told the Aba Commission of Inquiry.

> Women are very much annoyed. If I had a case with another in the Native Courts, that case would not be heard until I kept borrowing money, about £10 in all. If I do not borrow money, the case would be kept waiting for six months. That is what Chiefs do.... I want to tell you that these disturbances will go on perhaps for fifteen years unless these Chiefs are decapped.... Otherwise the trouble will go on.[96]

These testimonies were worded in such a manner as to demand change and not to attract sympathy. Initially, the women had directed their attacks against the warrant chiefs and their courts, but the revolt soon was directed against the colonial administration and the factories of European traders. Until this point, how to attack the native authority system was a perplexing problem for the local people. It appears also that the colonial authorities in Eastern Nigeria were remarkably ignorant of the level of corruption in the native authority system. During the rebellion, Native courts were destroyed or damaged and the chiefs were challenged because the native courts were seen as the outward symbol of the colonial government. The warrant chiefs, as a class

A court-house destroyed during the Women's Revolt. (Reproduced with the kind permission of the Bodleian Library, University of Oxford.) RH, Mss Afr. s. 1000, Edward Morris Falk papers.

Villagers gather at a court-house destroyed during the Women's Revolt. (Reproduced with the kind permission of the Bodleian Library, University of Oxford.) RH, Mss Afr. s. 1000, Edward Morris Falk papers.

Group of villagers standing near a court-house during the revolt. (Reproduced with the kind permission of the Bodleian Library, University of Oxford.) RH, Mss Afr. s. 1000, Edward Morris Falk papers.

Colonial troops used to suppress the revolt. (Reproduced with the kind permission of the Bodleian Library, University of Oxford.) RH, Mss Afr. s. 1000, Edward Morris Falk papers.

and as members of the local administrative system, were seen as the instruments of a European government. "The Court Members whether customary heads or not were assaulted or had property damaged without discrimination," the Resident for Owerri Province wrote in a memo to the secretary of the Southern Provinces. "Of approximately 150 Court Members in the Division the members who escaped some form of indignity could be counted on both hands," he remarked.[97] The women "only attacked houses belonging to court officials and people connected in some way with the court," the district officer for Ahoada wrote. "About 31 court members suffered damage," he concluded. And in Okigwe, the district officer related that several court members were attacked and their houses looted.[98]

Many chiefs were unjust. Overwhelming evidence was provided at the commission's hearings regarding the "persecution, extortion, bribery and corruption in the native courts."[99] The commission rightly believed that political discontent over the "persecution, extortion and corruption by the native court members (Warrant Chiefs) was [a] principal contributory cause."[100] The commission concluded that "although allegations of corruption and bribery were of a general nature, we heard enough to be satisfied that persecution by native courts members and corruption in the native courts are a source of very considerable discontent among the people."[101]

Overall, the immediate cause of the revolt has to be located in the economic conditions of eastern Nigeria, the depression in the economy from the beginning of the 1920s, the agrarian economy and link to a capitalist world market, which directly affected the lives of Igbo peasants. Indeed, the political grievances articulated by the peasants during the revolt were deeply rooted in the economic stress in the rural society in the 1920s. Although colonial officials admitted the existence of widespread economic distress due to the introduction of tax and the slump in produce prices,[102] they were not sympathetic to the communities that engaged in the revolt. For the colonial government, the imposition of a collective punishment would deter future revolts and perhaps teach the men a lesson or prevent their wives from engaging in such acts in the future.

A POUND OF FLESH: THE COLLECTIVE PUNISHMENT

In connection with the disturbances generally, I think a bit too much fuss has been raised by the fact that the victims were of the gentler sex. We are liable to forget that the Kings of Dahomey's Amazon bodyguard was not a fiction, but an unpleasant fact, and if a howling mob of excited female savages who would be quite ready to tear a man in pieces with their hands is about the place, the only thing to do is to take strong action. It is quite easy for us to criticize them here, but I wonder what anyone in this Office would do in a similar situation – *Colonial Office, London, 1930, PRO, CO 583/176/9*

The position of the Colonial Office, as expressed above, was probably shared by many officials in the colony. The protest was seen by British officials as a threat to authority and a disruption of economic and political life. Europeans in the colony had hoped that the revolt would end quickly. Mrs. Falk, whose husband was the district officer for Calabar, wrote in her journal on 8 December 1929 that steps were taken as soon as the outbreak of violence occurred "to frighten all the other grumblers sufficiently to keep the peace." The troops, she wrote, were "simply dying to be called in.... A few of the young officers ... are itching to go and get a chance to shoot. They fervently hope that the political officers will not be able to settle the affair with the help of the police only."[103] However, things got worse. A. B. Henderson, supervising agent of the United Africa Company noted that the attitude of the revolting women "was far from peaceful." According to R. L. Attwood of West African Motors, Aba, "the women were all in very aggressive mood, right from the start, and most of them were armed with heavy sticks, which they did not hesitate to use to damage property."[104] The assistant district officer for Bende, J. Cook, told the commission of inquiry that there was determined attitude of hostility towards Okugo as more and more women from distant towns gathered at Oloko market.[105]

As such, the government took drastic measures to suppress the revolt. The police and a detachment of the army were used at various hot spots to disperse the women. This became inevitable after the looting and destruction of European trading posts at Mbawsi and Aba and the threat of extending the rampage to the coastal trading stations. The colonial Resident addressed the European traders at Olomo about an impending attack on the trading posts in the Delta and the measures being taken to safeguard life and property. "Some show of force makes it clear that we shall have to take drastic measures to halt the spread of this astonishing hysteria," he noted. "We have information that the women are massing several thousand strong at various villages and markets in the forest behind this creek. So far, everything has been done to avoid conflict, but now the Government in Lagos has instructed us that a firm stand, with the use of force if necessary, must be made to bring the chaotic position to an end."[106]

The impact of the revolt was felt by all within the affected areas. The British made sure that the local population paid for the damages in cash. The revolt was estimated to have led to the destruction of goods and infrastructure valued at £60,000.[107] Estimates were based on the claims submitted by firms and private individuals and testimonies of witnesses and the value of the loot taken by each village was estimated by "dividing the total amount amongst the total number of women in proportion of each village incriminated."[108] This is perhaps an under-estimation, considering the destruction of personal property and the large number of police and army members that were transported and fed while the revolt lasted.

Throughout the affected areas, huge sums of money were collected by local district officers as compensation for aiding the revolt or for personal and government property destroyed by the rebels. The British idea of male complicity meant that men directly bore the cost of the revolt. In doing so, the British were hoping to teach the men a lesson and make them do a better job of controlling their wives in the future.

The Igbo have an adage, which says: *Otu aka ruta nmanu, ya ezuo ibe ya* (When one finger is dipped into palm oil, it smears the other fingers). The collective punishment imposed by the British on participating communities was their "pound of flesh," and as in the Igbo adage, it smeared all fingers. The amount collected from each community was based on the adult male population and a percentage of the tax rate and on the level of participation. The amounts imposed varied from a few pounds for communities that did

Table 4.1. Collective fines on Nguru area, Owerri Division.

TOWN	FINE IMPOSED		
	£	S	D
Inyogugu	626	11	3
Umunama Town	94	5	0
Nguru	774	1	3
Onicha	25	1	0
Umuhu	218	15	0
Lagwa	218	8	9
Avuvu Town	275	0	0
Ibeku Town	186	10	6
Azaraegbelu	10	6	3
Udo	96	8	6
Ahiara Town (Nguru Area)	1,000	0	0
Amuzi	171	12	11
Umuokrika	160	0	0
Amumara	420	0	0
Itu	258	2	0
Obizi	351	5	3
Eziborgu	150	8	9
Oboama	193	18	9
Ahiara	43	0	0
Akaba (Nguru)	79	1	3
Ihitte	280	0	0
Ogtuama	44	1	3
Eziudo	417	0	0
Ugiri	60	6	3
Amumara	50	0	0
Okpofe	215	0	0
Mpam	294	0	0
IhitteAfuku	320	0	0
Total	7,033	3	11

Source: NAE, UMPROF, file no. C.53/1929, vol. 26.

Table 4.2. Statement of deposits taken from towns.

TOWN	DEPOSIT (£)	INCIDENCE (S/D)	AREA	REASON
Nguru	250	1/8	Nguru	Deposit taken by Captain Wauton, reason not known.
Umuhu	30	2/–	Nguru	Took active part in disturbances hence incidence* 3/– more than 25% of tax incidence.
Ime Onicha	475	7/–	Nguru	Took active part in disturbances, truculent and a murder was committed in this town hence 5/3 over 25% of tax incidence.
Onicha Ama	375	7/–	Nguru	Same as in Ime Onicha.
Ngor	50	1/6	Ngor	Deposit taken by Captain Wauton; reason for incidence not known.
Ntu	30	1/–	Ngor	Same as in Ngor.
Obokwe	10	1/4	Ngor	Took no active part, hence –/5 under 25% of Tax incidence.
Umukabia	20	2/–	Ngor	Same as in Umuhu.
Nguru	50	1/4	Ngor	Took no active part, hence –/5 under 25% of Tax incidence.
Obike	80	1/4	Ngor	Took no active part, hence –/5 under 25% of Tax incidence.
Emweinwe	95	1/4	Ngor	Took no active part, hence –/5 under 25% of Tax incidence.
Umukam	50	1/4	Ngor	Same as above.
Orisa Eze	20	1/4	Ngor	Same as above.
Muokoro	5	1/4	Ngor	Same as above.
Elelma	25	1/4	Ngor	Same as above.
Ngwoma	30	2/6	Ngor	Took part in attack on Olakwo on 21/12/29 and in demonstration against Troops on 22/12/29 hence –/9 over 25% of Tax incidence.

THE LAND HAS CHANGED

Table 4.2. (cont'd)

TOWN	DEPOSIT (£)	INCIDENCE (S/D)	AREA	REASON
Loghara	90	2/6	Ngor	Same as in Ngwoma.
Umohiagu	100	2/6	Ngor	Same as above.
Umowa	60	3/–	Ngor	Took active part in closing main Owerri-Aba road, hence 1/3 over 25% of Tax incidence.
Ihitte	65	3/–	Ngor	Same as Umowa.
Isubiangu	120	3/2	Ngor	Lead attack on Olakwo on 21/12/29 and had demonstration against troops on 22/12/29 hence 1/5 over Tax incidence.
Obokwe	30	2/8	Ok-pala	Took active part in disturbances hence –/11.
Norio	55	2/6	Ok-pala	Took active part in disturbances, hence more than 25% of Tax incidence.
Eziama	300	3/9	Ok-pala	Spread disturbances area and thence into Ngor hence 2/– more than 25% Tax incidence.
Oboro	24	2/9	Ok-pala	Took active part in disturbances, hence 1/– more than 25% of Tax incidence.

Source: NAE, UMPROF 1/5/21, file no. C.53/1929, Vol. 21, district officer, Owerri, to Resident, Owerri Province, 19 January 1930.

*Incidence is probably the normal tax rate for the community.

Table 4.3. Obowo Court Area reasons for detailed statements of deposits.

TOWN	MALE POPULATION	AMOUNT DEPOSITED (POUNDS)	INCIDENCE/(PENCE)
Amumi	393	20	12.2
Alike	821	40	11.69
Okwohia	330	5	3.6
Avutu	657	30	11.0
Umuoke	472	10	5.1
Umilogro	276	5	4.3
Umunachi	365	5	3.3
Ehume	376	5	3.19
Odenkume	305	5	3.9
Umuarian	747	10	3.2
Atchara	151	3	8.0
Umuosochie	228	5	5.26
Umungwa	188	5	6.4
Amanze	216	5	5.5
Umuegehu	416	7	4.0
Amakohia	713	10	3.3
Nkumato	336	5	3.57
Umuihi	510	10	4.7
Amainyi	748	10	3.2
Umunakano	1,143	10	2.1
Lowa Onicha	645	10	3.7
Ikperejere	395	5	3.0
Abeke-uku	586	10	4.1
Nsu	1,346	10	1.78
Total	12,363	240	

Source: NAE, UMPROF, 1/5/21, file no. C.53/1929, vol. 21, Resident, Owerri Province, to the Secretary, Southern Provinces, 20 January 1930.

not take an active part in the revolt to £2000 for the hardest hit areas. In Nguru Court Area, Captain Wauton collected £1130 from Nguru, Umuhu, Ime Onicha, and Onicha Ama for "taking active part in the disturbances" and other unknown reasons.[109] The Obowo Court Area, with an adult male population of 12,363, provided £240. Obohia Native Tribunal Area, which colonial officials described as inhabited by sophisticated and comparatively wealthy individuals, was expected to pay a substantial amount of money into the colonial treasury. Women in these towns aided by the young men were said to have set colonial courts and native administration buildings on fire and rescued fourteen convicted prisoners after the Peace Preservation Order had been proclaimed. Houses belonging to members of the native courts were attacked and looted. Damage to property was estimated at £200, and property worth £550 was looted. For their severe crime, the towns of Akwete, Oham-bele, Obako, Obanko, Obohia, Ohanko, Ohuru, Ohanso, Obunku, Mkpo-robo, and Umuosi in the Obohia Native Tribunal Area paid a combined sum of £2,942. The fine imposed on each community was based on 25 per cent of the normal tax rate (7/-) plus or minus a varying amount depending on the degree of culpability of the town in the revolt.

The 1929 revolt has been presented as gender-specific and has often been portrayed as an anomaly. Why was the revolt dominated by women? What role did men play? What impact did it have on men, women, and the communities involved? If we consider the 1929 revolt as a feminist revolt, as some scholars have presented it, we overlook critically important dimensions of rural political activism, their gendered nature, and the broader context in which the women placed their demands with regard to the British colonial authorities and the few African men who served in local administration. Admittedly, women drew upon traditional forms of political language and discourse to articulate their demands. They used female-specific ideology in framing their actions, but they did not act in the interest of women alone. As one colonial official acknowledged, women suggested that even men "should not be taxed."[110] Therefore, the women's revolt reflects the collective experience of the rural population as a social group or class rather than experience along gender lines.[111]

Although opportunities for women remained limited, women did not often question the existence of the empire, as they had accepted it as an inevitable part of their lives by the 1920s. "We wish relations between us and government to be as cordial as those existing between us and the Reverend Fathers," a group of women wrote in a petition to the government. "If there is co-operation between us and government we shall be able to select new men to take the place of those chiefs who have been oppressing us."[112] So, the revolt was a struggle triggered by conflicts deeply rooted in the colonial extraction of peasant resources and declining incomes. A combination of economic concerns existed: the perceived unfairness of colonial trading patterns, price controls, and rising inflation underscored the revolt as a peasant protest that had significance in relation to subsistence and survival.

Women's domination of the revolt was a mark of their importance in the economy in general and the produce trade in particular. The women were fighting for the survival of the household. The comment of the district officer for Owerri reflects this view:

> The introduction of Tax and the consequent necessity of providing ready money has resulted in women having more work to do in preparation of produce. The fact that the fall in prices of produce has resulted in less money being forthcoming from this extra work than would have been the case had prices been maintained at their former level, has caused discontent among the women.[113]

Onwatugo of Akabo in Owerri Division told the commission that "we have no money to maintain our children, how much more then can we afford to pay tax? If a woman has four or five children, the first thing she does in the morning is to get money to buy food to feed the children."[114]

While we clearly hear the voices of the women who planned and took part in the uprising, it was the view of many officials that men helped the revolt actively through craven inaction. Colonial officials blamed the men for pushing the women into the open while lying low in the background. The Colonial Office concluded that there was apparently no seditious goal to "arouse the women to action; it was simply a case of the movement growing beyond the powers of the leaders to control the worst sections."[115] While African men remained aloof during the protest movement, where their loyalty lay was not in doubt. The domination of these protests by women raised a serious ideological dilemma for the British administration. Some officials propounded pseudo-scientific theories to explain the women's reaction. The secretary of the Southern Provinces in a memorandum opined: "In the dry season women are in a more neurotic condition than other seasons and consequently are more liable to break out in disorder."[116]

Although men had dominated the earlier conflicts and resistance to colonialism, the women saw this fight as their own. Indeed the British had curtailed men's ability to protest on a large scale by the time the women's revolt broke out in 1929. The brute force of previous British military expeditions was still fresh in the minds of many Igbo people. In the areas that became Mbaise, an important site of the women's revolt, for example, the Ahiara massacre of 1905 was fresh in the memory of many. The British expeditionary force had massacred many villagers for killing a white man, J. F. Stewart. In the area affected by the revolt of 1929, many other communities had witnessed the military power of the British during the pacification period. But there was a widespread perception, according to Onyegbule Korieh, "that the colonial officials would not use such force against women."[117] Apparently this perception was wrong, considering the number of women killed during the 1929 revolt.

Overall, Ogu Umunwayi bore a classic resemblance to social movements elsewhere. Sociologist Clifton A. Marsh has argued that "Economic inequality, denial of a voice in the political process, and a subordinate social status ... are breeding grounds for social conflict."[118] Yet the event of 1929 was defined by its agrarian roots and its mobilization of women that would irrevocably change British administration in Nigeria and the lives of many ordinary people. Frustrated by the low prices of palm produce and the treatment they received at the hands of the British-appointed native chiefs, peasant women turned their anger against political institutions in the effort to secure their rights. These women represented a new voice and a vanguard that would eventually force the colonial administration to rethink its administrative philosophy toward the Igbo people, whom most of the British administrators had come to regard as the most intractable of all Nigerian groups.

The character of the revolt and its widespread appeal to women over most of Owerri Province and Opobo exposed the growing chasm between local people and the colonial authority symbolized by the native administration system, which excluded the majority of the local people and their voices. But the dispute at Oloko and the wider crisis that followed had their roots in the structure of the colonial economy and the further integration of the Igbo into the world market. The local economy was under enormous stress during the depression of the late 1920s and early 1930s. The roots of the conflict in the economic decline of the late 1920s and the colonial government's imposition of direct taxation are underscored by numerous references to the low price of palm produce in the testimonies of men and women, some of whom had participated in the protests.[119] The event at Oloko revealed the tenuous nature of African-European relations under colonial rule and the attempt by rural farmers and traders to protect a local economy that had come to depend on the palm produce trade. The revolt in 1929 shared certain basic aims: economic emancipation, social freedom, and an improvement in conditions. Yet its agricultural roots, ideology, and domination by women set it apart from any previous social movements in colonial Nigeria.

Igbo peasants did not often achieve all of their objectives, but their actions ultimately forced reforms. Given the dozens of complainants from witnesses at the Aba Commission of Inquiry, colonial officials knew that allegations of judicial corruption could not be swept under the carpet. The response was drastic and came in the form of suspension from native courts, withdrawal of warrants, imprisonment, or all of these. Warrant chiefs like Ezewuro of Ahiara, Iwuala of Akpoku, Chiaka of Umunama, Chiaka of Umuokirika, Wachuku of Mbutu, Wachuku of Obokiri, Nwankwo of Aluru, Wigwe of Ife, Ihekoronye of Uvuru, Njoku of Oburu, and Nwachukueze of Umudimoka were suspended in February and March 1930 for various offenses including judicial corruption. Some like Ezewuro, Iwuala, and Chiaka of Umunama were imprisoned for six months.[120] The warrants of Chief Ezima of Ihie, Chief Ochinga of Obegu, and Chief Nwalozie of Umuaro, among many others, were cancelled for alleged misconduct during the disturbances.[121]

Major administrative reforms followed the revolt. The arbitrary appointment of warrant chiefs without consultation with the local people was scrapped. Local authority holders or *ezeala* were appointed to replace the warrant chiefs following the administrative reforms that were introduced in the 1930s. For the first time, women were appointed members of the native courts in Nguru Mbaise, Umuakpo, and Okpala.[122] Among the most prominent of these women was Ahebi Ugbabe of Enugu-Ezike, popularly called "Agamega" or "The Female Leopard."[123]

Although the reforms that followed the 1929 revolt have been seen as dramatic, they did not do much to calm official apprehensions about the African population or the agitation of the local people. Nothing could more clearly show that the British attempt to calm the Igbo did not succeed than the protests that continued in Owerri Province soon after the 1929 revolt. Like other protests before them, the rural protests that took place in the 1930s were defined by their economic/agrarian roots.

CONTINUITY OF RURAL PROTEST

F. B. Carr, colonial Resident in Owerri Province in the 1930s, wrote in his memoir that eastern Nigeria "had always been the trouble spot of Nigeria." Even though there was no major disturbance after the 1929 Women's War,[124] the unpopularity of tax and discontent remained high in the countryside more than a year afterward. This was particularly so in parts of Owerri Province. The police report in 1931 noted that all the inhabitants of most of the division in Owerri Province were "against both the rate and principles of tax," despite the propaganda promoting the benefits of taxation. Although the rate of income tax was reduced from 7/– for an adult male to 5/– in Nguru area and 6/– in the Isu area of Owerri Division, the people remained unsatisfied and demanded further reductions.[125] The report noted that the people in Nguru area "are very poor" and suffered from considerable hardship even in paying the 5/– demanded as income tax.[126]

Grievances remained. A police report in February 1931 warned: "On the surface the Division is quiet, underneath there is considerable unrest and discontent." A considerable police presence was required to maintain order and collect tax. In fact, F. W. Tristram, the assistant commissioner of police in charge of Okpala in Owerri, wrote in a memo on 15 February 1931 that the local native administration "exists in name only." The newly appointed chiefs "have as a rule, no authority whatever," and as a link between the district officer and the people "they are useless, as they neither pass on the messages sent out by the D.O. for the benefits of the people nor do they report matters of interest to the D.O."[127]

The "feelings of women against the old Court members is still high," a former official noted in the 1930s.[128] The same can be said of the endemic corruption and bribery within the native courts. While things had been generally brought under control by 1930, there was still, in the opinion of some officials, the danger of a general movement by women, and a significant increase in surveillance was implemented to monitor the activities of women. Meetings of women, aside from burials, marriages, and similar gatherings, the police advised, "Should not be ignored."[129] Officials were aware that real troubles could arise from small meetings, as the police observed:

> The women are just as determined as ever that tax must be abolished and the old court members removed.... The Organization of women known as 'OHANDUM' is well established in the Nguru, Ngor, Okpala and Isu areas of the [Owerri] division and if they receive recognition or encouragement in any way the work of building up a Native Administration may be seriously impeded or even rendered impossible.[130]

Post-1929 revolts were also rooted in the peasant economy. Some of the protests that continued sporadically in the 1930s came to centre on the introduction of produce inspection, although taxation continued to be an issue of concern for the rest of the colonial period. One such protest occurred in 1930 when a new system of testing palm oil, known as the "one-shilling test," was introduced.[131] In May and June 1930, the United Africa Company (UAC) complained that some shipments of palm oil to New York from Opobo had been shown to contain as much as 3.9 to 5.5 per cent of extraneous matter.[132] The UAC's complaints prompted the introduction of produce inspectors and a more rigorous inspection procedure. This was a departure from the guesswork that had characterized the previous inspection procedures. Overzealous produce inspectors introduced the "shilling test" in an attempt to remove the inconsistencies in palm oil inspection. Inspectors conceived the test as a process of determining whether palm oil contained as much as 2 per cent of impurities. Oil would be rejected if the residue covered a shilling piece.[133] The shilling test was introduced in Oguta after the UAC's complaints, but the issue of inspection had already emerged in the women's protest of the previous year. Many traders chastised produce inspectors for interfering in local trade.

The new inspection procedure, which one official described as an "unwise and unwarrantable action," had a significant impact on supplies. The effect, he argued, "has been not only to cause a drop in the quantity of palm oil brought at Oguta from about 200 tons in the week ending 20th September to nothing at the present time." He was concerned that the policy had started the women's movement again, as women were barricading trade routes and holding up trucks to exact tolls. He noted that the policy had "created widespread alarm in the Onithsa and Owerri Provinces" and necessitated the calling in of more police to the area and generally "added to the anxieties of an already sufficiently harassed administrative staff."[134] The shilling test was abolished on 7 December 1930, but disagreements continued over other regulations, including the rules governing the drying of palm kernels.

The problems that confronted the Nigerian oil palm industry in this period did not arise out of the depression alone. By the 1930s, the oil industry was facing increasing competition from the well-established plantations of Sumatra and Malaya. In 1933, the United Africa Company (UAC) wrote to the governor of Nigeria from London, stressing the need to improve production by adopting the plantation model. The UAC maintained that "the future of the Nigerian palm oil industry appears to us to be gravely compromised by the development of the industry in Sumatra and Malaya ... unless the African can be induced and enabled to adapt his methods of cultivation to modern methods the natural palm industry of Nigeria is in serious danger." Years of research and cultivation of selected varieties of palm had led to the expansion in production in Asia. The UAC was convinced that the "immediate future of the palm industry in Nigeria lies in the development of plantings of oil palm trees from selected seed properly cultivated and maintained."[135]

The unrest that occurred in 1938 in Okigwe Division was a constant reminder that the effects of the introduction of taxes were still felt in the countryside. Early in December 1938, crowds of men numbering about 400 in each case gathered at Isuikwato and Eluama, southwest of Okigwe, to express their grievances against high taxation and low prices for produce. A few days later, a crowd of women gathered at Okigwe demanding a reduction in tax. The women, according to the governor, dispersed after indulging in "frenzied outbursts of singing and dancing." However, armed with sticks, the crowd increased the next day at Isuikwato and threatened to destroy the livestock of those who paid their tax and to destroy the properties of the tax collectors.[136] The protest covered an area of approximately 545 square miles, including the Isukwuato, Uturu, Nneato, Isuochi, Umuchieze, Otanzu, and Otanchara communities of Okigwe Division and reaching the Alayi, Item, and Umuimenyi communities of Bende Division, attracting approximately 127,000 people. There was passive resistance to demands for tax payment in the affected area. A month after collection should have commenced in the area, the colonial governor reported that "no payment had been made and attempts by the administrative staff to reason with the people and persuade them to pay were unsuccessful."[137]

Aside from the issue of taxation, an important cause of the disturbance was the belief that the low prices offered for produce were artificially controlled. Although the government concluded that the tax rate was reasonable, it believed that the system of tax assessment and marketing of produce needed modification. While the scale and extent of this protest pale in comparison with the 1929 revolt, it nevertheless represented a familiar trend among peasants across much of southeastern Nigeria. As with previous revolts, the roots of the 1938 revolt lay in the peasant economy that continued to be under stress.[138] Indeed, the acting secretary of the Southern Provinces conceded that the administration was not able to make proper allowance for the effects of trade decline in adjusting taxes. In his view, "the assessment of the flat rate should be more scientific than it is now."[139]

Peasant protests continued in the 1940s in response to the introduction of innovations in the oil palm industry. Attempts were made in this period to introduce palm oil mills in eastern Nigeria. This was followed by widespread protests in parts of Owerri and Calabar provinces. Women were the most vocal opponents of the mills for a number of reasons. The introduction of the mills would and did certainly shift production from the household to the mills. There was a feeling among women that their husbands would sell palm fruits directly to the mills, thereby depriving the women of the income they derived from palm kernels.[140] Others perceived the introduction of these mills as a prelude to the takeover of their land and palm trees by the government.[141]

The evidence presented in this chapter suggests that there is considerable agreement on the origin of these revolts in the extension of the Native Revenue Ordinance into this part of the Southern Provinces, yet its agrarian roots have been under-estimated. Far from being a feminist revolt fed by female consciousness, the 1929 Women's Revolt and the others that followed emerged from genuine peasant consciousness in which women, as part of the peasant class, spearheaded the revolt and its framing. The chapter also restores women to their rightful place, reveals their human agency, and challenges the view expressed by officials like C. H. Ward, who argued that women in Owerri Province customarily "have no authority in their towns."[142] While the tax issue helped to shape the nature and scope of the rural response, the revolt itself was a synthesis of many factors. The decline in the price of palm produce played a significant role in the timing of the revolt. In addition, the fact that the focus of the debate often shifted from the tax issue to the native administration system and the blatant corruption of the African political class highlights the multifaceted factors that led to the revolt. Overall, the historical analysis presented in this chapter exposes the role of violence in maintaining colonial domination and provides important insights into the gender relations of production in an otherwise patriarchal colonial setting and into the rural agrarian roots of the revolt of 1929.

CHAPTER FIVE

THE SECOND WORLD WAR, THE RURAL ECONOMY, AND AFRICANS

We are in the midst of the most destructive war the world has yet seen, and it is the duty of every citizen of this country, as it is of every liberty-loving soul in every part of the world, to bear the greatest sacrifice ungrudgingly and contribute his maximum in every way possible, little or great to bring the success of the Allied forces nearer. – *West African Pilot, 12 February 1942*

... by the allocation of 4 bags monthly as compared to my previous shipment of 50 to 100 bags, my business will be very much crippled and the life of my entire family placed in jeopardy. – *Amos Okafor to District Officer, Aba, 8 July 1943*

We require 200 bags of gari to feed our selected laborers on the Tenti Dam Construction and shall be obliged if you will issue permit.... We certify that Tenti Dam is for our hydro-electric works generating power for the tin fields and can be justly described as a war effort. – *J.E.A. FitzGerald to Assistant Food Controller, Aba, 10 July 1943*

The advent of the Second World War and its immediate aftermath had an impact on several areas concerning the rural economy in colonial Africa. The war had "far greater local impact and indeed was to lead to far reaching changes," according to F. B. Carr, who served as the Resident for Owerri Province and later as the Chief Commissioner in charge of the Eastern Provinces in 1943.[1] The conditions of rural life in the East were less than satisfactory on many fronts before the war began. The new Native Administration system, which was still on trial, had not improved local access to resources. Carr noted that there were "many problems of more material nature which had to be tackled" and many demands for "improved material conditions many of which though far beyond resources led to heightened interest in progress and highlighted the urgent need for development on a vast scale."[2] This was the state of the local society when the war broke out. Still, there was the desire to pitch in, no matter how little, on the part of the local population. Carr recalled the desire to support the war effort by the African population:

> At the outset raising money for war purposes became a dominant feature of daily life and the response was quite astonishing. "Win the War Fund" and "Spitfire Funds" were fully supported and even the poorest – and none was particularly well off in those days – gave their bit. The salaried classes, clerks and the like with a meager average of, say £50 a year volunteered a monthly deduction from their pay.... Indeed, a wave of loyalty seemed to sweep through the country and even in the remote villages all seemed to want to help.[3]

This chapter examines how the wartime mobilization of African labour affected Igbo villages, towns, and cities from 1938 when the mobilization began to the end of the war in 1945. It outlines the key changes in the colonial agricultural policy in relation to the production of much-needed raw materials such as palm oil and the increased mobilization of the local population for food production. It also examines the new regulations and laws introduced during the war to control the local agrarian economy and commerce. The chapter links the political and economic landscape of the war era and their impacts on African population to the unique forms of protest that occurred in response to British wartime policies as reflected in the petitions they wrote to colonial officials during the war.

CHANGES IN AGRICULTURAL POLICY

Until the Second World War, the colonial department of agriculture focused on the expansion of palm oil production, improvement in the quality of produce, and the efficient marketing of agricultural produce.[4] Progress was made in the export and commercial sectors of agriculture, but the government faltered on the development of the subsistence sector. Nigeria, like other colonial territories, experienced rising food prices and increased importation of food items during this time. The impact of past policies became evident toward the end of the 1930s and government became increasingly aware of the need to encourage the production of food. The government remarked in the 1938 report of the department of agriculture: "The production of export crops, important as this is to the wealth of the country and to the revenue of the government, must not be subordinated to the production of foodstuff for local consumption, for those who are underfed cannot do the maximum amount of work."[5] While the government recognized the need to improve local food production, it did not provide any direct incentive to farmers until the outbreak of the war. On the outbreak of war, colonial officials were asked by the imperial government to carry out the task necessary to secure the local resources needed from Nigeria to support the war effort. Thus, officials embarked on a double strategy of encouraging more export production and a more aggressive drive to increase local food production.

THE EXPORT SECTOR

The colonial government initiated a broad range of measures designed to increase the supply of palm oil and kernels and commodities such as wild rubber, which were desperately needed during the war. Palm oil was particularly important in war production and the manufacture of cooking oil and margarine for British citizens. Indeed, Carr described the production of palm oil and kernels as "a matter of first priority."[6] The loss of the British Far Eastern colonies increased Nigeria's strategic importance as a supplier of palm oil. The Ministry of Food in London was vested with the power to purchase Nigeria's palm produce as part of the measures to ensure an efficient supply system. The prices of palm oil and kernels were raised to encourage production for export.[7] In a dispatch to the colonial administration in Nigeria, the secretary of state for the colonies urged it to "exert every possible effort to obtain maximum production of export crops."[8] The war also exposed the need for new export crops independent of the market for oils and fats.[9]

The strategic importance of Eastern Nigerian peasants during the war is reflected in the reorganization of the colonial administrative personnel to ensure maximum mobilization of the local population. Despite their already depleted numbers, administrative officer were appointed, "for the sole duty of urging and supervising maximum production."[10] As a result, all of Eastern Nigeria witnessed an extensive demand for cash crops, exploitation of forest resources, and forced labour because of British demand during the war. Colonial officials campaigned through the press and placed posters at strategic locations, calling on farmers to harvest and process their palm produce. Schools and churches were incorporated into the campaign through the organization of palm kernel cracking competitions.[11] A palm produce drive team was formed in 1943 to stimulate production, collect information on production and marketing, and examine ways to improve both.[12] Additional produce buying centres were established at Owerri, Okigwe, and Oloko Item. Palm nut cracking machines were installed at various locations in the region.[13] The export duty on rubber was abolished to encourage increased harvesting of wild rubber.[14] The government also encouraged internal migration from Abaja in Udi Division, the migrants being hired in Awka Division to harvest oil palms. The deputy controller of palm produce, Mr. L. T. Chubb, advised the government to exempt these migrant villages from recruitment for the army and the coal-mines until the government achieved its objectives.[15] Significant progress was made towards export production, but not always to the satisfaction of imperial officials like Lord Swinton, who was appointed Resident

Table 5.1. Palm produce export data, 1939–46.

YEARS	TOTAL PALM OIL EXPORT (TONS)	TOTAL PALM KERNELS EXPORT (TONS)	PRODUCER PRICE/ TON PALM OIL	PRODUCER PRICE/TON PALM KERNELS
1939–40	157,970	342,580	£5:19s: 3d	£5: 2s: 9d
1940–41	141,703	262,575	6:3:4	4:11:6
1941–42	147,678	344,820	6:3:3	4:13:0
1942–43	153,537	323,555	9:4:6	5:14:4
1943–44	134,664	330,647	10:1:0	7:16:4
1944–45	139,464	320,764	12:6:6	8:14:0
1945–46	110,242	283,471	13:1:6	9:4:0

Source: Calculated from Report of the Mission Appointed to Enquire into the Production and Transport of Vegetable Oils and Oilseeds in West African Territories (London, 1947), 57.

Minister during the war. "With characteristic energy and drive he spurred on everybody and demanded greater and greater efforts," wrote F. B. Carr.[16]

The demand for palm oil and kernels from the region helped to facilitate further developments in the rural economy; farmers responded positively and increased production.[17] In 1941, for example, the department of agriculture acknowledged that the demand for all the principal crops and products of Nigeria had increased on an unprecedented scale and had led to an increased output from peasants.[18] Nevertheless, export figures fluctuated, despite marginal increases in prices from 1942, due to the internal market for palm oil in Northern Nigeria.

The state did not always garner popular support from farmers, because the structural changes implemented were not always accompanied by higher prices for peasant produce. Additionally, some wartime and post-war developments affected the export of palm produce from Nigeria. These included a significant increase in the export of oil from the Belgian Congo, which rose from 60,000 metric tons during 1934–38 to 118,000 metric tons between 1948 and 1950. This trend would continue. Exports would rise to 140,000 metric tons by 1953.[19]

Measures adopted by different European nations to protect their own economies directly affected African farmers who depended on the export of primary produce. Germany, for example, was a major importer of Nigeria's palm kernels in the interwar period. Trade with Germany became impossible. When the Depression hit Germany, the country imposed very high tariffs on foreign sources of oil and subsidized domestically produced mustard seed and linseed. The country's importation of vegetable oil, which was valued at 70 million marks in 1929, fell to 27 million in 1935 and continued to slide thereafter.[20] Before the Second World War, the United States imported about 8 per cent of its oil from Nigeria. The imposition of high tariffs on imported oil raised domestic oil prices by approximately 100 per cent. This measure significantly reduced U.S. oil imports from 164,000 metric tons in 1936 to an annual average of 70,000 metric tons in the following years.[21]

In the United Kingdom, war hampered trade and supplies and led to a lower demand for imported goods, including palm produce, as the war progressed. The Ministry of Food in Britain called for a reduction of palm kernel production because of unfavourable conditions in Britain and the austerity measures implemented.[22] Although the restriction was imposed on most parts of the western provinces in August 1940, Eastern Nigeria was spared because the region had no alternative export. To avoid what officials described as "undue hardship" for the people of eastern Nigeria, the department of agriculture allowed the region to produce kernels on a quota basis. The quota was based on the average quantity purchased in the previous three years.[23] The restriction on palm kernels was lifted in May 1941, when demand increased again in Europe.[24]

Indeed, there was no consistent policy. By October 1940, considerable debate was going on between the department of agriculture and officials at the Nigerian Secretariat, Lagos, over the provision of the necessary fund required by the department to reorganize production to meet its objectives. J. R. Mackie's frustration was evident when he wrote on 19 October 1940:

> In the course of the very frank discussion which I have had recently with you and your staff I have gathered that, as has so often happened in the past the instructions which I have received from the Government and the policy laid down by it for my Department

are incompatible with the financial resources of the country. As such a state of affairs is liable to misunderstanding between your branch of the Secretariat and myself and to cause a great deal of unnecessary work, before submitting to you further proposals and before preparing my estimates for 1941–42. I should be most grateful if you would be good enough to give me a frank and clear answer to the following questions:– (i) is the work of my Department still considered to be essential for the prosecution of the war? (ii) If so can I expect to be supported by sufficient funds to enable my officers to work to their absolute maximum?[25]

While the government considered the work of the department of agriculture essential to the war effort, the desire of the government to maximize production in Nigeria was going to be based on increasing the level of peasant production with minimum additional resources.

The first response of the British was to increase palm oil production. As soon as it became clear that palm produce in particular was not required in the quantities originally anticipated, they tried to reduce or increase it depending on demand in Europe. But what actually determined local producers' response to demands was the price of produce. Many farmers were indifferent to the call for increased production of palm oil when produce prices were low. The production of kernels in Onitsha, Awka, and Agwu Divisions and Nnewi District fell from 14,359 tons in 1939 to 10,100 tons in 1942.[26] Eleazer Ihediwa recalls that "Low prices forced many people to abandon the harvest of oil palms."[27] An agent of the United African Company at Ogrugru in Onitsha Province reported in 1939 that "little produce was coming in."[28] After a meeting with middlemen and producers in 1939, the Resident for Onitsha Province agreed: "There is no doubt whatever that the people are holding up production – and if we are going to consider extended palm produce production the question of a guaranteed price must be answered." In his view, "even without increased production, something needs to be done and the guaranteed price or government 'subsidy' or other expedient should be adopted."[29] The government was willing to adopt measures to compel farmers to expand production.

COERCION AND RESTRICTIONS

The war was a catalyst for new forms of colonial control and major changes in imperial policy in Africa. The new forms of control – or what John Iliffe calls "new colonialism"– was marked by direct intervention in the local economy. In the words of Basil Davidson, the pressure brought on the local peasantry in the form of forced and selective production of crops and marketing reforms "upset rural stability," and according to David Anderson and David Throup, this period marked an "important transition in British attitude to Africa."[30] The Nigerian experience of this period supports this analysis. In Nigeria, officials restricted the movement of food items, especially gari (the most common local staple produced from cassava tuber) from one part of the country to another. They controlled prices on other local and imported food items and initiated an unprecedented mobilization and control of peasants for food and export production.

The colonial approach to extraction in this period is reflected in the laws and regulations implemented to control peasant production, marketing, and accumulation, and in the local reactions. Enforcement through Nigeria Defense (Oil Palm Production) Regulation No. 55 of 1943 began to chip away at peasant autonomy. This regulation made it compulsory for farmers to harvest and process their oil palms, or face incarceration. The regulation also empowered the deputy controller of oil palm production to order the harvesting, processing, and marketing of palm produce. Furthermore, Defense Regulation No. 89 of 1945 compelled native authorities to ensure the implementation of Regulation No. 55 or face possible prosecution. Through this regulation, the colonial state increased its pressure on the peasantry. In Owerri Province, where the regulation resulted in prosecutions, the palm production officer, P. L. Allpress, noted that: "Palm production has greatly fallen off, and from the unharvested areas I have found in the Aba Division one must conclude that this is to some extent due to the dilatoriness of the people. No amount of talk has any effect on the people unless one's threats are backed by action now and then."[31]

The colonial authorities doubled the tax rate as a penalty in areas where oil palm owners refused to harvest their crops. The government was committed to the recruitment of labour for the harvesting of oil palms where the rightful owners failed to do it. To increase its resource base further, the colonial government increased the exploitation of other forest products, such as wild rubber, wild silk, honey, beeswax, gum copal, charcoal fuel, Calabar ordeal beans, raffia, and bamboo. There was initial enthusiasm for wild rubber exploitation in parts of Ogoja, Onitsha, and Owerri Provinces, although the harvesting was discontinued in 1945 because of the high cost of extracting the rubber.[32] And by this time, the war was essentially over.

The failure to meet official expectations led to the prosecution of many local farmers. Between 1943 and 1944, for example, available records show that over a thousand persons were prosecuted in Obudu District for failing to crack palm kernels or harvest oil palms.[33] The crackdown associated with the export drive in the period was widespread.[34] Fines ranged from £1 to £5 or terms of imprisonment of one month or more. The low price of export commodities and the forced labour policy imposed on rural peasants provided the context in which many rural dwellers shifted attention to food production, which attracted high returns.

FOOD SECTOR

The greatest problem facing the colonial administration was how to curb the rising cost of living that became more pronounced when the Second World War broke out in 1939. The British mounted a vigorous propaganda campaign, pointing out the importance of increasing the amounts of food and export products to support the British war effort. There was a tremendous outpouring of patriotic passion on the part of the British. They expected Africans to react in a similar way by producing more food and raw materials. This did happen sometimes, even though the local people had other interests and were more concerned with matters of immediate and practical nature – their own subsistence. The *West African Pilot* wrote in an editorial of 23 February 1942: "if we cannot produce munitions, we can certainly produce raw materials and food products which are equally important for the successful prosecution of the war."[35]

The war made the colonial authorities nervous about reliance on imported food. Furthermore, the constraints on world shipping made it extremely difficult to rely on imports.[36] As the war went on, it created an increasing number of immediate problems. These included low levels of imports and exports, shortage of food, and rising prices for locally produced food items, such as rice, yams, and pepper, as well as high prices of imported items. There was a perception that Nigeria was capable of producing enough food to feed the country's population and to meet Britain's import needs in wartime.[37] Indeed, Nigeria occupied a unique position among British colonial possessions in West Africa with its enormous agricultural potential and population.

Beginning in September 1939, the agricultural officers were fully occupied with the effort to mobilize human and material resources to achieve a level of local self-sufficiency and increase food production.[38] In a dispatch to the colonial administration in Nigeria, the secretary of state for the colonies urged it to "exert every possible effort to obtain maximum production of export crops."[39] Sir Frank Stockdale, who was appointed to inquire into the production and transportation of vegetable oils and oil seeds produced in the West African colonies, advised the director of the Agricultural Department, J. R. Mackie, after his visit to Nigeria in 1938, that Nigeria, like other colonial territories, should endeavour to be self-sufficient in food supplies.[40]

The department of agriculture made plans to deal with the uncertainties of war. In a circular issued to all agricultural officers, the Agricultural Department summarized its goals to meet the needs of the war as follows:

   (a)   To be ready to help the Imperial Government by producing such crops as it may ask;

   (b)   To ensure that Nigeria is, as far as possible, self-sufficient in foodstuffs including those that are normally imported from elsewhere; and

   (c)   To do what we can to make the West African colonies as a whole self-supporting.[41]

Officials pursued these goals with great zeal. The view of one such official, J.A.G. McCall, who was appointed the controller of palm oil production, perhaps summarizes the prevailing view: "The production business is our particular war effort, and surely that should come first, even if other duties of

Table 5.2. Imported foodstuffs for the year 1938 for African consumption.

COMMODITY	QUANTITY	VALUE (£)
Fish	212,000	396,000
Salt	1,000,000	267,000
Rice	186,000	96,000
Sugar	140,000	105,000
Bread and Biscuits	20,000	47,000
Flour	50,000	46,000

Source: L. Wale Oyemakinde, "The Pullen Marketing Scheme: A Trial in Food Price Control in Nigeria, 1941–1947," *Journal of the Historical Society of Nigeria* 6, no. 4 (1972): 414.

Table 5.3. Imported foodstuffs for the year 1938 (mainly for European consumption).

COMMODITY	QUANTITY (LBS)	VALUE (£)
Milk	1,205,000	35,000
Butter	256,000	15,000
Lard	153,000	4,700
Meat	12,000 ctw	47,000
Bread and Biscuit	20,000	47,000
Vegetables (fresh)	1,603,000	7,8000
Tea	226,000	14,000
Jam	220,000	7,2000
Confectionery	332,000	13,200
Fruit (dried)	97,000	1,500

Source: L. Wale Oyemakinde, "The Pullen Marketing Scheme: A Trial in Food Price Control in Nigeria, 1941–1947," *Journal of the Historical Society of Nigeria* 6, no. 4 (1972): 414.

administration have to suffer thereby."[42] By 1941, the importation of rice into the country had virtually ceased, falling from 14,900 tons in 1936, thereby creating a food crisis during the war.[43] The frantic urgency of the food production drive contrasted with the laissez-faire pace of pre-war agricultural planning. The department intensified experimentation in wheat, potato, and rice production and established a fisheries section to replace the supplies formerly obtained from overseas.[44] The Agricultural Department was particularly anxious to introduce upland rice, a new and unknown crop in the area.[45] The colonial administration increased the budget allocation to agriculture from £136,726 in the 1939–40 fiscal year to £397,260 in the 1945–46 fiscal years.[46] Although research continued to be an important item on the agenda for the Department of Agriculture, the war era marked the beginning of direct intervention in the local agrarian economy.

In further pursuit of its wartime goals, the colonial government set up a supplies department in September 1942 under Dr. Bryoe as chief supply officer and four administrative assistants. Earlier, in November 1939, the department of agriculture had appointed Mr. E. McL. Watson as marketing officer. The creation of this post anticipated increased production and the need to develop both the internal and international trades in foodstuffs. The duty of the supply unit was to organize grain supplies for the army, the mines, and government reserves. By the end of March 1943, the government had purchased 10,300 tons of millet, 2,700 tons of rice, 7,000 tons of yams, 18,200 tons of guinea corn, and 3,400 tons of maize.[47]

The food scheme was not very successful. The government's effort to improve food sufficiency concentrated on the cultivation of rice, and the Department of Agriculture was directed to slow down or discontinue other lines of agricultural production, including the oil palm planting program.[48] The department asked for the authority to convert funds allocated for the palm oil program to food production and propaganda, although it was not clear if it received this authority. The cultivation of rice received a boost in some parts of Igboland, although local people preferred imported rice to locally produced rice because of its higher quality. In Onitsha Province, the acreage devoted to rice rose to 5,000 in 1943 due to the war.[49] However, a combination of drought and poor management led to a yield of only 2,000 tons in 1943. Ironically, the colonial emphasis on the production of exotic food crops, such as rice, potatoes, and various vegetables, did not improve food security. The emphasis on vegetable production, for example, catered to the needs of European residents and the army, while the production of rice was meant to reduce dependence upon imports. The failure of the production drive, however, led to increased control of locally produced food items, especially yams and gari, which were the most important food items for both the rural and the urban population.

Mobilization for the war created an opportunity for African entrepreneurship and facilitated the rise of an enterprising class of traders cashing in on the increasing demand for local foodstuffs, especially gari. By the 1930s, a lucrative market in locally produced food items had developed. Farmers expanded food production due to increased demand for gari and palm oil in Northern Nigeria. In the Ozuitem area, for example, farmers concentrated on growing yams and, partly or completely turned to palm produce when the European traders paid high prices.[50] Significant quantities of gari were railed to Northern Nigerian towns from the late 1930s. The quantity of gari railed to Northern Nigeria increased progressively from 1,414 tons in 1937, worth about £4,989, to about 5,428 tons, when the war began. In the first half of 1942, the quality railed to the North had reached 6.804 tons, worth about £57,661.[51] Growing concern over the threat of urban and rural food insecurity prompted the colonial authorities to regulate the distribution of both imported and locally produced foodstuffs.

FOOD RESTRICTIONS AND RATIONING

On 22 April 1941, S.A.S. Leslie, the colonial food controller, issued a circular that outlined the general scheme for food control in Nigeria. In a seemingly arbitrary manner, the food controller was empowered to direct the distribution of available supplies while deputy food controllers and their assistants initiated and administered rationing in local areas. Although the supply of food items in this period, except for milk and flour, which were already rationed in Lagos, were not seriously threatened by the war, the government was making plans to ensure that rationing would follow periods of severe shortfall.[52] Importing firms in Lagos, Nigeria's largest city, were required to ensure that consumers purchased all of their supplies from a single source in order to regulate the distribution of essential commodities. Consumers were required to nominate the firm with which they wished to deal. This enabled the government to compile a list of customers, which would then be circulated among the firms.[53]

Control was extended to the rural areas. By 1942, rural farmers and traders were cashing in on the increased demand for local foodstuff in Northern

Nigeria. Large quantities of yams, cocoyams, coconuts, and maize were being transported to the North for sale. The colonial administration imposed strict restrictions on the movement of food items, such as yams, cocoyams, and gari, within the Eastern Region. District officials, alarmed by the quantity of food items moving out of the region, began to implement restrictions in order to stave off crisis within their districts. Some district officers regulated the quantity of local food items, such as yams, cocoyams, and gari that could be moved from particular districts to other areas by invoking the Nigeria General Defense Regulations (Law No. 75 of 1941), which came into effect during the war.

In 1943, the transportation of these food items from the Aba railway station to the North had increased by a very large amount from the previous year (see Table 5.4). J.V. Dewhurst, district officer for Aba, was so alarmed at the rate of outward movement of these food items that he suggested the need to prohibit the movement of yams from the district. Writing to the Resident, he noted: "The increase in the export of yams is very great indeed and, in view of the fact that Aba normally imports yams, disquieting. So too is the increase in the export of maize and I am not certain that the export of this also ought not to be prohibited."[54] In June 1945, the acting district officer for Ikon Division issued a memorandum that stated: "no person shall export yams and cocoyam from the Ikom Division except under permit from the District Officer."[55] Farmers in most parts of Abakiliki Division faced difficulties in obtaining seed yams due to a combination of drought and the scarcity of seed yams in the market.[56] In a similar fashion, the Resident for Ogoja Province prohibited the movement of yams from the province. According to his order, "no person shall export yams by rail from the Ogoja Province except under permit from the district officer in Afikpo."[57] In Awgu Division, the resident, Dermot O'Connor, issued an order restricting the movement of yams outside the division.[58] In Onitsha Province, the colonial resident issued an order restricting the movement of gari and yams from Udi Division "except under permit signed by 'Competent Authority.'"[59]

The intervention of the government and the controls imposed could be justified as food prices rose beyond the reach of the average household. Although many farmers and traders were unhappy with the regulations, the reaction of district officials made a certain kind of sense. The *West African*

Table 5.4. Quantities of food items railed from Aba Station in 1942 and 1943.

FOODSTUFF	NO OF BASKETS RAILED IN JULY 1942	NO OF BASKETS RAILED IN JULY 1943
Yams	35	573
Cocoyams	23	96
Coconuts	523	1589
Maize	3	143

Source: NAE, ABADIST,1/26/958.

Note: The weight of a basket varied between 1.5 and 2 cwts.

Pilot agreed that "war conditions and profiteering" caused a rise in the cost of essential goods, including food.[60] The district officers in the food-producing areas were reacting to potential tensions that could arise with severe food scarcity. They did not want food riots on their hands. But they were also interested in playing their part in the war effort by ensuring that local food products were fairly distributed, particularly in this period of low imports. Indeed, food producers and marketers were cashing in on the scarcity of imported food and increased demand for local products. By 1944, the price of gari had increased considerably from about 1/- 6d to about 9/- in urban markets. In a report to the district officer, Aba, in May 1944, the Nigerian Police wrote that gari producers "are making an exorbitant profit when one considers that they found it worth while to make gari up to 1940 and sell at 1/6d – 2/- a bag." Prices were even higher in places such as Uyo, where it was suspected that large quantities of gari were being exported to Fernando Po, where a large number of Igbo and Ibibio migrant labourers were working in Spanish plantations.[61] Such an escalation of food scarcity and high prices was considered a threat to political and social stability.

The restrictions imposed by officials had enormous effects on the local population. Restrictions disrupted food security arrangements and the internal trading network that linked the region into a form of economic commonwealth. Women farmers could not sell excess produce, such as cocoyams and cassava, to their traditional trading partners, and many Igbo men and women expressed concerns over the difficulties they faced in carrying out local trading. Colonial officials were inundated with petitions, complaints, and requests, as local people faced increasing difficulty in meeting their subsistence needs.

The low prices of export produce, the high cost of imported goods, and the government-imposed restrictions exacerbated discontent and disillusionment. In particular, the government's control of locally produced foodstuffs and distribution implemented during the war left little room for dealing with mounting food scarcity. It also created an atmosphere of crisis and conflict. The records do not tell exactly the level of the food crisis in the region and the interruption of local and regional trade, but by 1943, several petitions and appeals from local traders to colonial officials had asked for permission to transport food items within the region or to the North. Yet the colonial officials provided few satisfying measures even as peasants were called upon to work much harder than before to support the colonial state.

The food supply problem remained the most pressing issue during the war. Both the imperial government and governments in the individual colonies provided some incentive for peasants to increase production, but the authorities also used coercion to ensure that peasants complied with their demands. Colonial officials, despite their attempt to encourage free trade and uninhibited access to the market for local people, were also noted for restrictions and intervention in production and marketing of local produce. The war helped to create barriers and officials enacted legislation, not only to prevent the free flow of local products such as gari and yams, but also to determine the price at which they could be sold.

CONTROLLED PRICE

By 1941, the colonial government was forced to impose price controls to deal with the looming food crisis. Shortages of food had been anticipated right from the outbreak of the war in 1939, which prompted the directive that provincial and district officers should watch food price trends and adopt measures to conserve available food.[62] The purchase of locally produced food items by the government at a fixed price was common during this era. But colonial price controls extended to both imported and locally produced foodstuffs.[63]

Although the agricultural department noted in its 1940 report that Nigerian peasants regarded the restriction and loss of income "as a contribution towards the prosecution of the war,"[64] many peasants were visibly discontented with the restrictions. Opposition to price controls and regulations was not confined to inter-regional traders and urban associations. The long-standing tradition of women's opposition to state intervention in the rural economy arose again in the rural areas and markets. Their acts of protest gave women a voice. The opposition to price controls, championed by rural women, swept through most of southeastern Nigeria. The Omuma Native Court, Aba Division, reported women's protests against the imposition of price controls on gari by government officials in November 1944. Some of the women who attended protest meetings, he wrote, "do not make gari," but they followed others in these meetings.[65] As in previous protests, the participation of non-gari producers is a mark of the solidarity and consciousness that existed, and still exists, among Igbo women.

In Owerri Province and in the Bende Division in 1944, women protested against the colonial government's attempt to control the price of gari, the most important staple in the area. The purchase of gari by officials at a controlled price threatened the income of the women who controlled the production and sale of the product.[66] To the women who produced the gari, the action of the government officials was seen as an attempt to take over their cassava farms. As in earlier years, women fought against this threat to their livelihood in an already precarious agricultural economy. On 1 July 1942, rice and bean traders in the city of Onitsha petitioned the resident for Onitsha Province regarding the imposition of price controls carried out in the market by the police on 16 June 1942. In a carefully detailed calculation, the traders estimated the losses they incurred due to the price controls.[67] On 17 July 1942, an Onitsha gari trader and twenty others petitioned the district officer for Onitsha, suggesting that the authorities reconsider the price control imposed on gari. In their petition, the traders noted that the recommended price of 1d for 5 cups and ten shillings per bag was below the estimated delivery price of twelve shillings per bag of gari from Aba, from which Onitsha traders got their goods.[68] Protests and petitions remained a means that peasants used to confront the intervention in the rural economy in this period.

Table 5.5. Price control of foodstuffs at Enugu Market, 16 February 1942 to 25 August 1942.

ITEM	PRICE NOT EXCEEDING	QUANTITY
South African salt	3 d	per lb.
Boneless beef steak	5 d	per lb.
Other beef cuts	41/2 d	per lb.
Shoulder or leg of mutton	6 d	per lb.
Shoulder or leg of pork	9 d	per lb.
Ox tongue	4 d	per lb.
Gari	1 d	6 cigarette tins
Eggs	1/2 d	Each
Yams	11/2 d	2 lb.
Cassava (prepared)	2 1/2 d	Per lump
Groundnuts	2 1/2 d	per lb.
Beans	1 1/2 d (or 3 d)	per lb. (2 lb.)
Millet	1 1/2 d	per lb.
Guinea corn	2 d	per lb.
Rice (Nigerian grown)	6 d	per lb.
Egusi	1 d	per lb.
Pepper (dried)	1 1/2 d	per cigarette tin
Chickens	8 d–1/- 2 d	according to size
Palm oil	2 1/2 d	per gin bottle
Potatoes	14/- 4 d	56 lb.

Sources: NAE, EP, OPC, 122, vol. vii, ONDIST, 13/1/2, "Public Notice," B. W. Walter, Local Authority, Enugu, 28 October 1942.

Table 5.6. Control of local foodstuffs in Abakaliki Division.

ARTICLES	QUANTITY	PRE-WAR PRICE	QUANTITY	PRESENT PRICE	QUANTITY	CONTROLLED PRICE
Yams: September–March	12 large 15 medium 30 small	1/– 1/– 1/–	6 large 10 medium 20 small	1/– 1/– 1/–	9 large 12 medium 17 small	1/– 1/– 1/–
April–August	6 large 15 medium	1/– 1/–	3–4 large 8–9 medium	1/– 1/–	5 large 9 medium	1/– 1/–
Rice (local)	24	1/–	8 cups	1/–	10 cups	1/–
Gari	12	1 d	6 cups	1 d	8 cups	1 d
Egg (hen)	4	1 d	2	1 d	4	1 d
Egg (duck)	2	1 d	2	1 d	2	1 d
Pepper	2 Cigarette cups	1 d	1 Cigarette cup	1 d	2 Cigarette cups	1 d
Plantain	10	1/2 d	6	1/2 d	8	1/2 d
Banana (ripe)	12	1/2 d	8	1/2 d	12	1/2 d
Oranges	10	1/2 d	5	1/2 d	6	1/2 d
Okro	40 Capsules	1/2 d	15	1/2 d	20	1/2 d
Cassava	large basket	3 d	large basket	6 d	Large basket	5 d
Coco yam	8	1 d	8	1 d	8	1 d
Groundnuts	24 cups	1 d	4 cups	1 d	6 cups	1 d
Palm oil	1 bottle	6 d	1 bottle	9 d	1 bottle	8 d
Palm wine	Calabash	2 d	Calabash	3 d –4 d	Calabash	2 d
Ideal Milk	6 oz tin	4 d	6 oz tin	8 d	6 oz tin	6 d

Source: NAE, CALPROF, 3/1/2329, District Officer Abakiliki, 9 November 1942.

AFRICAN RESPONSES TO COLONIAL CONTROL: PETITIONS AND SUPPLICATIONS

> The power of the written word, the "book" juju, is taking a hold
> on Nigeria, a sad but inevitable consequence of the spread of edu-
> cation. – *A. E. Cooks, District officer, Owerri Province*

The restrictions imposed on food items were the most stressful of all the demands made from the local population. It was evident to the rural and urban
population that they would have to adopt desperate measures to survive. But
their strategies did not often succeed. The British, on their part, were even less
successful in handling such hard times. For the societies of Eastern Nigeria,
the dependency on palm produce and the vulnerability of the rural population
to the slightest change in market prices created a much more hostile relationship between colonial officials and the indigenous people. This period was the
beginning of the consolidation of peasant consciousness as well as a period of
awakening political consciousness.[69]

Petitions and supplications were Africans' preferred method of appeal-
ing to colonial authorities regarding their conditions or protesting against
policies.[70] The personal – and often intimate – letters of African traders
and farmers paint a unique portrait of a rough-and-tumble time. The danger posed by the restrictions was made explicit in the lives of small traders,
farmers, traders, and the urban population. On 11 July 1943, for example, a
local trader, Mr. O. O. Muoma, wrote to the British district officer for Aba.
In his petition titled "Injustice: Gari Railing to the North," he told the district officer that his name had been deleted from the list of traders permitted
by the government to export gari to Northern Nigeria. Muoma considered
this development "abnormal" and an "injustice." Cutting him out of the gari
trade, he argued, deprived him of his livelihood and threatened the lives and
subsistence of his two sons, who lived in the northern city of Kano.[71] Another
trader, J. O. Okorocha of Mbawsi, wrote to the District Officer, protesting
the allocation of a meagre quota of ten bags of gari to him. Mr. Okorocha,
who had exported 209 bags of gari to Northern Nigeria in 1942, stated: "my
quota is too poor considering my intensive trade last year.... I beg of you to

remedy the situation and award me what is due."[72] For Muoma and Okorocha, like many others, the war economy was not only the cause of hardship and despair but also a time of opportunity and entrepreneurship that expanded as the scarcity of imported food items provided opportunities for the expansion of trade in locally produced foods. Still, the war revealed the reality of daily life for ordinary people in colonial Eastern Nigeria.

The case of a certain Mr. Udeh, who lived at Abalikiki, is typical of a problem that many farmers faced. Mr. Udeh had applied to the District Officer for Abakiliki Division on previous occasions for permission to transport seed yams to Awka for planting. On 19 May 1945, he wrote again, after his previous requests had been ignored: "I am still asking you about a permit which I will use in carrying down seed yams down to Nawfia[,] Awka. The reason why I say very much about it is that the time of planting is passing.... I beg your Honour that you may consider about it."[73] With the farming season ending, Mr. Udeh wrote the District officer again. He had become desperate. "In my first and second letters that I wrote to you, I begged you to give a special permit to transport one trip of seed yam to Nawfia in Awka District. I told you that I have lived [in] Abakiliki for 25 years and I have not get any yam to plant in my town. Now this season of yams plantation is coming to end.... Yam plantation remains not more than 20 days now."[74]

While the outcome of Mr. Udeh's petition is not known, his was not just an isolated case. J. E. Akajiofo of Mbawsi wrote to the district officer for Aba on 1 September 1943 requesting a permit to rail his yams to Northern Nigeria for sale. Mr. Akajiofo wrote:

> Sir, I have the honour most respectfully to forward this humble application to your worship. My request is that your servant bought 16 tons of yams during the month of July. The yams should have been railed that month, but unfortunately for me, one of my sons died. Then I had to leave the yams in the railways station, and go to my town. I returned on the 11th of August, and on the 14th of the same month a consignment was issued me by the station master, Mbawsi. The next morning when I came to weigh the yams the station master informed me that he received a letter from the District officer, Aba restricting the exportation of yams to the north.

I respectfully pray that your worship may render me the necessary help to rail the yams.[75]

An application written by an Umuahia trader shows that the effects of the restrictions on the movement of goods were widespread. Mr. Eze, who had been a long-time yam trader, wrote to the district officer, Aba, for permission to transport yams to Kano and Jos in northern Nigeria. He had yams ready to be railed up north but was unaware of the new requirement for a permit. He pleaded:

> I humbly beg to state that on absent of the previous knowledge, that yams shall be under permit, I t[a]ke the liberty to ask your worship to grant me a special permit to rail out those baskets I have already got at the Station.... Therefore I humbly crave for your mercy consideration and attend to this matter immediately by granting me the permit as requested otherwise my said food stuff will rot due to long stay in the shed.[76]

"I regret I cannot accede to your request," the District Officer wrote to Mr. Eze.[77] This was a familiar response.

The appeal of an Aba trader, A. Jamola, to the District Officer, Aba, for a reconsideration of an earlier petition that was denied, like many others, is an excellent indication of the suffering of rural peasants. Jamola wrote asking for a sympathetic reconsideration of his humble request for permission to rail gari to the North. "I am a stranger in this community with a family of wife and children who are dependent upon me for subsistence.... Since the early part of this year I was engaged in gari trade with the north and our lives depended entirely upon the small profit occasionally derived therefrom."[78]

Like many others, Jamola was caught in the colonial attempt to manage the deteriorating economic conditions that confronted the imperial as well as the colonial governments. The tearful and emotional petitions by Akajiofo and others were striking evidence of the increased intervention of the colonial administration and the effect on their lives. David H. Kubiri, when he sought a permit to sell gari in the north, wrote:

I with all civility prostrate to ask earnestly, in the name of your families and home, in the name of British justice, which makes an Englishman superior to other races, in the name of all [unclear word], and in the name of your past honour and trust, which has designated his worship as a capable and able rule[r]. [T]hat the permit of gari might be rendered to your humbly servant the families may not die away for starvation.[79]

These emotions were expressed in many petitions and reveal the levels of dissatisfaction within the local society, but the humble phrases used in these petitions remind one of the paternalistic structures of the colonial society. In the experiences of Udeh, Ajajiofo, Jamola, and others, one sees a persistent marginalization of local interests in this period and the intensified insecurity created by the colonial economy. In some ways, it seems to have been the traders who suffered the most.

Urban areas were not spared the agony and hardship of the war. Significantly, the rural areas played an important part in the supply of food to the urban areas. Urban households that depended on food supplies from the rural areas faced both physical food shortages and restricted access to food. The disruption of local trade was felt in areas such as Aba and Onitsha. *The West African Pilot*'s editorial of 5 June 1942 wrote of the city of Aba: "Black market masters seem to have things all their own way and, of course, the people suffer." The paper, like the average person in the urban areas, was concerned about the rising cost of local food materials. The Aba Community League wrote to the District Officer in August 1943 regarding the effects of food restrictions on the residents of the town:

I am directed to bring to your notice the very grave danger of famine which is threatening this township consequent upon the present restriction of the railment [*sic*] of gari to the North which has led to the adversely affected traders seeking other avenues of business and have tapped the scanty resources of the township of yams, plantains, cocoyams, corn and other articles of food. Time was when these lines of foodstuffs came into the local market from Onitsha, Itu and other places which in addition to local resources barely met the requirements of the township.[80]

While reliable data on urban employment and welfare during the colonial period are scarce, wages for non-agricultural labour were very low.[81] In 1933, for instance, Native Administration road labourers received ten shillings per month. Railway track gangs received a wage of eight pennies per diem. Casual labourers received between three and four pennies per diem without food.[82] The difficult economic situation that culminated in restrictions on food was felt in the cities in the form of dwindling food supplies and a rising cost of living.

The letter from the Aba Community League quoted above demonstrates that concerns about the regulations were not expressed by individuals alone. In a petition to the Resident, Owerri Province, the Aba Gari Traders Association expressed its dissatisfaction with the quota system implemented by the District Officer for Aba. The association argued that "the method will annihilate the gari trade, and undoubtedly impoverish the average trader involved and render life 'not worth living.'"[83] The association further noted: "As free citizens of the Empire," the association argued, "we have a right to live, and this right [sic] we pray for an amendment to the method or system of control of the railment of the commodity as advanced by the District Officer, Aba."[84]

Transportation also presented problems and restricted the ability of producers to move products. As the Aba Gari Traders Association indicated in its petition to the Resident for Owerri Province, the restrictions imposed on the gari trade created additional burdens for rural traders who already suffered from the restrictions on transportation because of the war. The petitioners noted: "We are ever willing to speculate on the produce trade, but transport forms a great barrier. During those days of unrestricted transport facilities, produce flowed into the township from Okpala, Ulakwo, Owerri, Nguru, Okigwi, Ife etc. Today, the damage of Owerrinta Bridge and motor transport restrictions have barred that trade."[85]

The efforts to implement colonial restrictions and control were not always successful. Archival records describe an informal market system that began to emerge because of government control. In August 1943, the secretary of the Aba Community League wrote to the district officer, Aba, regarding what he described as "unauthorized markets outside the township."[86] In July 1944, the government estimated that about 200 tons of gari, more than the permitted quota, were exported to the North under illegal conditions.[87]

Nevertheless, the war had some positive effects on the internal market for foodstuffs. The market for food items created an incentive for farmers to increase production. Despite the problems generated by the war, urban demands for food supply from the hinterland provided opportunities for African producers, most of whom had been supplying foodstuff from the time of the emergence of urban centres in the 1920s. The increased demand for gari and palm oil in northern Nigeria worked against external export. J. S. Harris, in his study of the local economy of the Ozuitem Igbo noted that farmers concentrated on yam production but turned to palm produce when prices were high.[88] Men and women reacted in many parts of Igboland, moving from export crop production to food production, depending on which attracted higher price.[89] Between June 1943 and May 1944, 5,825 tons of gari were railed to northern Nigeria from Ogwe, Aba, Omoba, Mbawsi, Umuahia and Uzuakoli railway stations.[90]

Control was probably enforced on the railways because they could be policed. But traders were able to circumvent officially approved channels of trade. Some traders moved between different buying stations to maximize their profits and to increase the quota allocated to them. This prompted the Resident for Owerri Province, Mr. A.F.B. Bridges, colonial Resident for Owerri Province, to invoke the "Nigeria General Defense Regulation 1941 (Food Control)." Bridges was within the authority conferred by the Defense Regulations when he implemented the "(Gari – Owerri Province Non-Removal Order 1944)," which stated: "No person shall export Gari by rail from the Owerri Province to any station north of Enugu except under permit signed by the competent authority, or by a person authorized by the competent authority."[91] A personal allowance of 8 lbs. per person for travellers was allowed. The threat of famine and the breakdown of existing trading relationships forced the most vulnerable districts to introduce measures to preserve food.[92]

In September 1944, Mr. H.L.M. Butcher, who had become the District Officer for Aba, introduced measures that were even more radical. In a circular directed to native councils and Courts Clerks, he directed that all gari permits granted to middlemen and traders be cancelled with effect from 1 September. Butcher regretted that: "traders have not adhered to the conditions of the permit issued to them, but have induced the Railway staff to send up far more that they were allowed." He accused the local agents of gari dealers in the North of selling gari there at "grossly excessive prices," which resulted in shortages in the Aba Division and unduly high prices in the North.[93] He proposed that Mr. Bleasby, a European and manager of Gibbons Transport Aba, act as the government's agent for buying gari at designated centres in the District and arrange for its distribution to Native Administrations in the Northern Provinces. He further proposed to set up a committee made of middlemen and producers who would sit monthly and fix the price of gari for the following month to ensure that gari was available and cheap locally.[94] Similar action had been taken in Calabar, where African traders were forced to supply gari to Mr. Nicholas, another European.

The responses of African traders were predictable. In a resolution passed at Aba on 23 August 1944, an association of gari traders noted that its members' role as traders contributed toward "the defeat of Hitlerism and all that it stands for in order that Democracy will rule the world and all forms of man's inhumanity to man be wiped off the face of the earth." The association described the colonial government's restrictions as "definite discrimination against the Africans."[95] It appeared to the association's members that the government failed to recognize African contribution to the war effort.

Although the petitions mostly came from men, suggesting their domination of the lucrative long distance trade between the eastern and northern Nigeria, some came from associations that may have included both men and women. Yet there were those written by women which brought their own perspective to the events of this period. Their letters shared the common concerns but also drew upon their femininity. Agnes Garuba opened her letter of petition with "Your maid servant humbly and respectfully begs to submit this my humble petition to you as a father." Her unique perspective is worth quoting:

> I am now as a widow, the Government only is my husband since my husband was transferred from Hausa down to Enugu and from Enugu now to overseas, I depend upon nothing than gari trading, buying from the natives and selling in the station, now many people had been stopped not to buy, and therefore I see no way of selling and maintain my life [sic] with my families which my husband left to me and sign soldiers work and went away. If [am] not

trading my former trade in gari then sure we must die for starvation.… May our heavenly father help my Lord the D.O. to put this my supplication in different consideration and to fulfill my wish, for we all are in the same flag that's why we don't care to send our husbands, brothers and young fathers to fight for the same flag.[96]

These traders did not fit into a typical profile. There were those like S. O. Enyiomah of Aba who had been trading in gari since 1927. He and four other traders requested the District Officer for Aba to consider their longevity in the gari trade and grant them permits. They stated: "we, the said petitioners are the only first and original people who engaged in the railing of gari from Aba to Northern Nigeria since the year 1927 and 1928 respectively up to the present moment without any confusion and trouble in our own part."[97] Others like Ikebudu Nzekwe, who started his gari trade in 1942, had rushed into the trade on account of the crisis created by the war.[98] There were women who retailed gari in northern Nigeria and were increasingly making a profession out of it. Agnes Garuba, like other women, had taken up the gari trader because her husband, the family breadwinner, had gone to fight in the war.[99] Many young men were sent to the north to represent major traders who regularly railed gari to them from Aba and other railing stations such as Umuahia and Omoba. C. O. Muoma put it well when he argued that the restriction of the internal export trade was not just an injustice to him but would threaten his two boys in Kano who will "lack maintenance as they will not get any supply of even their own food."[100]

Whether the petitioners were men or women, they worded their petitions or appeals in such a way as to gain as much sympathy or redress as possible. Still, the most vocal criticisms of colonial control in this period came from women, most of whom protested the imposition of price controls by local officials. The ability of Africans to write or to hire professional letter writers gave local people the ability and opportunity to speak back to power. It gave local people not only the opportunity to speak about their concerns but to establish a dialogue with colonial powers – a dialogue that received some level of acceptance because it fit into the structure of European habits of expression.

We can read beyond African voices in these letters and petitions. They offer a lens through which we can understand the broader attempts by colonial administrators to control the colonized through the restriction of movement, through surveillance, and through the extraction of resources. Yet these letters and petitions also place people in the colonial context and present the diverse life-stories and points of view of African individuals. The letters and petitions of this lowly class of peasants and traders not only defy the perceived social and political order of the colonial state but show the unappreciated contribution of Africans to the Allied war effort. Indeed, the demand for local foodstuffs imposed a heavy burden on the local population due to the scarcity and high prices created by the war.

POST-SECOND WORLD WAR DEVELOPMENTS

The end of the war was followed by unparalleled social and economic transformations. The immediate impact was created by the demobilization of African soldiers in 1946. The returning solders brought significant amounts of money into the economy. One of the most significant social and economic impacts was on bride price. The cost of marrying a wife became so high that the government was forced to intervene and regulate the cost of bride-price.

The problem of post-war reconstruction in Europe led to changes in colonial policies. It was envisaged that the colonies would play a significant role in post-war reconstruction through a more efficient exploitation of local resources.[101] The welfarist nature of development planning in this period was rationalized by the argument that greater expenditure on the welfare of colonists was a recognition of their contribution to the war effort.[102] Government assistance under the Colonial Development and Welfare Act of 1945 provided funds for long-term economic development. An offshoot of the act was the Nigerian Ten-Year Plan for Development.[103]

These innovations were introduced within the context of an economically weak British Empire, rising nationalism in the colonies, and a welfare-minded Labour government. The envisaged changes were often not reflected in the actions and opinions of the men on the spot. The government was not particularly enthusiastic about a proposal in September 1945 to grant credit facilities to farmers, peasant industries, and demobilized soldiers. L. T.

Chubb, secretary for the Eastern Provinces, argued that such schemes would likely fail in the Eastern Provinces. "With the difficult agricultural problems existing in the Eastern Provinces," acting assistant director of agriculture for the Eastern Provinces wrote in the *Report of the Mission Appointed to Enquire into the Production and Transport of Vegetable Oils and Oilseeds in West African Territories*, "I think that opportunities for constructive and economically sound assistance to peasant farmers in the form of credit facilities are likely to remain very limited until agricultural research has made more progress. We have no straightforward and proved system of agriculture such as mixed farming in the Northern Provinces."[104] On 1 November 1945, the government in the Eastern Provinces disbanded the palm production team, which was set up during the Second World War, and the Residents were requested to give a month's notice for the termination to the African staff.[105] Indeed the enthusiasm for agricultural development died down quickly after the war.

The most important structural change in the post-war period was the establishment of produce marketing boards. The origin of the marketing board system can be traced to the harsh global economic condition of the late 1930s, which had triggered a wave of anti-colonial protests.[106] These protests, it has been argued, led "British officials to consider greater state involvement in colonial economies."[107] With the outbreak of the Second World War, European businesses, including the United African Company (UAC), pressed for controlled marketing to "reduce the riskiness of the West African trade, and the Colonial Office acceded."[108] The West African Cocoa Control Board and the West African Produce Control Board, set up in 1940 and 1942 respectively, formed the roots of the marketing board system and the large surplus accumulated by the boards played an important part in the decision to continue statutory marketing of export products after the war.[109]

The Oil Palm Marketing Boards was incorporated in 1949.[110] The emergence of these boards led to greater colonial involvement in local economies. The boards were required to cater to palm produce and maintain an efficient organization of licensed buying agents to undertake the handling of produce and its delivery to the boards. Like other marketing boards, the Oil Palm Marketing Board was required to maintain legally prescribed grades and standards to improve the quality of export produce and allocate funds in the form of grants, loans, investments, and endowments for the purposes of economic development and research.[111] The actual purchasing of produce was left to licensed buying agents.[112]

The Eastern Nigerian Marketing Board was charged with the task of ensuring the maximum production of palm oil and palm kernels.[113] H. A. Oluwasanmi has described the system of licensing buying agents as a continuation of the *"status quo ante bellum."*[114] In the case of oil palms, prospective buying agents were required to show evidence that they could purchase, with reasonable regularity, at least 400 tons of palm kernels and 200 tons of palm oil in their first year of operation. In addition, licensed buying agents were required to produce "acceptable evidence of their ability to provide the necessary capital to finance their purchases."[115] This limited the ability of Nigerians to become licensed buying agents. Therefore, expatriate firms dominated the produce trade until the 1960s, when Nigerians became important players.[116]

Although the boards were supposed to ensure the stability of producer incomes and act as buffers for African peasants, they diverted a large portion of the gross income of farmers to finance development in infrastructure, plantations, farm settlements, and research.[117] The differences between producer prices and market prices, P. T. Bauer noted, left the primary producers at roughly the same income level as they had been before the Second World War.[118] Still, post-war conditions generated a certain measure of socio-economic progress and development. Higher prices introduced by the government as well as improved processing techniques accounted for the boom in peasant production in the 1950s.[119] The period witnessed an increase in the role of indigenous entrepreneurs in the economy. Investments in transport and the distributive trade were also made in this period. This era also witnessed a remarkable number of co-operative societies catering to the interests of rural farmers.[120]

SOCIOCULTURAL DEVELOPMENTS

The Second World War accelerated the rate of social change, and this created uncertainties and reduced the motivation of young men in particular to remain in the villages. A young man, Gilbert Uzor, discouraged by the poverty

Table 5.7. Provincial details of palm oil target figures and actual tonnage passed for export, April 1945–46.

PROVINCE	PALM OIL TARGET (TONS)	TONNAGE FOR EXPORT	SHORTFALL
Onitsha	9,104	4,655	4,449
Ogoja	797	423	365
Calabar	41,451	32,014	9,437
Owerri	38,663	32,679	5,984
Total	90,015	69,771	20,235

Source: NAE, file no. 1642/vol. 11, ABADIST, 1/26/908, "Palm Produce Production," M. E. Broughton to Director of Supplies, Nigerian Secretariat Lagos, 6 March 1946.

of rural life in his village of Umunomo, was encouraged by the financial benefits that would accrue to him to join the army during the Second World War. Gilbert shared the view of many young men in the villages that the war created an opportunity to escape rural poverty. The army provided him the resources he needed to marry his wife Nwaeke and establish a home in Port Harcourt at the end of the war.[121]

The war changed the way people lived their lives. In fact, the war accelerated the increase in migration that had developed earlier, forcing more young people to leave their towns and villages. Linus Anabalam, who left his village with many other young men after the war, remarked:

> After the war, the eye of the world was opened and people began to migrate out of the rural areas. Many of us wanted to copy the life style of the returning soldiers, including the way they dressed. This encouraged many young men and some women to leave the villages because there was money to be made outside the village and quickly too.[122]

The implications of colonial policy for the people of Nigeria went beyond food security problems. Colonial policy also affected the people's everyday lives. Informants emphasized that many adult males could not fulfill social obligations, such as the payment of bride-wealth.[123] Young men were forced by such economic and social obligations to migrate from the villages as the monetization of the local economy increased. Anex Ibeh, who migrated as a young man, notes: "When European money was introduced, one could not live on farming alone. I walked on foot to the mid-west and worked in the timber industry to accumulate enough money to start a family."[124] For many men in the Mbaise area, for example, wage labour and migration took them away from their own fields. Subsistence, marriage, and conformation to a new life style drove many to migrate out of Igboland.[125]

In some parts of the region, however, the rural economy benefited from the growing expenditures on transportation and communication, rail and road development.[126] Between 1943 and 1946, the government allocated a substantial sum of money for the construction of rural roads to facilitate the evacuation of palm oil and kernels. The development of local feeder roads further integrated rural households into the colonial and war economy.[127] The development of roads and other forms of communication increased the ability of rural peasants to leave the villages. They migrated to the towns, taking advantage of the opportunities in the construction industry and in the expanding bureaucracy. Large numbers of Igbo labourers were attracted by the opportunities offered by plantations in the mid-western area of Nigeria and the timber industry in Benin. The increase in the migration of men and young boys tended to heighten frustration with the local economy, but migration provided a ready source of capital accumulation, a break from rural life, and opportunities that were absent from rural Igboland.

The Igbo formed the bulk of new migrants into the emerging cities throughout Nigeria in the post-war period. Their rate of migration was stimulated by the "increased income which migrants earned from abroad," Chief Eneremadu recalls.[128] The scarcity of agricultural land, the declining soil fertility, and the high population density of parts of the Owerri, Okigwe, Orlu, Awka, and Onitsha areas increased the rate of migration from these regions in the post-war period.[129]

Still, the developments in Eastern Nigeria in this period must be analyzed in relation to the environmental and demographic conditions of the region. Unlike many other parts of the country, the extensive exploitation of natural resources and the expansion of cash crop production took place in an increasingly fragile environment overburdened with high population density. Eastern Nigeria, like many other parts of tropical Africa, experienced a dramatic rate of deforestation during the colonial period. Although deforestation had a deep history, most of Igboland lost its forest cover in the first half of the twentieth century due to the expansion of cash cropping and agricultural intensification. An informant recalled the disappearance of different species of mushrooms, edible plants, and roots in the last fifty years.[130] Informants remember when people could make a meal out of collecting wild vegetables and tubers from the forest.[131]

Overall, colonial policies in this era demonstrated a serious anti-peasant bias and threw the burden of the war on the African population. Farmers did not have the capacity to meet the challenges imposed on them, and few colonial administrators had the capacity to relate production to local conditions or to peasant strategies. Intervention in the rural economy disturbed production and reduced the incomes of farmers.[132] The contradictory policy of the colonial authorities created the reality faced by the indigenous farmers who produced according to the demands of the European market.

The need for food and raw materials was central to British policy during the Second World War and it significantly altered peasant-state relations. Some peasants withdrew from the local economy and many migrated out of the region, due in part to the drop in the prices of palm oil and kernels and in part to the coercive nature of colonial policy in this period. While the prices of food items rose in the 1940s, the prices of palm oil and kernels did not. The instability in farmers' incomes contributed to the rate of agricultural involution after the Second World War.

The Second World War was a major cause of change for men and women from Igbo societies as they sought to negotiate their livelihood within an increasingly intrusive colonial society. As farmers and traders, they came to witness the increased intervention of the colonial state in their affairs through new market regulations, restrictions, and price controls. With the demands of the war, rural peasants were called upon to contribute to the British war effort through the production of palm produce and other export goods in addition to food items. In some cases, Igbo peasants capitalized on the opportunities provided by the war to produce more palm oil and food to meet increased demands. At other times they resisted the intrusion of the colonial state and its regulatory powers. They did this by subtle and less subtle forms of resistance, including the use of petitions and supplication.

The post-war period witnessed major attempts by the colonial authority to transform the colonial economy and increase the welfare of colonial subjects. These reforms were not always successful as the policies were at times at odds with the realities in the colonies. Yet government policy under the Colonial Development and Welfare Act of 1945, which provided funds for long-term economic development, formed the foundation of the economic development policy of the Eastern Region of Nigeria after political independence. The nature of agricultural policy and peasant response in the post-colonial period is the focus of the next chapter.

CHAPTER SIX

THE AFRICAN ELITE, AGRARIAN REVOLUTION, AND SOCIO-POLITICAL CHANGE, 1954–80

In 1954, the Lyttelton Constitution, adopted as part of British political reform in Nigeria, transferred aspects of economic planning to the regional governments.[1] By this time, agriculture was still contributing about 60 per cent of national income and subsistence for over 70 per cent of the population.[2] In the Eastern Region, agriculture was seen as the most important route to economic development and increased welfare for the population.[3] The palm produce trade still dominated the rural economy. Key policies focused around cash crop production on plantation models, incorporation of rural farmers into the government's agricultural programs, and the expansion of food production as a corollary of the government's overall agricultural program. However, the post-independence boom was short. By the mid-1960s, political and structural problems were holding back the progress and any gains made. Economically, this set the stage for the agricultural crisis that engulfed most of rural Igboland from the 1970s. This chapter examines the attempts made by the indigenous elite to expand agricultural production and the successes and failures of the agricultural programs implemented under the supervision of local state actors. It also explores how the unintended consequences of major political and structural changes in Nigeria, i.e., the Nigeria Civil War (1967–70), and later, the expansion of the petroleum industry from the late 1970s, affected the trajectory of agriculture. These two factors would drastically reduce the contribution of agriculture to the total export earnings of the region in particular, and the country in general, and threaten the subsistence and ability of the rural population to survive through agricultural production.

PRE-INDEPENDENCE INDIGENOUS REFORMS

The establishment of the Eastern Nigerian Development Corporation (ENDC) and Eastern Regional Development Board (ERDB) in 1954 marked an important step in the government's attempt to accelerate the pace of agricultural development, especially the oil palm industry. This pilot scheme, which began in Abak and Uyo Provinces, was extended to Owerri Province in 1954. It also marked the beginning of the indigenization of agricultural policy in Eastern Nigeria and the transformation of the nature and scope of government support for agriculture.[4] The mid-1950s also witnessed further development of earlier programs. Extension services and demonstrations to improve the skills of local farmers were introduced.[5] Practical school farms, field days, and agricultural shows were instituted to stimulate farmers' interest in new farming techniques.[6] Extension work became available to all the divisions of the region with the opening of the new School of Agriculture at Umuahia in 1955. The Eastern Regional Production Development Board also encouraged peasant participation by paying a subsidy of £5 to individuals or communities for the rehabilitation of palm grooves.[7]

The desire to improve quality led to the introduction of innovations in palm oil production methods. According to Eno Usoro, the survival of local producers in the industry "depended upon changes in processing to meet the marketing board's export quality requirements, especially since low grade oil was no longer purchased by the boards."[8] One of the significant innovations introduced was the hand press for extracting oil. Oil press operatives often carried out digesting and pressing procedures, after which the farmers took away their oil and kernels. Indeed, hand presses were much more successful than the manual method, extracting about 20 per cent more oil than the manual method.[9] The traditional method was also more laborious. Owners of hand presses often provided their facilities to farmers for a fee. It was the farmer's responsibility to strip the bunch and cook the fruit. The use of the hand press gained currency as a cheap competitor with the pioneer oil produced in the mills established by the government. There were over 3,200 hand presses operating in the Eastern Region by 1959.[10]

Another innovation introduced by the government was the installation of palm oil mills following the establishment of pioneer oil palm projects. The mills involved a simple factory process designed to process about three-quarters of a ton of fruit per hour. Between 1949 and 1954, the Board of the ENDC spent about £925,200, or 64 per cent of its expenditure, on the erection of oil mills.[11] By 1954, the 56 oil mills operating in the region employed a labour force of about 1,368 people.[12] The mills processed a total of 33,609 tons of fruits, producing 5,716 tons of oil and 2,331 tons of kernels, in the first quarter of 1953.[13] The export of palm products gained momentum from the mid-1950s onward, with the Eastern Region producing over 40,000 tons of palm kernel and over 170,000 tons of palm oil per quarter by 1959.[14] The numbers of mills would increase to more than 200 in 1962, producing about 25,000 tons of oil.[15]

However, there was resistance to the introduction of oil mills in some parts of Igboland.[16] Women resisted the introduction of both hand presses and pioneer oil mills in Ngwa and neighbouring areas because these innovations threatened their control over palm kernels.[17] They particularly resisted the attempts by European firms to buy the uncracked nuts.[18] The innovations were resisted for other reasons. According to Chief Eneremadu, "There was widespread mistrust of the government's real intentions and fear that the introduction of these mills would result in loss of land and oil palms to the government."[19] In addition, the substantial cost involved in acquiring hand presses discouraged many. The price of a press rose from about £45 in 1945 to an average of £65 by 1953 due to increased demand.[20] Overall, the 1950s witnessed significant increase in innovation and a corresponding increase in levels of production. The pace of government participation would accelerate after Nigeria gained full political independence from Britain in 1960.

INDEPENDENCE AND THE REGIONAL
AGRICULTURAL INITIATIVE

"Agrarian revolution" became a household slogan in Eastern Nigeria from 1960 onward. The population was still overwhelmingly rural and the economy was based on the production of palm oil and kernels and on services related to the bulking and transportation of the products to major produce-buying centres. Therefore, agricultural development was seen as an essential part of the process of modernization after independence. The region's agricultural revolution was shaped by the vision of the first premier of the Eastern Region, M. I. Okpara, and the minister for agriculture, P. N. Okeke-Ojiudu. Okpara was a strong advocate of what he called "pragmatic socialism." Okpara believed that the region's development lay in an agricultural revolution that created wealth for both the state and peasant farmers. As a pragmatist, he owned a large farm in his hometown. His proactive stance on agriculture was inspired by the desire to transform the countryside by creating an ideal agricultural economy that would embrace peasants.

The trust of the Eastern Region's agricultural development strategy was the creation of large-scale state-run farm projects under its tree crop program. The tree crop initiative was a composite of three main programs: the Oil Palm Rehabilitation Scheme (OPRS), aimed at replanting 24,000 hectares (60,000 acres) with new hybrid palms,[21] community plantations, and farm settlements. During the First Development Plan (1962–68), the government allocated £30.4 m, or about 37 per cent of its capital expenditure, to agriculture.[22] Approximately 62 per cent of the investment in agriculture was allocated to tree crops.[23]

Dr. M. I. Okpara: The architect of the agricultural revolution. (Reproduced with the kind permission of the National Archives, London. PRO, INF 10/253.)

OIL PALM REHABILITATION SCHEME (OPRS)

The OPRS began in 1962 as part of the first six-year development plan and was arguably the most import project under the tree crop program in terms of its scope. The government aimed to tie the government's agricultural programs to peasant producers through intensified oil palm rehabilitation. The major objective was the improvement and replacement of old oil palms with seeds capable of improved yields per acre, producing fruits with increased oil content.[24] The department of agriculture supplied free seedlings and fertilizer and offered extension services for the first five years to participating farmers. Participating farmers were also paid up to $28.00 per acre over the five-year period in which crops were expected to mature.[25] The incentives offered by the government generated enthusiasm on the part of rural farmers. Philip Njoku, who started commercial farming in 1961, recalls: "We were given free advice by extension officers as well as £5 per annum in cash from the government."[26] The OPRS accounted for 8.4 per cent of the total proposed expenditure on tree crops and by the end of 1966, about 50,000 acres had been planted with oil palm out of targeted acreage of 60,000, while over 4,000 farmers participated in the scheme in its first five years of operation.[27]

EASTERN NIGERIA DEVELOPMENT CORPORATION (ENDC) PLANTATIONS/ESTATES

The Eastern Nigeria Development Corporation plantations represented the most ambitious of the agrarian development programs.[28] In principle, the ENDC plantation project represented a major step towards the achievement of a comprehensive agricultural development policy. A total of 148,930 acres was acquired for plantation development in twenty-two locations in the first five years after independence.[29] By the end of 1965, 67,000 acres had been planted with cash crops, with 85 per cent coverage projected by the end of 1968.[30]

However, the conditions in most parts of the Eastern Region posed a challenge to the project. While some of the plantations were located in areas of relatively "low" population density in the Calabar and Uyo Provinces, the rubber plantations at Ameke in Umuahia Division, Emeabiam and Obiti in Owerri Division were located in high-density areas. By national standards, even areas of "low" population density in Igboland have historically been relatively high in density.[31] People living in these highly populated areas suffered the effects of large-scale land alienation. In addition, the establishment of plantations created land tenure problems and weakened local food security.[32] Disputes between village groups over the ownership of particular tracts of land occurred frequently. This continues to generate tension between communities and the government to the present day. Furthermore, loss of peasant lands forced many to depend largely on the market for subsistence. Oral sources from Emeabiam, where a rubber plantation was established under the government's plantation scheme, emphasize that local peasants

Table 6.1. Eastern Nigeria Development Corporation, Agricultural and Plantations Division Situation Report, 1963.

NAME OF PLANTATION	DATE STARTED	ACREAGE ACQUIRED	ACREAGE PLANTED BY 1963
Calaro Oil Palm Estate	Prior to 1960	11,000	7,700
Kwa Falls Oil Palm Estate	Prior to 1960	3,724	3,563
Ikom Cocoa Estate	Prior to 1960	4,707	3,089
Oghe Cashew Industry	Prior to 1960	1,870	1,870
Bonny Coconut Estate	Prior to 1960	1,000	860
Elele Oil Palm Estate	1960	7,500	3,100
Abia/Bendeghe Cocoa Estate	1960	6,000	3,514
Umuahia Cocoa Estate	1960	2,000	1,541
Arochukwu Cocoa Estate	1960	2,000	1,591
Eket Oil Palm Estate	1961	8,965	85
Elele Rubber Estate	1961	5,500	1,200
Obubra Cocoa Estate	1961	3,000	1,605
Obubra Rubber Estate	1961	10,000	380
Ibiae Oil Palm Estate	1962	12,000	120
Emeabiam Rubber Estate	1962	8,000	800
Amaeke Abam Rubber Estate	1962	8,000	300
Biakpan Rubber Estate	1962	8,000	300
Etche Rubber Estate	1962	15,000	500
Obrenyi Cocoa Estate	1962	4,000	103
Boje Cocoa Estate	1962	2,500	637
Nsadop Oil Palm Estate	1963	18,000	Nil

Source: NAE, ESIALA, 64/1/1.

resisted the alienation of their land for the rubber plantation. According to a former employee of the Emeabiam Rubber Estate, "the government has never compensated us adequately for our farm land which they took for the rubber estate."[33]

The plantation projects faced other problems, including inaccurate feasibility studies and topographic, pedologic, and cadastral map surveys. The lack of soil analysis and the planting of crops on unsuitable soils led to crop failures.[34] The Ubani section of the Umuahia Cocoa Estate and much of the

Onitsha Agricultural Show, c. 1962. (Reproduced from the library of the National Archives, Enugu.)

Tractor laying plastic water pipe at the Umudike Agricultural Research Station, 1963. PRO INF 10/250. (Reproduced from the library of the National Archives, Enugu.)

THE LAND HAS CHANGED

Arochukwu and Obubra Cocoa Estates failed to produce healthy trees, despite the heavy cost involved in the initial establishment of the plantations.[35] The Ubani and Arochukwu plantations were replaced with coffee and oil palm respectively because cocoa could not do well in these estates.

The mechanized nature of their operation did not generate the desired employment. It was estimated that the plantations would employ 80,000 elementary school graduates, 15,000 school certificate holders, and 2,000 university graduates.[36] Judging from the employment figures in 1966, the plantations did not create the desired employment.[37] In addition, agriculture by this time had become unattractive for young school leavers.[38] Planners also ignored questions of gender as officials followed a technocratic approach to agricultural development.[39] Zebulon Ofurum, who worked at the rubber plantation, located at Emeabiam near Owerri, remembers, "Women were not employed in the rubber plantations because the work was seen as men's work." They were employed occasionally to "weed the plantation," he recalled.[40]

COMMUNITY PLANTATIONS

The community plantation program was an important component of the first six-year development plan. The scheme aimed to change and modernize "village life in *toto*," based on communal self-help, economies of scale, and crop specialization, and to maximize marketed output.[41] There was the desire also to change the traditional land tenure patterns in order to guarantee larger land holdings to practising farmers and use local resources, especially "abundant labour, as a substitute for scarce capital."[42] Selected farmers received free seedlings and fertilizer to maintain their crops on land leased to individual members of farmers' cooperatives by the government.[43] In addition to the main crop of oil palm, each farmer was required to grow food crops to support himself and his family. By 1965, there were twelve community plantations with 970 participants, occupying 11,750 acres of land.[44] The establishment of these plantations marked the beginning of what Floyd called the "plantation decade."[45]

Nevertheless, there were obstacles to the successful implementation of the community plantation projects. A key requirement for participation in the new community farms was a farmer's ability to provide a minimum of five acres. But very few farmers had access to five contiguous acres that could be devoted entirely to oil palm. Indeed, over 85 per cent of the region's farmers cultivated less than 2.5 acres in 1963.[46] Most of central Igboland, including Mbaise, Owerri, Mbano, and Obowo, were already facing severe land scarcity. The communal land-tenure system in most parts of Igboland also precluded any large-scale organization of land to accommodate those who might have had an interest in participating in the program. The modernizing model adopted ignored the realities of rural life in Igboland. There were men and women who could have taken up cash crop farming on the scale envisioned by the government but abandoned such ideas. Substantial capital was required to employ farm labour and to rent or lease land. Farmers who did not have access to their own or family land could spend as much as £300 to purchase land.[47] Women faced even more obstacles than men did. They could not rely on family, friends, cooperatives, and *isusu* (thrift societies), among other sources, for loans as did men.

Although labour was readily available, it was expensive. Most of the labour was not family labour, and farmers hired men and women to clear the land, plant seedlings, and maintain groves on a daily basis. In the 1960s, farmers paid high wages to labourers in a combination of cash and provision of meals. Male labourers earned an average of 3 shillings and 7 pence with food and 4 shillings and 3 pence without food per day. Women earned about 10 or 11 pence less than men did. Wages for both male and female labourers were about 8 pence per day less when they were provided with meals.[48] In addition, the patriarchal assumptions about farmers and land tenure systems, which effectively vested land rights in men, limited the opportunity for women to participate in official agricultural programs. Philip Njoku, a beneficiary of the government's support for local farmers, recalls that men dominated oil palm production on the plantation model. He remembers, "Our oil palm cooperative had a membership of 100 men and two women."[49] Since women did not generally belong to the agricultural cooperatives, male-dominated cooperative societies defined the community agricultural schemes.

FARM SETTLEMENTS

The most elaborate of the government's agricultural programs was the farm settlement scheme. In 1961, the premier of the Eastern Region, Dr. M. I. Okpara, travelled on an economic mission, which took him to Israel, Malaya, Ceylon, India, West Germany, and the United Kingdom.[50] His visit to Israel, in particular, played a critical role in the development of the settlement schemes in the Eastern Region. On his return with a team of Israeli experts, the premier announced in a speech in 1961 that a number of farm settlements would be established at a cost of £500,000 each. Each settlement was expected to employ about 400 young settlers and their families on individual farms and villages with government financial and technical support.[51] The government's vision of settler life was informed by an ideology referred to as "pragmatic African Socialism."[52] P. N. Okeke, the minister for agriculture, noted that the farm settlements would serve as a model for the masses of the peasant farmers to emulate. This was the most important long-term aim of the scheme.[53] The government believed that the trickle-down effects of the scheme would ultimately improve peasant production, provide career opportunities for school leavers, and stem rural-urban drift.[54] The achievement of these objectives was predicated upon the availability of boys who would make careers in modern farming.[55]

The scheme was based on the Israeli model of the "smallholder village" (*moshavin*), where settlers have secure title to their holdings as part of a larger cooperative.[56] The settlement scheme was expected to provide an alternative to the system of extension services, as a means of disseminating new techniques to farmers throughout the region, as well as to promote the social integration of different ethnic groups and communities.[57] The government enlisted the help of international agencies and organizations to assist in the realization of its agricultural development program.[58] In 1963, U.S. Agency for International Development (USAID) provided technical assistance in the form of a poultry specialist and an irrigation engineer.[59] During the same period, Israel provided technical assistance to the region for the planning and implementation of the farm settlement scheme.[60]

By 1965, the government had acquired about 148,930 acres in 22 locations for various schemes under the plantation development program.[61]

Table 6.2. Farm settlements in Eastern Nigeria, 1962–66.

SETTLE-MENT	TOTAL AREA PLANNED		AREA PLANTED DECEMBER 1966		TYPE OF CROP	SETTLERS BY DE-CEMBER 1966	DATE STARTED
	Acres	Hect-ares	Acres	Hect-ares			
Boki	11,541	4,616.4	528.65	211.5	Oil palm, citrus	240	Nov., 1962
Uzou-wani	10,562	4,224.8	619.00	247.6	Rice	190	Mar., 1965
Igbar-iam	6,560	2,624.0	1,775.00	710.0	Oil palm, citrus	350	Nov., 1962
Erei	10,385	4,154.0	1,338.00	535.2	Oil palm	360	Oct., 1964
Ulonna South	2,018	807.2	892.00	356.8	Oil palm, Rubber	240	April, 1964
Ulonna North	5,780	2,312.0	623.00	249.2	Oil palm, Rubber	120	Jan., 1965
Ohaji	14,929	5,971.6	2,053.40	821.4	Oil palm, Rubber	360	Nov., 1962
Total	61,775	24,710.0	7,829.05	3,131.7		1,860	

Source: H. I. Ajaegbu, *Urban and Rural Development in Nigeria* (London: Heinemann, 1976), 65.

The government set up six farm settlements covering 61,775 acres at Ohaji, Igbariam, Erei, Boki, Ulonna South, Ulonna North, and Uzouwani.[62] With 1,070 settlers in 1966, about 7,829.05 acres were planted with various cash crops, including oil palm, rubber, and citrus fruit.[63] The settlers were required to cultivate food crops, such as yams, cocoyams, and maize on their compound plots for their own subsistence. Equipped with modern facilities including maternity homes, schools, and modern farming equipment, these plantations represented the most ambitious of the government's agricultural development programs and its attempt to create "modern," self-supporting farming households.

The scheme did not work in practice for several reasons. The government hoped that there would be a steady inflow of foreign capital, to the tune of

Prospective settlers being interviewed for admission to a farm settlement. (Reproduced from the library of the National Archives, Enugu.)

nearly half of the total outlay of the budget. This was not realized.[64] In addition, conditions in Eastern Nigeria were different from the conditions in Israel upon which the government modelled the settlement project. The success of farm settlements in Israel is related to the peculiar historical circumstances in which the members of the Jewish Diaspora found themselves. With little agricultural experience, the scheme offered them the opportunity to acquire agricultural skills in a new environment.[65] On the other hand, the Eastern Nigerian farmers already had centuries of agricultural knowledge and many years of commercial agriculture experience behind them, so the new system operated under different socio-economic conditions.

Other factors worked against the settlement scheme. The policy of collectivization was implemented under unfavourable social conditions and was imposed on an unwilling rural population. The settlement projects, which amounted to forced villagization and relocation of peasants, was not very successful in attracting settlers because the project was adapted in a capitalist system that lacked the force involved in most socialist societies where collectivization had been adopted. The settlement schemes often took settlers away from their immediate localities, which tended to sever kinship-based labour networks. The official report of the ENDC in 1963 acknowledged: "more attention needs to be given to the social implications of plantation development and to the impact on society and the social structure."[66] Subsequent studies by a social anthropologist attached to the Uzouwani settlement agreed that the whole idea of a farm settlement was "strange."[67] He noted that dormitory life, communal feeding, and the separation from home were like "taking a plunge into the unknown."[68] Floyd and Adinde noted that sociologists or human geographers who studied the farm settlements could detect the phenomenon described as "settler shock," and without difficulty.[69]

Furthermore, the management of the settlements was not in line with indigenous ideas. The activities of government supervisors, the over-centralization of decision-making, and the bureaucratic approach of the civil service made work frustrating for those who had to do it.[70] In addition, many settlers saw themselves as civil servants working for the government and did not regard settlement farms as their own private, profit-oriented enterprises.[71] This was more so for those settlers from the land-owning communities. Eugene Nwana, who was the administrator of the Ohaji Farm Settlement during the first five years of its establishment, explains:

> Settlers from the land-owning communities were less committed to their work than their mates from other parts of the region. They were more readily diverted by events in their native homes, such as a festival, death of a relative or friend and the mourning engagements that followed, meeting friends outside the settlement, and general laziness.... It appears that what attracted them was the stipend that settlers were paid during the period when the food farm and plantation were established.[72]

The desire to create employment was not very successful. The farm settlements were semi-mechanized and capital-intensive rather than labour-intensive and, therefore, employed very little labour. The number of settlers (1,860) in 1966 was very limited compared to the number of unemployed people in the region. Intakes of settlers continued to remain low, and, by 1970, the settlements had only 3,350 families, which represented what Floyd called "a drop in the bucket" compared to the employment needs of the region.[73] Similarly, the system did not do as much to relieve congestion in high-density areas as had been envisaged. The Igbariam scheme drove a number of tenant farmers who had historically survived by working the land in the area to scatter. Since these tenant farmers were forced to forfeit their livelihoods in the Igbariam area, many moved to other parts of Onitsha Province, where they added to the rural over-population and under-employment.[74] The growth of dependency on imported food items was an indication of the failure of that settlement and other agricultural schemes.

There was much government control of settlers' lives. The contracts which settlers were forced to enter into made them tenants-at-will to the government of the region for a period of thirty-five years, which was considered the length of the active working life of the first settler.[75] Settlers undertook to work hard each day in accordance with a regimented timetable. Absence from work except in cases of certified illness incurred a fine of £1.[76] A settler could not sublet or fragment his holdings. A tenancy could change hands under three conditions. First, a son could inherit his father's holding on the death of his father. Second, the settler could be evicted if the conditions of entry into the settlement were violated. In this case, the new settler inherited the capital liabilities involved in establishing the holding. Third, a settler could leave the settlement voluntarily, after which the holding would be transferred to a new settler under the same conditions as above.[77] It is doubtful that settlers could easily break away from their contracts since they were required to repay subsistence loans of three shillings per day provided for the first two years. Besides being forbidden to form trade unions, settlers were forced to belong to farm settlement co-operative societies and to sell their produce to the co-operative societies for wholesale marketing.[78] This organizational structure, the insecurity of tenure, the fear of eviction, and the regimented settlement life could not have made for optimum production. In actuality, the settlers were mere labourers instead of owner-operators.

Acquisition of land was not easy. The alienation of large tracts of land created problems in some of the settlement communities. In the Ohaji settlement, for example, Nwana noted that different communities owned the 6,085 hectares of land acquired for the settlement. Since settlers were supposed to enjoy the right of ownership, tension arose between local and "foreign" elements in the settlement communities. Some villagers opposed the recruitment of non-indigenes, in particular Igbo, into the settlements located in the non-Igbo areas. Many in these areas viewed this as the colonization of their land by the Igbo, since each settler was perceived as enjoying rights of ownership of his plantation land in perpetuity. Others saw the project as "land snatching and rather than co-operate with it, even by copying good methods of agriculture, tried as much as possible to get even."[79] One village group remarked: "a time will eventually come when a large proportion of our productive land will pass into the ownership of non-indigenes. It is a dangerous threat to our interests and those of our posterity."[80]

There were demographic issues. About 7,828.05 acres of land had been alienated from rural communities for the settlement schemes by 1966.[81] With population densities far above the national average in a land-hungry society, land alienation remained a potential source of conflict.[82] In addition, farm settlements, the activities of migrant farmers, and the expansion of land-hungry communities took a "heavy toll on the natural vegetation of the region."[83] The entire program was far too costly, and any achievement came at a high environmental and ecological cost, and as had been the case with previous state-sponsored agricultural projects, the government was mainly concerned with export crop development. Few farm settlements cultivated food crops such as yams, cocoyams and maize in their compound lands, but all the farm settlements grew export crops.[84] Because the mono-cropping system of agriculture practised in the settlements was based, for the most part, upon the practices of other societies, the issue of local conditions and sustainable practices was not fully considered by the government.

Life in the settlements and the contractual arrangements there revealed an implicit and explicit bias against women. The notion of a self-supporting farming household inevitably drew women into the farming settlements, but the program emphasized their reproductive role as child bearers and their productive role as subsistence producers. Men were responsible for cash cropping, while women provided domestic services, raised children, and tended food crops in the farm plot.[85] Ony male settlers cound enter into a contract with the government for the duration of the contract.[86] The conditions under which tenancy could change hands also favoured male children over a settler's spouse or female children. Tenancy changed hands under three conditions – by a man's son inheriting his deceased father's holding, including assets

THE LAND HAS CHANGED

Table 6.3. Marketing board purchase of palm oil and kernels, 1960–66 (in thousands of long tons).

	PALM OIL		PALM KERNEL	
Year	Nigeria	Eastern Region	Nigeria	Eastern Region
1960	190	170	423	208
1961	173	161	430	208
1962	128	121	362	169
1963	149	139	414	197
1964	148	139	401	203
1965	164	158	449	n.a.
1966	130	128	415	n.a.

Note: One long ton is the equivalent of 2,240 pounds.

Source: Federation of Nigeria, *Federal Office of Statistics, Annual Abstract of Statistics, 1964* (1965) and Central Bank of Nigeria, *Annual Report and Statement of Accounts, 1966* (Lagos, 1967).

and liabilities, by the eviction of a settler if he violated the conditions of his contract, and by the voluntary relinquishment of a holding at the settlement. The settlement scheme was structured to benefit male farmers, and passing land and property from father to son(s) was the norm.

The objective of social integration and improvement in the economic status of settlers could not be attained. The scheme operated as a state enterprise in a region where peasants were historically the backbone of export production. Although Floyd identified what he characterized as positive gains, reflected in neatly kept farms and modern houses as opposed to the "primitive" dwellings of the farmers,[87] such a view is reflective of the official idea that the key to agricultural development lay in the "transformation of peasant life."[88] Officials ignored indigenous knowledge and failed to recognize that peasants had other interests and motivations that often did not fit into the official concept of development. The removal of settlers from their roots and communities and the setting up of settlements in new, unfamiliar locations was contrary to basic Igbo ideals that emphasized the primacy of locality and space as a source of individual and group identity. Moreover, the young boys who were supposed to be drawn into the program did not show interest in farming. By the 1960s, the civil service provided what most of them perceived as a more prestigious form of work. Yet the regional government saw its mission as a noble one. It increased its hold on the peasants through market intervention and the extraction of peasant surplus and land.

Overall, there was progress in the cash crop sector during the first few years of independence. By 1963, the government had made significant progress in the palm rehabilitation scheme, the rubber planting scheme, the cocoa improvement scheme, extension services, and in the extension of agricultural credit to farmers.[89] Agriculture still contributed about 80.3 per cent of the total value of Nigeria's exports in the first few years after independence with a significant percentage coming from the palm oil sector in Eastern Nigeria.[90] Although, Okpara's policy began as a revolution in favour of rural farmers, the agricultural programs were state-centred and continued to be guided by government "experts." But, the post-independence era differed in some respects from previous years.[91] There was recognition of the need for a balance between export crop production and the production of food crops for the population. The incorporation of rural farmers was deemed essential to achieving the dual goal of expanding both cash crops and subsistence production.

FOOD CROP DEVELOPMENT

Significant effort was made to improve food production through both direct and indirect support for peasants from 1958. The inclusion of foodstuffs and poultry production was to reduce the dependency on imported food. Although 75 per cent of the capital outlay for agriculture in this era was allocated to tree crops, an important aspect of the regional agricultural program was intercropping food crops in oil palm plantations. About 67 per cent of the land under the rehabilitation scheme was intercropped with food crops. Indeed, the allocation of 7.6 per cent of the agricultural budget to food crop production was a significant improvement over earlier years.[92] Cassava accounted for 85 per cent of the cultivated acreage. Yams accounted for 5.8 per cent, and maize, cocoyam, bananas, plantains, okra, pepper, and other vegetables made up the remaining proportion of crops.[93]

The cultivation of rice, for example, continued to expand into the mid-1960s. The government distributed improved rice seeds (B.G. 79) to farmers. It increased the distribution of rice to farmers from 3,799 lbs. in the 1958/59 planting season to 47,809 lbs. during the 1963/64 planting season.[94] Rice cultivation in Abakaliki, Ogoja, and Enugu Provinces received a major boost because of favourable growing conditions. In 1964, 145 of the 225 privately owned rice mills in the Eastern Region were located in Abakaliki.[95]

The regional government also aggressively pursued cassava improvement. The Ministry of Agriculture developed a new strain of cassava, which doubled previous yields. In 1964/65, the government distributed 1,214 bundles of 50 cuttings to farmers. This was one of several attempts to improve agricultural productivity by rural farmers.[96] The popularity of gari and fermented cassava (akpu), the low cost of production and the ease of storage contributed to the extensive cultivation of cassava throughout the region.[97] These advantages compelled increasing numbers of farmers to embrace cassava cultivation.[98] The expansion of cassava cultivation led to an increase in the quantity of gari exports to northern Nigeria.[99] Once gari became a popular staple, Sarah Emenike of Item recalled, "Every man and woman turned their attention to cassava."[100]

In 1966, the Ministry of Agriculture published a guide as part of its overall development program, which included provisions for the extension of rural education to women. The department of agriculture argued that rural women need "an opportunity to develop their potential" and to "broaden their experience to meet and bridge the demands of a growing Nigeria."[101] Thus, the department sought to provide rural women with "an informal educational service in home economics and related and pertinent agricultural interests." The specific areas of interest to women under this program were listed as food preparation, nutrition, clothing, home improvements, child care and care of family members, and garden and livestock production (for use in the home and marketing).[102] This line of thinking was at the centre of the agricultural development strategy of the region.

Overall, the state's agricultural policies in the early post-independence period were pragmatic, but the broad contours of development ideology did not change radically. Indeed the revolution was incomplete. Onyegbule Korieh recalled that many rural farmers had hoped that prices of palm oil and kernels would improve under the new government. According to him, "We hoped that our own people will provide better prices than the European offered us. We thought that our conditions would improve, but this was not always really the case."[103] At the same time, the government's policies developed bureaucratic and anti-peasant aspects that ignored the social and cultural dynamics of the region's people. The top-down approach that marked colonial agricultural development and the emphasis on cash crops continued in the post-colonial era.

The government's extension services did not reach the majority of rural farmers. This was especially the case for those who did not engage in cash crop production. Participation in official agricultural schemes was not often possible for women or the average rural farmer. A survey in 1968 indicated that the participants in government schemes were "not typical farmers." The average participant was "a married man about 40 years old with some formal education and some non-farm and commercial experience."[104] Sybilia Nwosu, like many others, could not recall ever encountering an extension officer or receiving any support from the government with regard to her farming.[105] Despite the rhetoric, the exclusion of women from state-supported agricultural programs, which began in the colonial era, continued. As Jerome Wells has noted, the agricultural policies of the post-independence state were patriarchal, paternalistic, and reflected the anti-peasant stance of local bureaucrats.[106]

In spite of the temporary boom of the early years of independence in 1960, the agricultural sector was in decline by the late 1960s. The most devastating impact on agricultural and rural life were the outcomes of political crisis linked to the Nigerian civil war and the structural changes in the overall economy rooted in the emergence of petroleum as a major source of revenue for Nigeria. These two issues and the responses of the state will be the focus of the rest of this chapter.

THE NIGERIAN CIVIL WAR AND THE
AGRICULTURAL CRISIS, 1967–70

Emergent Africa has known more than its share of strife and bloodshed, from the Mau Mau terror in Kenya to the carnage of Congolese secession. But in scope of suffering, in depth of bitterness, in the seeming hopelessness of any solution short of wholesale slaughter, there is no parallel to the tragedy that has been gathering force the past 14 months in Nigeria – once Africa's brightest hope for successful nationhood. One of the opposing forces, wielding a full array of modern weapons from Britain, Russia and much of Europe, is the federal government of Nigeria. It is determined to crush a rebellion that it feels will destroy its republic. On the other side, armed chiefly with determination, stands the secessionist state of Biafra, the home of Nigeria's Ibo tribe. The Ibos [sic] are convinced that they are fighting not only for independence but for their survival as a people. – *Time, 23 August, 1968*

The story of the Nigeria-Biafra civil war is frequently told in many Igbo families. While the stories of the horrors visited on individuals are well known, the economic and social effects of the war, particularly for agriculture, have not been studied. The memories of men and women reveal that they starved and were desperate for food. The level of food insecurity and starvation was something that had not confronted the region before. "My son, do not remind me of that war. It is something you want to forget,"[107] was how my own mother, Amarahiaugwu Korieh, began when I asked about her experiences during the war. Personal and community life was transformed. It is estimated that in the years of the war, 90 per cent of the Igbo population lived below starvation levels because of the disruption of agriculture and trade. "The horror of the war remains indelible," notes Onyegbule Korieh, concluding, "I have never seen human suffering and death on such a scale before."[108] Alpelda Korie recalls, "The war and the disruption it brought created the kind of poverty and hopelessness never experienced in the region before."[109] Charity Chidomere described it as a period "characterized by pain and hopelessness."[110] Most households were not able to farm during the war. "I did not farm a lot under such insecurity ... no one was sure of living to harvest the crop," Susan Iwuagwu explained.[111] These memories provide an important entrée to an analysis of the social and economic changes that occurred among the Igbo because of the civil war.

By the mid-1960s, the Nigerian federation was in a political crisis that led to a civil war between the predominantly Igbo-speaking people of eastern Nigeria and the rest of the federation.[112] In January 1966, the civil government of Nigeria was overthrown in a coup carried out by army officers and replaced by a military government under Major General Aguiyi-Ironsi. In July, a second coup by Hausa officers from the Northern Region removed the government, which was headed by an Igbo. What followed were the massacre of Igbo men, women, and children in the north and the mass migration of many living elsewhere to their homes in the Eastern Region. Following the massacres in northern Nigeria and the failure of negotiations, Lt. Col. Chukuemeka Odumegwu Ojukwu proclaimed the secession of the Eastern Region from Nigeria on 30 May 1967 and declared the establishment of the Republic of Biafra. The federal government declared war and imposed economic sanctions on secessionist Biafra. Despite initial successes in the war, the Biafran army suffered heavy losses, and there was large-scale disruption of economic life, including agriculture. Severe food shortages followed until the war ended in 1970. The severe nature of the economic crisis has to be understood in the context of the structural weakness of the prewar agrarian economy – an economy that did not have the capacity to withstand any significant disruption of its fragile base.

WAR AND A FOOD-RESERVE-DEFICIT ECONOMY

> I do not want to see any Red Cross, any Caritas, any World Council of Churches, any Pope, any Mission, or any United Nations Delegation. I want to stop every single I[g]bo being fed as long as these people refuse to capitulate. I do not want this war. But I want to win this war. – *Major Benjamin Adekunle, August 1968*

Most parts of Igboland were what could be characterized as "food-reserve-deficit economies." As a period of "food reserve-deficit," the war, despite its relatively short duration, left many households vulnerable and exacerbated the agricultural crisis that parts of Igboland were already facing because of population and ecological factors. Igbo subsistence farmers, in most cases, were unable to produce enough to meet their food requirements exclusively from their own farms. Subsistence levels varied widely and were "nowhere very high," W. T. Morrill noted for the Igbo.[113] Food intake in most parts of Igbo territory was highest in November and December, but the hungry season or period of low food intake *(unwu)* occurred after the planting season, between February and June. People relied more on the market and wild food items during these annual periodic famines.

Although the region was not prone to persistent drought, high population and poor soil led to chronic food shortages. Low or late rains often upset the food security of the rural population. Early colonial reports mentioned the vulnerability of some areas to food insecurity. The district officer for Owerri had reported as early as 1928 that "looking back over a period of years I cannot recollect seeing in the market any quantity of yams for sale."[114] By the middle of the twentieth century, these ecological variables had made it imperative for most parts of Igboland to depend on yams, rice, onions, meat, and other consumables from outside the region. However, most parts of Igboland could support themselves in basic carbohydrate foodstuffs before the war. Kenneth Lindsay, former professor of history at the University of Nigeria (renamed the University of Biafra for the period of the war), writing in the *Globe and Mail* in 1968 noted that the areas under Biafran control in September 1968, with a normal estimated population of two and a quarter million, produced about 2 million tons of yams and a substantial amount of cassava and plantains in peace time.[115] So even in peacetime, production was less than needed. The disruption of regional trade made matters worse when the war broke out for many parts of Igboland that were already food-deficient.

Although the unfavourable food outlook already endemic in some parts of Igboland did not lead to a food emergency as soon as the war broke out, the civil war fell upon the Igbo like an eclipse, judging by the experience of rural households. The Biafran government had some time to prepare for war after the initial massacre of Igbos in 1966. In the months following the massacre, the government turned its attention not only to building the necessary infrastructure for a possible war but also to increasing the food supply. An American CIA report stated that farmers were encouraged to plant more and "with additional labor available from the refugees, they almost certainly did so."[116] While cassava and yam production in the region was expected to alleviate the carbohydrate need, the CIA concluded that a "large segment of the population, however, will remain in need of protein from outside sources."[117]

Some Igbo areas faced a serious food crisis one year after the war began as the previous year's harvest depleted. In addition, the use of food as a weapon of war by the federal government was quite effective.[118] A *Time* report on 23 August 1968 describes the food crisis faced by the Igbo:

> Crowded into hardwood forests and mangrove swamps that cannot possibly support them, Biafrans are starving to death, by a conservative estimate, at the rate of 1,000 a day. Most of the 4,500,000 refugees from all corners of Nigeria who returned to the Ibo heartland live in makeshift camps, totally dependent on scanty government and missionary rations. The price of staple foods has risen fantastically (cost of a dozen eggs: $4), and salaried work is almost nonexistent.[119]

In October 1968, UNICEF predicted a "serious famine crisis in Biafra in December 1968, when it was expected that supplies of yams and *garri* in the enclave would be exhausted." An ICRC agricultural expert reached the conclusion in October 1968 that a "catastrophe of enormous dimensions was on its way due to a forthcoming carbohydrate shortage."[120] Pacificist Lord Fenner Brockway, head of the Committee for Peace in Nigeria,[121] in a speech at a Washington public meeting on 11 January 1968 said that the carbohydrate shortage and the lack of seedlings for the new farming season could mean 25,000 deaths daily "in a few months."[122] Mr. Nordrum, an ICRC agricultural expert, stated in an article in *Aftenpost* on 18 November 1968 that hunger had "forced a good many farmers to start harvesting too early with the result that total yields of foodstuffs were a good deal below that of a normal year." At the same time, there was double the normal number of inhabitants. "It was therefore realistic that the fenced-in population in the enclave had only 40 percent of the amount of food which would have been consumed normally to last them until next harvest,"[123] he concluded. In November 1968, the Biafrans were already "consuming next year's seed crop, and once that had gone the area would be faced with famine on a massive scale," a visitor to Biafra quoted the relief agencies.[124] The threat to household and farmers' security was not limited to crops alone. Several families were killing off their domestic animals because they could not move with them as they left their homes to escape soldiers who ravaged villages seizing whatever valuable crop or animal they encountered.

The food requirements within Biafra were not precisely known because of inadequate information about both population and food production. The population estimates ranged from 4 to 7 million. According to a British report, the area under Biafran control supported perhaps about 3.5 million people before the second influx of refugees in 1968.[125] With diminishing areas under Biafran control, much of the population required full feeding from various humanitarian agencies. The large number of returnees to Igboland exacerbated the food crisis. Refugee figures varied widely. Kenneth Lindsay estimated the number of Igbo refugees in Biafra at 4 million, of whom he thought 571,000 were in refugee camps.[126] The International Committee of the Red Cross (ICRC) based its November 1968 appeal for food in aid of Biafra on an estimated refugee population of 3 million by February 1969.[127]

The crisis was clear enough to attract considerable international attention, most of it not supporting the breakup of Nigeria but attempting to save Biafrans from genocidal starvation. The Catholic Church was particularly important, probably because of the role of Irish Catholic priests in Biafra. In North America, the Igbo Diaspora played a major role. They created aware-

Starving Biafran children. (Reproduced with kind permission of the Bodleian Library, University of Oxford.) RH, Mss Afr. S. 2399, Britain – Biafra Association.

ness and ran circles around the better funded representatives of the federal government of Nigeria. The members of the diaspora wrote articles, talked to newspaper editors, spoke to any group that would listen to them, and organized many propaganda efforts.[128] "In the shadow of the Nazi Holocaust, and recalling the profound Jewish anxiety over the Arab threat to massacre Jews in Israel last year," Marc H. Tanenbaum of the American Jewish Committee wrote, "clearly the Jewish conscience ought not to permit us to remain silent in the face of such an incredible tragedy.... We have an obligation as Jews and as human beings to help alleviate the suffering of so many men, women and children."[129] The American Jewish Committee combined efforts with Catholics and Protestants as well as other organizations to raise money to defray the cost of airlifting food and medicine to Biafra. Each flight carried 13 tons of food and medicine at a cost of $6,300 to São Tomé (off the coast of Nigeria), from where it was transported to Biafra. It was hoped that such joint Christian and Jewish efforts would help sensitize the conscience of

THE
REFUGEES
ARRIVED

BY ROAD..

Biafran refugees returning from northern Nigeria. (Reproduced with kind permission of the Bodleian Library, University of Oxford.) RH, Mss Afr. S. 2399, Britain – Biafra Association.

America and other countries to respond more generously. The Catholics and Protestants made it clear that they would welcome "even a single symbolic effort for its important publicity value."[130] This effort yielded significant results, including donations of medical supplies by Dr. Richard Hahn, director of the Alliance for Health, San Rafael, California.

The large influx of returnees and displaced persons into the region created social and economic problems never previously experienced in the area. In December 1968, a British agricultural expert estimated that there were two and a quarter million refugees living with relatives and 3 million refugees without relatives to help support them. Other reports supported this estimate. By October 1968, about 3 million people who had no family connections with local farmers were being fed at feeding centres, according to an ICRC agricultural expert.[131] The forced repatriation of Igbos into the Biafran territory put increasing pressure on the already fragile environment. Stealing of farm produce increased astronomically. Mbagwu Korieh recalled, "Refugees ravaged farms and pulled out freshly planted seed yams from the ground for food."[132] This type of abominable act that would have attracted severe sanctions and ritual cleansing in peacetime was a way of life during the war. "People had to survive at all costs," Jonah Okere of Umuekwune, Ngor Okpala, remarked.[133] This was the beginning of a serious, prolonged agricultural crisis for many peasant farmers, as many never recovered from their losses. Many farmers related the experience of losing their yams during the war as marking the end of their farming occupation and role as breadwinners. It was difficult to rebuild decimated villages and the rural economy at the end of the war.[134]

Federal policies from 1968 crippled the region by starving it of food supplies. Forced displacement and indiscriminate violence left hundreds of thousands of rural farmers in a precarious situation. The federal government blockaded the considerable interregional and intraregional trade in foodstuffs that had taken place before the war.[135] Food imports including cattle, soya beans, and corn from northern Nigeria and practically all fish and European products that came through Port Harcourt were blockaded. This eliminated the principal sources of high-quality protein that had added significantly to nutritional balance given the high carbohydrate diets derived from local food.

The price of essential food items rose beyond the reach of even wealthy Biafrans. The greatest need was primarily salt, high protein foods, and baby food, which often came in the form of powdered eggs and fishmeal. *Ofe mgbugbu* or soup with no salt was normal for many families. Gari, a staple food among the Igbo, "rose in price by 18–36 times during the period [1968] to 3–18 Nigerian shillings per cup; four bananas increased by 40 times to 10 shillings; a pound of salt rose over 1,000 times to £N14-16."[136] By mid-1968, many parts of Biafra were in dire need of food as result of the federal blockade, a situation exacerbated by a high population density that was two and half times the national average.[137]

Insecurity prevented any meaningful farming or internal trade. Yam, cassava, and rice production in the Abakiliki and Ohaozara areas was severely disrupted. The supply of cassava and fish from Etche and Ikewere in the south faced similar disruption.[138] The occupation of much of the east by federal troops by late 1968 led to the resumption of some trade, particularly along the borders of the region. Yet Biafra "remain[ed] almost totally cut off

THE LAND HAS CHANGED

from its normal sources of high-protein foods," according to a CIA report.[139] Hunger continued to be widespread throughout the federally controlled portions of the former Eastern Region because of unsettled conditions in these areas, the lack of central government authority, and the low level of cultivation.[140] Toward the end of the war, however, USAID began a discussion with the Red Cross to purchase seeds and yams in the Abakiliki area for distribution to farmers in northeastern Igboland. Plans also were made to purchase 100,000 machetes for farmers who had "resorted to sharing tools."[141]

FOOD CRISIS AND *"AHIA* ATTACK"

"The last days of 1969 were the worst time we ever had," recalls Chilaka Iwuagwu, of Umunomo, Mbaise. "We just wanted the war to end; we scarcely could get enough to eat."[142] The Biafran leader, Colonel Ojukwu, in a broadcast on 18 January 1969, launched an emergency food production program. Each community, he said, should be able to produce enough food for its people. Ojukwu enjoined every family to maintain a small vegetable garden. Farmers were urged to lend to others any land that they might not be able to cultivate themselves in order to utilize the "vast areas of uncultivated land" in the region. The Biafran leader emphasized that "Biafrans should not sit back and expect to be fed by relief organizations."[143] He urged individuals to keep small poultry farms and children to collect yam heads for planting.

The Biafran government responded to the crisis by changing the priority of the former regional government, which had emphasized the production of cash crops for export.[144] The Biafran government established the Biafra Development Corporation (BDC) and Food Directorate to oversee and coordinate food production in the territory.[145] From 1967 onward, the BDC utilized some of the forestland acquired by the government for the plantation programs for the production of maize, rice, onions, tomatoes, groundnuts, pigs, and poultry.[146] By 1969, about 15,906 acres of land had been cultivated in this way. The BDC hoped to achieve three main objectives: to commence the production of food for the population immediately; to raise immediate cash revenue; to plough back any surplus into the tree crops program; and to convert available

Col. Chukuemeka Odumegwu Ojukwu, head of state of Biafra. (Reproduced with the kind permission of the National Archives, London. PRO, INF 10/253.)

land under the plantation program into farmland in order to increase food production and increase the potential for export.

Biafrans did not wait to be reminded to be resourceful despite the hopelessness of their situation and the damage to the local economy. Both men and women went to war, but they fought on different fronts. One of the immediate and enduring impacts of the war was the significant transformation in men's and women's roles in the economy. New doors opened to women in transregional trade as they struggled to feed their households in a time of war. The channelling of men's labour to the war effort left women to support the household, with the result that women bore the burden of the crisis in the agricultural sector. The Biafran government capitalized on the availability of female labour, incorporating women into the food production drive. Women organized food campaigns, assisted in the food production effort, and supported the Land Army Program.

The destruction of the rural agricultural base and the food crisis in Igboland forced many women to enter into long distance and cross-regional trade. Although rape was a constant threat, especially for young women and girls, they foraged for food and traded in areas where men could not. This was popularly known as *ahia* attack (trading on the war front). Those who survived the horror reminisced about this form of trading. Maria Gold Egbunike was a thirty-one-year-old school teacher at the beginning of the war in 1967. Maria, who traded salt and other goods during the war, recalls: "Many women traded during the war. Women would go to the relief centers and get stockfish, salt and other goods, and sometimes they will trade it with other people or sell it. This trade is what sustained the Biafran economy."[147] Ezenwanyi Anichebe of Eziowelle town, Anambra, recalls:

> Igbo men, who were not soldiers, were afraid of both the Biafran soldiers as they were of the Nigerian troops, so trading across the war zone was essentially women's business because of the potential danger men could face if caught by the Federal troops or the Biafran soldiers. Women therefore took advantage of this situation to dominate the frontier trade.... The profit we made was enormous, and served in saving the families during this critical period.[148]

The resourcefulness of the Igbo of both sexes and ages was important in their survival. Young children learned to survive on their own by collecting wild fruits, vegetables, and palm kernels.

The women's war effort was especially channelled toward cassava production. The importance of cassava, previously regarded as a reserve against hunger and as food for the poor, increased tremendously during the war. This drastic change was induced both by the hunger that many families faced and by the policy of the Biafran government, which compelled rural people to cultivate the soil to ensure their own survival. The total acreage of cassava was estimated to have increased and *gari* production intensified to supply the "refugees and ... the army."[149] A Reuters report of 15 January 1969 spoke of forest areas being "hacked down in late 1968 to grow cassava."[150] This crop had several advantages over yams under the war conditions. Amarahi-augwu Korieh recounts that early maturing species known as *Ofomi iwa*

[Cameroon cassava], *Nwa ocha* [White child], which could be roasted and eaten, and *Nwayi Umuokara* [Umuokara lady] were introduced during the war to stave off hunger."[151] Women and children were resourceful in other ways. Snails, rats, ants, and grasshoppers collected from the forest provided much-needed protein supplements. Palm oil, the normal cooking ingredient, became a good supplement to the nutritional deficiencies of the available starchy food.

The war also had a liberating effect on women, as Oruene Olaleye pointed out. The war forced them to assume new responsibilities as breadwinners of the family but at the same time enhanced their status.[152] Women increased their participation in trade during the war to meet household food needs and to fill the void created by the recruitment of men into the Biafran army. Dependency on the market for household subsistence created avenues for women's mobility on a scale unprecedented in Igboland.[153] The war also broke the social barriers that had constrained women's mobility in the past. Onyegbule Korieh remembers, "the survival of the family depended largely on the resourcefulness of women who did not stand the risk of being conscripted into the army. Men allowed their wives to travel more freely than before. This did not stop at the end of the war."[154] Women's changing roles eroded the patriarchal assumptions about the male as breadwinner, thereby introducing changes in gendered relations of production and power. The Igbo accepted, it seems, a reconfiguration of male and female roles in trade and a greater concentration of female labour effort on food production.

At the end of the war, many rural people went back to their farms to eke out a living, but many women continued with the regional trade that they had started during the war. Nwadinma Agwu recalls how women sought to help each other through collective farming and pooling of resources to acquire land and seedlings, and the proceeds were shared among the contributors after the harvest. Thus, "women were able to raise money to start petty trading and sustain the household."[155] Others, like Edna Okoye of Umudunu, Abagana, left her village as a migrant farmer to escape "poverty and infertility of soil."[156] Migrant farming became a major source of income, which helped to sustain the domestic economy. Margaret Nwanevu of Amumara Mbaise, who had lost her home during the war, was able to rebuild it with the proceeds from her palm oil trade. "I was able to build a mud house with zinc roof where the entire family lived after the war.... I sustained the family for a long time after the war because my husband had nothing and had to begin all over again in life."[157] Still others formed trading guilds to bring in essential commodities into places like Mbaise. Chinyere Iroha of Uvuru Mbaise recalls:

> Postwar hardship and poverty necessitated the formation of women trading guilds in our area, which brought fish and other seafood from the Rivers area into Mbaise. These were sold in local markets here as well as in neighboring towns by the women in the guild. With this, we were able to assume the role of providers for the families when our husbands were still disillusioned at the loss of a war. We really moved the economic fortunes of this area after the war.[158]

The Nigerian Civil War had the most important long-term effect on agriculture among the Igbo. The civil war ruined the Eastern Regional government agricultural programs and destroyed the optimism of the early 1960s. With the outbreak of the Nigerian Civil War in 1967, palm oil exports, the major cash crop for the Igbo, declined to 16,000 tons, compared with 165,000 tons in 1961. Palm kernel export, a major source of income for women, also declined, to 162,000 tons in 1967 as against 411,000 tons in 1961.[159] The trajectory of economic policy and governance followed a new path leading to the abandonment of the agricultural projects. Thus, the potential ability to revolutionize agriculture and the rural landscape may never be known, as the war sowed the seeds of the agricultural crisis that was exacerbated by the development of the petroleum industry.

THE PETROLEUM ERA AND THE AGRARIAN CRISIS IN THE 1980s

Before the 1980s, Nigeria was the world's leading exporter of palm oil and peanuts and a major producer of cocoa.[160] Most of Igboland was also largely dependent on palm oil and kernel production for income. Beginning in 1970, however, the Nigerian economy was significantly transformed by the expansion of the petroleum industry. The struggle over control of the petroleum resources in the eastern part of Nigeria seems to have played a significant part in shaping policies before, during, and after the Nigerian Civil War. Many foreign countries recognized the important role Nigeria's oil would play as the demand increased from the United States and the emerging nations such as China. A CIA intelligence memorandum rightly noted: "The brightest feature of Nigeria's postwar economic scene is the rapid rise in oil revenues. After being hit hard in the early years of the war, the industry since has expanded dramatically ... making Nigeria one of the ten largest oil-producing countries in the world."[161] "Paradoxically," the CIA wrote, "the civil war played a major role in stimulating new production" and increasing federal revenue. "With the known oil-producing areas of eastern Nigeria cut off, the oil companies intensified production in the mid-western region and offshore, and these areas currently are more important than the east." Production increased significantly, reaching an estimated 1.4 million barrels per day in January 1971.[162]

The formation of the Organization of Petroleum Exporting Countries (OPEC) led to sharp increases in oil prices. Prices dramatically increased by 130 per cent in 1979 with the formation of OPEC. The Arab-Israeli war in 1973 and the oil embargo on Western countries that supported Israel caused dramatic increases in oil prices. The Iranian Revolution of 1979 led to another oil shock, and prices rose by over 130 per cent. Similar occasional crises continued into the 1990s. These crises brought huge revenues to Nigeria. Oil prices rose from $3.78 per barrel in October 1973 to $14.69 per barrel by the beginning of 1974.[163] The export boom led to an increase of nearly 10 per cent in GDP annual growth during the 1970s and early 1980s.[164] Revenue went from $411 million in 1970 to $26.62 billion in 1980.[165]

The dramatic rise in global oil prices coincided with the Second National Development Plan, 1970–74. Agriculture, mining, and manufacturing were projected to contribute 44.2 per cent 13.4 per cent, and 12.4 per cent, respectively, to GDP. The actual contribution to GDP, however, was only 24.7 per cent for agriculture, 45.1 per cent for mining, and 4.8 per cent for manufacturing. Following the expansion in petroleum production, Nigeria's GDP grew at 6 to 8 per cent, while non-oil exports declined by about 60 per cent between 1964 and 1980. The importance of crude oil to the economy in this

Table 6.4. Production and resulting payments, 1965–70.

YEAR	PRODUCTION (MILLION BARRELS)[a]	PAYMENTS TO THE GOVERNMENT (IN MILLION US $)
1965	99.4	36.0
1966	152.4	53.2
1967	116.5	75.6
1968	51.9	44.8
1969	197.2	78.4
1970	395.7	280.0
1971	620.5 [b]	600.0

a. Excluding production from the eastern states for the period April 1967–September 1968.

b. Estimated.

Source: Central Intelligence Agency, "Nigeria: The War's Economic Legacy," 10 May 1971.

Table 6.5. Nigerian federal government revenue from crude petroleum, 1970–80.

YEAR	OIL REVENUE	TOTAL CURRENT REVENUE	OIL REVENUE AS % OF TOTAL REVENUE
1970	166.4	633.2	26.3
1971	510.2	1,169.0	43.6
1972	764.3	1,404.8	54.4
1973	1,016.0	1,695.3	59.9
1974	3,726.7	4,537.0	82.1
1975	4,271.5	5,514.7	77.5
1976	5,365.2	6,765.9	79.3
1977	6,080.6	8,080.6	75.2
1978	4,654.1	7,371.1	63.1
1979	8,880.9	10,913.1	81.4
1980+	9,918.6	11,859.8	83.6

+ Nine-month period.

Source: Toyin Falola, *Economic Reforms and Modernization in Nigeria, 1945–1965* (Kent, OH: Kent State University Press, 2004), 222.

Table 6.6. Sectoral composition of Nigerian output for selected years, 1960–75 (in %).

SECTOR	1960	1963	1970	1975
Agriculture	64.1	55.4	45.8	28.1
Oil and Mining	1.2	4.8	12.2	14.2
Manufacturing	4.8	7.0	7.6	10.2
Building and Construction	4.0	5.2	6.4	11.3
Others	25.9	27.6	30.0	36.2
	100.0	100.0	100.0	100.0

Source: M. Watts and P. Lubeck, "The Popular Classes and the Oil Boom: A Political Economy of Rural and Urban Poverty," in *The Political Economy of Nigeria*, ed. I. W. Zartman (New York: Praeger, 1983), 110.

period is reflected in the significant rise in revenue from $189 million in 1964 to $25.5 billion in 1980 – that is, from 1.3 per cent of GDP to 24.4 per cent.[166]

As government revenue increased, the state adjusted to the new conditions that emerged from the expanding oil sector. Indicative of the new power at the federal level was the allocation of a major part of revenue to the federal government. Allocations to the states also grew, from 323.8 million naira in 1974 to 2,534 million in 1979–80.[167] In general, the growth in the Nigerian economy was rapid, with an annual GNP growth of 7.4 per cent between 1970 and 1979. This period, however, was qualitatively different from the pre-1970s. The growth in the oil sector was accompanied by a significant sectoral transformation. These changes in sectoral composition reflected "not simply a growth in non-farm activities but a stagnant agrarian economy."[168] Investment in agriculture fell from 7 per cent of the budget in 1971 to 4 per cent in 1981.[169] While the tide had turned in favour of the state in the form of huge revenues from oil, the rural population was frustrated as the state turned its back on agriculture. Peasant production and dependency on agriculture went into serious decline.

The Nigerian government's interest in agriculture flagged considerably, shifting to the more lucrative oil sector.[170] A number of scholars have linked the agricultural crisis in Nigeria to the emergence of petroleum as Nigeria's main export product. Sara Berry suggests that, by the 1970s, farmers in western Nigeria, for example, were too busy scrambling for a share of the oil

An abandoned oil palm mill at Owerrinta, Abia State. (Photo by author.)

wealth in Nigeria to have time or energy to invest in expanding or upgrading their cocoa farms.[171] The neglect of agriculture and dependence on oil combined to expose the fragility of the Nigerian economy and heightened class contradictions.[172] As Nicholas Shaxson has recently shown, the paradox of African oil is the enormous wealth it generates for a few, and the poverty and political and economic insecurity it brings for the majority in oil-exporting countries such as Nigeria.[173]

As the government wavered between encouraging food production and importing food cheaply from abroad, massive importation of all kinds of foodstuffs, including those that could be produced locally, struck at the roots of the rural economy.[174] In 1961, for example, the value of food imports into Nigeria was about 45.44 million naira. By 1974, Nigeria's food import bill had tripled, and it increased still further to about 1.8 billion naira in 1981.[175] Toward the end of the 1980s, Nigeria's food import costs had increased from

Table 6.7. Share of food in total import value, 1971–85 (million naira).

YEAR	TOTAL IMPORTS	FOOD IMPORTS	% SHARE OF FOOD
1971	1,069.1	88.3	8.2
1972	990.0	95.8	9.7
1973	1,241.1	128.0	10.3
1974	1,737.3	154.8	8.9
1975	3,721.5	297.9	8.0
1976	5,148.5	440.9	8.0
1977	7,093.7	786.4	10.4
1978	8,217.1	1,020.7	12.4
1979	6,169.2	952.4	15.4
1980	6,217.1	1,049.0	12.8
1981	12,602.5	1,820.2	14.4
1982	10,100.2	1,642.2	16.0

Sources: Government of Nigeria, *Know Nigeria Series No. 1: Towards Self-Sufficiency in Food* (Lagos: Federal Ministry of Information, 1991), and Central Bank of Nigeria, *Economic and Financial Review* (various years).

509.79 million naira in 1964 to 9,658.10 million naira.[176] Consequently, between 1973 and 1980, there was an overall annual decline in agricultural production, while the GDP growth rate was more than halved.[177] Revenue slumped to $13.1 billion between 1981 and 1982 and was reduced to approximately $7 billion in 1988. While the country gained from the expanding oil wealth, rural people faced a crisis that permeated every facet of their existence. The change in the accumulative base of the state had a direct impact on rural production, and increasing pauperization emerged as the economic fortunes of the country changed.

The Igbo found themselves in a unique position because most of the oil fields were located in the southeastern region of Nigeria. They were able to deal with decreasing land and low productivity in rural agriculture by diversifying household incomes. But the emphasis on the non-agricultural sector continued at all levels of the economy, as governments and individuals made choices that intensified the agricultural crisis at the national level but also ameliorated its effect for some rural dwellers. In rural Igboland, the average holding declined considerably. The average farm size in most of Igboland in 1974 was under 0.10 hectare per household. Only 5 per cent had between

THE LAND HAS CHANGED

2 and 3.99 hectares, which represented the largest holdings.[178] However, the economic opportunities the non-agricultural sector offered to peasants brought them some relief.

The increase in revenue from oil led to massive infrastructural development, an extensive growth in the urban population, and the rapid growth of an industrial labour force. It was followed by the commoditization of urban social relations, a sharp upturn in the size of a disenfranchised and militant "floating population," and new waves of migrants from rural to urban areas.[179] At the same time, the expanding urban sector demanded semi-skilled labour for the construction industry and other service jobs. This development was important in two ways to the Igbo countryside, which experienced a population growth rate of over 3.0 per cent but produced few jobs. It created a favourable environment for rural urban drift and the loss of potential agricultural labour. From the 1980s, rural peoples sought new opportunities as the oil boom fuelled the explosion in infrastructural development. In most parts of Igboland, the rate of movement into the major cities in Nigeria and beyond was enormous. People migrated out of rural Igboland in large numbers. Lagos, Port Harcourt, Aba, and Enugu, as well as many cities in other parts of the country, became important destinations for many migrants from southeastern Nigeria. Lagos, for example, with a population of 665,000 in 1960, grew to 1,153,000 in 1985. Port Harcourt, which attracted a large number of Igbo migrants, had a population of 315,000 people in 1985, rising from a population of 180,000 in 1960.[180]

The contradictions of an urban-biased development policy and industrialization and their effects on local agricultural sustainability left the rural population few choices. Many sought other sources of income. The most important development in this transformation was the high level of small-scale and large-scale trading activities that became the hallmark of the Igbo economy – a new economy that became dominated by rural women. Many people responded to the circumstances of the oil boom era and the various constraints in other innovative ways. The lure of the city remained strong. Even with few skills, some found it easiest to migrate to the booming towns to find various forms of employment. Many moved to the cities without the promise of a job. Alban Eluwa, who left his village in 1974 for Lagos, recalls: "You know in those days, if you did not have the money to attend a secondary school, your next options were to become an apprentice of some sort or move to the city sometimes with no specific aim. You have to leave when your mates have all left the village."[181]

Others tried to improve their conditions within the rural areas by producing foodstuffs to feed the expanding urban population. Others found employment in the expanding service-oriented society that emerged with the increased importance of petroleum to the national economy and the huge expenditure on infrastructure in the urban centres. Onyegbule Korieh, a former migrant, noted, "When I arrived at Obigbo in 1973, I found that there was money to be made from selling garri and part-time farming. I could make enough money out there to feed my family well and send my son to a secondary school in 1973."[182] For Onyegbule, like many others, survival in the rural areas was becoming outdated for many by the mid-1970s.

Nigeria emerged in the 1980s as a robust semi-industrial economy from its prewar mercantile basis and dependence on the export of agricultural commodities.[183] As the petro-economy expanded, a new entrepreneurial class of "contractors" emerged more clearly than before the 1980s. They engaged in the booming supply business and used their profits to invest in trading. Others used their salaries to invest in agriculture. In Imo State, they leased land and hired labour to produce cassava in areas such as Ohaji and Egbema.[184] Yet the condition of many peasants remained precarious because they lacked the land, labour, and cash to invest in agriculture. The general outlook was one of decreasing food and agricultural productivity. For the overnight petro-contractors, their livelihood became very unstable as the oil revenue declined and the construction boom dried up from the 1980s onward.

Concomitant with the increase in oil revenue was the expansion of social services such as education. By the end of 1959, the enrolment figure in the Eastern Region was 1.4 million[185] out of an estimated total population of 8.1 million.[186] By the 1970s, school enrolment in southeastern Nigeria was above the national average. This trend continued in the oil boom era. The high enrolment figures meant a significant reduction of available household agricultural labour and an increase in the tasks of women, who had to take on farm tasks previously performed by children. Although school attendance was encouraged for two main reasons – social and economic mobility and status enhancement and the introduction of universal free primary education – the result was a labour crisis, since most educated household members never returned to farming. But young people who got good jobs elsewhere earned more money than they could have earned on the farm.

The ruinous inflation that accompanied the oil wealth was reflected in the inability of farmers to survive on what they produced. As subsistence production became radically undermined by the petroleum economy, they and a large segment of the overall population suffered dire consequences. Andrew Ibekwe, a retired bank employee who now farms in his village, attributed rural poverty to the fundamental changes in the economy that have taken place in recent times. He states, "The village landscape has changed to the extent that farming has become very insignificant in rural livelihood."[187] Onyegbule Korieh attributed the difficulty of rural life to the cost of living, which had gone beyond the reach of many rural dwellers. He says, "Things are tough presently because of overpopulation, scarcity of farmland, and very high inflation. But our sources of earning a living have not changed at all."[188] Despite the boom, the rural and agricultural landscape and its associated problems remain relatively unchanged.[189] As Michael Watts notes, some classes benefited materially from the commodity boom, as measured by the consumption of purchased imports, "but the majority of the urban and rural poor found any hard-won gains rapidly eroded by inflation."[190] Indeed, the post-1970s developments weakened the essentially agriculture-based economy. The petroleum boom brought new opportunities, and the wealth created by the oil industry made the state and some people wealthy. It was also a period of ambivalence. The inequality and inflation that it generated in the 1980s led to the devastation of the economy and to rural and urban poverty.

The development of the petroleum industry and its emergence as a major revenue earner for Nigeria was perhaps the most profound structural change that shaped the countryside from the 1970s onward. As petroleum became a major source of state revenue, the agricultural orientation of both the government and rural communities was faced with powerful trends that drew both the population and the state away from agriculture. The intermittent attempts to reverse these trends have not been successful.

THE STATE ATTEMPT TO COME TO THE RESCUE

Given the declining contribution of agriculture to the national economy at the end of the Second National Development Plan period, the Yakubu Gowon administration (1966–75) recognized that dependence on oil led to economic vulnerability. For this reason, agriculture received high priority during the Third National Development Plan period, 1975–80.[191] About 57 per cent of the allocation to agriculture remained unspent in the previous development plan, and the Gowon administration interpreted this as indicative of a fundamental defect in the design and implementation of agricultural programs. The gross under-spending of the allocation to primary production revealed fundamental problems in the implementation of agricultural programs at the federal level.

A substantial part of the remaining money was allocated to government projects such as farm settlements, irrigation schemes, and plantation projects for cash crop production. The new plan also recognized the need to check rural-urban migration through a balanced development agenda for both the rural and the urban sectors, but the continued expansion of the oil sector and the opportunities it provided for employment and trade continued to undermine the agricultural sector. No direct support was provided for peasant farmers, who contributed over 95 per cent of both the export and domestic productions. The government's support for farmers in the form of credits went to a few commercial farmers and bureaucrats disguised as farmers. Under these conditions, many peasants abandoned their farms to seek other forms of employment.

The major intervention to deal with the agricultural crisis in the country came in 1976, when the Olusegun Obasanjo administration launched the Operation Feed the Nation (OFN) program. The objectives of the OFN program included the mobilization of the nation toward self-sufficiency and self-reliance in food production. However, these objectives were not achieved. The program collapsed for various administrative and logistical reasons. Political expediency inhibited its successful implementation. As a program designed to make fertilizers, in particular, available to farmers through the various state ministries of agriculture, the scheme faced many logistical problems. Fertilizers often arrived so late in some areas that they could not be applied to crops. Storage facilities provided under the program were grossly inadequate. In many cases, bureaucrats hijacked the fertilizers and tried to resell them to peasant farmers at prices that many could not afford.[192] Unfortunately, the OFN did not target the peasants who had been the backbone of the country's agricultural production. The political elite usurped the gains that could have been made. Because the government did not deal directly with peasants, the peasants did not comply with the wishes of the government. All this needs to be seen against the massive importation of food by the end of the OFN program in 1979.[193]

The government also intervened in the customary land tenure systems in Nigeria through the implementation of the Land Use Decree in 1978.[194] The decree sought to eliminate the problems associated with traditional land tenure systems.[195] It also aimed to create a uniform tenure system and to eliminate any tenure arrangements that inhibited large-scale agricultural development.[196] Farming became the favourite part-time occupation for the military elite. The decree created opportunities for the military and bureaucratic elite to take land from peasants. It vested authority over land in the governor of each state. Although most rural areas remained relatively unaffected by the decree, communities located on the periphery of urban centres lost their land to urban development, with peasants continuing to be squeezed off their land as the cities expanded their housing and industrial projects.

The creation of new states in 1976 was accompanied by more reforms at the state level. Like their predecessors, the new states continued to attach a great deal of importance to agriculture. At the top of the policy-making apparatus was the Ministry of Agriculture and Natural Resources. One of the most important projects for agricultural development in Nigeria was the setting up of agricultural development programs (ADPs). The ADPs, which started in 1974, were federal programs implemented at the state level. These corporations were charged with the production and processing of agricultural products.[197] In Imo State, for example, the ADP produced maize, cassava, and horticultural crops, including citrus and pineapple, all of which had become highly commercialized.[198] The ADP also engaged in export crop production. The responsibility of the ADP was to cater to the needs of the small-scale farmers. However, Gavin Williams has noted that the benefits "accrued to the rich rather than the poor," and that "some projects have excluded the poor from access to productive resources and redistributed the assets and incomes to the rich."[199] These projects included oil palm projects at Ohaji, Ozuitem, Nkporo, and Ulonna North and South, and rubber plantations at Obitti, Emeabiam, Ameke Abam, and Ndioji Abam.

The corporation's projects did not revolutionize agriculture in the region. The commission of inquiry set up to review the activities of the ADP in 1980 found that the cashew plantation at Mbala was unprofitable as an economic venture.[200] Only 10 per cent of 4,092 hectares of mature rubber were being tapped in 1980. The oil palm development projects did not increase overall production from the region. The Pioneer Oil Mills (POM), located in Imo State, for example, could not break even. The commission of inquiry set up to examine the activities of the project observed that "the project cannot be a viable venture because of its structural rigidity, paucity of palm fruit supply and the salary/wage bill that has been too heavy for the small volume of business the POM handles."[201] The food crop project, which was fully funded by the state, did not fare better than earlier projects. The establishment of the rice project at Ugwueke, the Commission of Inquiry argued, did not lead to profits because it appeared to be "politically motivated."[202]

The Fourth National Development Plan, 1981–85, sought to expand local food production and the production of basic raw materials for industries under President Shehu Shagari's Green Revolution program launched in May 1980. The Green Revolution program called for an accelerated increase in agricultural production through the removal of the constraints to increased production and the provision of agricultural input and extension services to farmers.[203] The government established eleven river basin development authorities and nine integrated agricultural development projects. The projects were attempts to increase the production of rice, sugarcane, millet, sorghum, maize, wheat, cassava, and yams. The irrigation policy aimed to develop a system of multiple cropping in the northern arid zones of minimal annual rainfall. While the government allocated the substantial amount of 8.828 billion naira to agriculture in this plan period, the legacy of incompetent management, corruption, nepotism, and lack of adequate feasibility studies hampered the chances of success.[204] Like the OFN, the Green Revolution program ignored rural farmers and the program achieved very little due to weak and corrupt leadership.[205] The government's "quick fix" attitude toward agriculture did not reflect its overall development ideology of rural development. The inherent contradiction in the state-peasant relationship was reflected in the failure to provide farmers with adequate incentives and support to permit the widespread adoption of improved techniques.[206]

Huge agricultural projects and schemes generally did not succeed, and irrigation projects in particular did not revolutionize agriculture or increase food production. Government incentives benefited only commercial farmers, who often diverted agricultural credits to other uses.[207] In addition, the location of agricultural projects was often politically motivated. For example, the building of irrigation projects in parts of northern Nigeria was motivated by the need to distribute the benefits of the oil boom of the late 1970s.[208]

Overall, the decades after the civil war were landmark years for many rural dwellers. The peasant class largely disappeared from the southeastern Nigerian agrarian scene during the period after the end of the war. Despite the increased involvement of the state in agriculture, rural peasants did not respond favourably. In Imo State, for example, the total area under cultivation fell from 203,000 hectares in 1976 to 52,000 in 1981, representing an annual decline of 32 per cent. The output of yams in 1981 was 22,000 tons, representing a 39 per cent decrease from the 1976 output. Likewise, the output of cassava fell by 78 per cent between 1976 and 1981.[209] On the individual level, it was difficult to rebuild decimated Igbo villages and the rural economy after the civil war. Poverty struck all parts of Igboland, but it was worse in central Igboland with its high population and land scarcity.

The 1960s were years of optimism for many African societies in both political and economic terms. This certainly was the case for Igbo society. However, by the 1970s, the euphoria of the early years of independence had turned to frustration. The agricultural revolution of the Okpara government had not borne much fruit before the Nigeria civil war broke out in 1967. The war changed Igboland and its economy irrevocably. Furthermore, the expansion of the petroleum industry changed the economic landscape fundamentally. The massive infrastructural development in the urban sectors with no corresponding development in the rural areas exacerbated the crisis already faced by rural dwellers. The government's reforms in agriculture were a brutal failure as the problem of agrarian opportunities that they tried to address was exacerbated as inequality grew. The structural changes that emerged because of the war undermined the institutions and practices that governed the local community, including agricultural practices and labour arrangements, and eroded rural identity. While the demand for palm oil continued within the domestic economy, the returns from sales have not kept up with the massive rise in the cost of living. This has left the rural population poorer, despite the rise in real income since the 1980s. As before, the greatest asset of the rural population has been their resilience in the face of major crises in the rural economy. The final chapter in this book sorts through the spirited attempts to deal with the crisis that seemed to engulf the rural population in the 1970s and 1980s, putting the events of these years in historical perspective.

CHAPTER SEVEN

ON THE BRINK: AGRICULTURAL CRISIS AND RURAL SURVIVAL

> When both the identity of self and of community becomes in-
> distinguishable from that of the land and its fabric of life, adapta-
> tion follows almost instinctively, like a pronghorn moving through
> sagebrush. – *Donald Worster, Dust Bowl, 164*

By the end of the 1980s, low agricultural productivity, food insecurity, and environmental degradation became apparent in many parts of Igboland. Indeed,
most of central Igboland struggled to feed a growing population on a diminishing area of farmland under the impact of years of neglect of agriculture
in favour of a national economy dependent on the petroleum industry. These
trends threatened the ability of farmers to increase productivity and to practice sustainable agriculture.[1] The crisis in the rural areas forced many rural
dwellers to rethink the approach to their livelihood and survival strategies.[2]
This chapter goes beyond the focus in the colonial and early post-colonial periods to outline the condition of agriculture since the 1980s and to examine
the way in which the Igbo, especially in central Igboland, have sought to deal
with issues of livelihood historically.

Although Nigeria experienced a degree of prosperity from oil revenue
before the mid-1980s, this prosperity was short-lived. The dramatic fall in
oil prices from more than $35 a barrel in 1980 to as low as $10 per barrel in
1986 affected the economy at all levels.[3] The changing fortunes of the state
affected the rural economies as the country faced a recession and increased

debt burden. The introduction of painful austerity measures culminated in the implementation of World Bank and International Monetary Fund macroeconomic policies in the form of a structural adjustment program (SAP) by the Ibrahim Babangida administration in 1986. The aim of the SAP was debt recovery in the short term and poverty reduction through economic growth in the long term. The liberalization of the economy involved the elimination of stimulus programs, abolition of subsidies, reduction in price control, export promotion, devaluation of the currency, privatization of state owned industries, and reduction in public spending.

The structural adjustment program affected rural life. It led to an increase in rural poverty and in the prices of all necessities of life. Although the cost of living increased enormously, real wages actually fell under the SAP and massive lay-offs of workers and overall economic problems followed. Household consumption data collected between 1980 and 1996 and agricultural census information collected in 1993 and 1994 show an increasing level of poverty in the agricultural sector.[4] In the mid-1990s, about 67 million people, or about 65 per cent of the population, were identified as poor. A poverty assessment (PA) study carried out in parts of Nigeria revealed that 87 per cent of the core poor in 1985 were members of rural agricultural households.[5]

Raluca and John Polimeni's study of the impact of the globalizing trends of the 1980s on Igbo society show that the restructuring of the public sector diminished the capacity of rural people to cope, even through traditional institutions, including the extended family, attempted to help communities to "maintain cultural traditions."[6] Eugenia Otuonye, a rural dweller and mother of five children, reflected on the impact of the SAP era: "We have not had a good life since the government brought *ota na isi* [knock on the head]; life is much harder than words can describe."[7] A school teacher recalled how she embarked on backyard farming to help sustain her family. The case of Isidore, who lost his job as a construction worker in a Port Harcourt shipyard, was not an isolated case. "I have never held another paid job since I left Port Harcourt in 1989, and it has been a struggle to support my family."[8] Many young people faced similar uncertainties and experiences. *Ota na isi* as a metaphor for hardship became part of the political and economic discourse for many rural dwellers. Songs reflected the hardship of the era.

At the state levels, government agencies encouraged farmers to diversify, with emphasis on food production. The federal and state governments throughout Nigeria, with the support of the World Bank, set up the Accelerated Development Agricultural Programs (ADAPs). This was the outcome of the Nigeria Food Strategies Mission, which had been concerned with the deterioration of the food situation in the country.[9] The Imo State Accelerated Development Agricultural Program (ISADAP), which covered most of central Igboland, was established in September 1982 to capitalize on the food production functions of the Ministry of Agriculture and, at the same time, minimize the protracted problem of red tape in public sector activities. The reduction of government's hold on the operational mechanism of ISADAP was a tacit acceptance of the constraints that state interventions had imposed on agricultural development in the past.

The first three years of ISADAP witnessed improvements in extension and agronomy services. These involved about 707,983 farming families in Imo State, from whom 28,077 contact farmers were selected.[10] The program also embarked on the provision of high-yield cassava cuttings and seed yams to farmers. While this represented an improvement on past years, only 380 extension agents were provided, a very small number in relation to the number of farmers in the region. However, the production data from ISADAP in 1983 showed an improvement in food production. The total area under cultivation went up from 52,000 hectares in 1981 to 88,000 hectares in 1983. With an estimated increase of 16 per cent in cultivated area for 1984, ISADAP appeared to be the only state-owned agricultural project that made some gains. Cassava output in 1983 was 293,000 metric tons. This represented a 24 per cent increase over the 1977 figures. Rice production increased from 5,000 metric tons between 1980 and 1983 to 55,000 metric tons in 1985. The prices of major food items, including yams and gari, fell by between 29 and 52 per cent in different towns in the state.[11]

ISADAP made some gains, but the fall in staple food prices may not have been directly related to its activities. The crisis of the 1980s brought both the rural and urban population face to face with a major rise in the cost of living. Both rural and urban dwellers adopted a self-help strategy of producing more of their own food. In many Igbo households, the proverb, *aka aja aja wetara onu nmanu nmanu* (it is the soiling of the hands that brings about the oiling of the mouth), came to be strictly applied in daily living. Open spaces in towns and backyards were suddenly converted to farms. The general economic decline in the country compelled many rural and urban dwellers to engage in some production.

Yet these gains did not translate into an overall improvement in agriculture. Government programs remained largely inefficient while rural farmers did not always respond favourably or put government advice into effect. In fact, the total area under cultivation had fallen over the longer term, from 203,000 hectares in 1976 to 52,000 in 1981 in Imo State, representing an annual decline of 32 per cent.[12] The output of yams in 1981 was 22,000 tons, representing a 39 per cent decrease from the 1976 output. Likewise, the output of cassava had also fallen by 78 per cent between 1976 and 1981. While these data may not be very reliable, they indicate an increasing crisis in the rural sector.

The ADAP programs did not produce any significant change in the gendered pattern of previous policies. Women were ignored as independent farmers. The perception of women as "homemakers" excluded them from agricultural credits and other forms of support for farmers. The home economics centres proposed by the federal government under the ADAP scheme focused on nutritional education for rural women.[13] Moreover, women's inability to provide collateral such as land worked against them in obtaining credits and loans under the program.[14]

Rural life continued to go through a severe crisis, despite attempts to revitalize agriculture. In most of Igboland, where the average holding had declined considerably, the economic opportunities provided by agricultural pursuits offered rural dwellers little relief. The average farm size in most of Igboland in 1974 was under 0.10 hectare per household. Only 5 per cent had between 2 and 3.99 hectares, which represented the largest holdings.[15] It was the emphasis on the non-agricultural sector that often ameliorated the effect of agricultural decline on rural dwellers.

While these problems persisted, the expanding urban sector demanded semi-skilled labour for the construction industry and other service sector jobs. This development was important in two ways for the Igbo countryside, which experienced a population growth rate of over 3.0 per cent but produced few jobs. First, the Igbo responded to the urban economic growth and the

THE LAND HAS CHANGED

Table 7.1. Average size of farm and holding, 1984–85.

STATE	NO. OF FARMING HOUSE- HOLDS (IN THOU- SANDS)	TOTAL NO. OF FARMS (000)	TOTAL AREA OF FARMS (PER THOU- SAND HA)	NO. OF FARMS PER HOUSE- HOLD	AVERAGE SIZE OF HOLDING (HA)	AVERAGE SIZE OF FARMS (HA)
Anambra	565	1,363	137	2.41	0.25	0.10
Imo	634	1,529	109	2.41	0.18	0.07
National Average	6,066	12,141	6,608	2.00	1.14	0.57

Source: *Rural Agricultural Sample Survey, 1984/85, 1985/86* (Lagos: Federal Office of Statistics, 1990), 16.

opportunities it afforded them. It was by diversifying household incomes that the Igbo were able to deal with decreasing land and low productivity in rural agriculture. Second, the expansion that occurred in the petroleum sector stimulated new forms of economic activities outside agriculture, especially an expansion in the service sector and a booming supply business.[16] Others leased land as absentee farmers in areas such as Ohaji and Egbema and hired labour to produce cassava. But this was not the story of the majority of the population, who continued to struggle and to adapt to the changing landscape of rural Igboland.

Given the role that migration and the adoption of crops such as cassava have played in recent years, the rest of this chapter will examine the resiliency of the rural population, the changing cropping patterns, and how many have coped with the changing rural landscape through migration. How these changing livelihood strategies have affected rural identity and rural gender roles will be examined.

CONTINUITY AND CHANGE IN RURAL LIFE

The ability of rural Igbo, like many Africans, to eke out a living in the face of economic adversities beyond their control is remarkable. Even after all the structural changes that have occurred, a large portion of the Igbo population has remained in the rural areas. Igbo tenacity can be linked to a sense of place that comes from having a link with the land – *ala* – even as survival strategies have shifted enormously from dependency on farming to dependency on a variety of non-farm income-generating activities, including wage labour and migration. This sense of place defines rural adaptation in ways that reflect the important link between the land, farming, individuals, and group identity.

Two conditions have influenced the way in which people have responded to the conditions in the rural economy: The first is the tendency for some rural dwellers to reinvest excess income in other income-generating activities. The second is what Bongo Adi identifies as the push factor. Here, some rural dwellers have no option but to diversify in response to declining agricultural productivity, land scarcity, and population pressure.[17] Bryceson has argued that the attempt to eke out a living in some rural economies has led to de-agrarianization, in which most rural farmers shift away from agriculture to non-agricultural income-generating activities.

Most of central Igboland have exhibited such tendencies since the 1970s. The scarcity of land, poor soil, and population pressure has made diversification inevitable for communities such as Mbaise, Mbano, Etiti, and others in central Igboland. Nwanyiafo Obasis, a rural farmer in Mbaise, explains: "We combine farming and trading in order to survive. If one is a trader without a farm, one is taking a risk because the market could fail."[18] In Nguru, Mbaise, since the 1970s, Adi found that poor soil, scarcity of farmland and a very high population density has left the people "little option but to move away from agriculture as a significant source of income."[19] Here, as in other parts of Mbaise, the number of landless people has increased significantly while fallow periods have been eliminated altogether. The significant shift to non-agricultural activities came in response to the crisis in agriculture, and people responded to both the internal and external structural changes through their social institutions. According to Eugenia Otuonye, "We are left with no option but to do other things in order to survive. I sell *akara* [bean cake] and do other odd jobs to provide food for my children."[20]

Yet we cannot generalize about the nature of the agricultural crisis because conditions differ markedly even within one region. While most of central Igboland faced a severe agricultural crisis, agriculture remained an important source of income for some parts of the region. Umuagwo, Ohaji, and Oguta have continued to engage in significant commercial and subsistence food production that supports the local population as well as the urban population of Owerri. Here, relatively smaller population and the rich soil support both commercial and subsistence agriculture. Farmers' incomes have remained relatively competitive because farm products have continued to attract high prices due to the demand from other parts of the region and the urban areas. As Adi noted in a study of livelihood in Umuagwo, the average fallow period is between five and ten years because it is relatively abundant in comparison to the rest of the region, where the average land holding is about 1.2 hectares per household. Umuagwo remains a significant source of cassava and gari for parts of central Igboland, an important food item among the Igbo today. The adoption of cassava has radically transformed the agrarian culture of the Igbo, their dietary habit, and rural identity, as they eat more cassava than the traditional favourite – yam.

CASSAVA PEOPLE AT HEART?

Until the mid-nineteenth century, cassava was confined to the root crop belt of West Africa, primarily in the lowland tropical forests.[21] As noted in chapter 1, the Igbo were a "yam people at heart," but this icon of Igbo agriculture, masculinity, culture, and identity has been put on a back burner in Igbo agriculture. Indeed, since the 1970s, the Igbo can be aptly described as a "cassava people at heart." The adoption of cassava, as Simon Ottenberg argues, is an important measure of the "index of the level of change," among the Igbo.[22] Susan Martin has stressed the major transformations that followed the introduction of capitalist agriculture (palm oil exportation) and how male control of the product and low-priced palm oil products sometimes forced women to divert their labour to cassava production. The cultivation of cassava led to changes in food production methods, gender, and intergenerational relationships as rural families confronted the agricultural crisis of later years.[23]

Nevertheless, the changes began much earlier. European accounts show that cassava reached Onitsha about 1857.[24] Recollections of rural people confirm that the Igbo initially looked down upon cassava. An evangelist in Onitsha in 1863 noted that those who grew cassava were the poor who could not "afford to plant yams."[25] The early rejection and skepticism were partly associated with the poisonous characteristics of poorly processed cassava.[26] However, it began to "defuse less tardily" during the influenza pandemic of 1918–19.[27] The First World War and the famine that followed the pandemic, Don Ohadike had noted, increased the importance of cassava in parts of Igboland. By the 1920s, Ormsby-Gore observed that cassava had become a major supplement to the native food crops.[28] In addition, a district officer in Owerri described cassava as the main supplementary foodstuff. He noted that the average amount of cassava planted was about the same as that of yam in 1929.[29] Its importance would increase in later years.

Cassava gained prominence in Igboland for a number of reasons. Clearly, it provided a suitable alternative source of cheap carbohydrate and soon became a famine-relief crop that alleviated the traditional hungry period (unwu) preceding the yam harvest. Luke Osunwole remembers: "In those days there was famine, usually after yams had been planted. June and July were the worse months. Our people would do their best to survive on cassava, cocoyams, and yams set aside for eating (ji njakiri) that had been stored in the storage house (mkpuke)."[30]

Advances in the production and utilization of the cassava tuber made it a popular food item in the urban areas. Processed cassava in the form of gari became very popular among the expanding urban and working class population, which was dependent on a cheap source of food. Morgan observed that cassava was a cheap, easily transported food of increasing popularity among the majority of the employed population and had the "advantage of harvest in May and June when no other fresh food is available."[31] Often referred to as nri okopkoro (food for spinsters/bachelors), gari was a very convenient food because it was very easy to prepare – by simply pouring boiled water over the flour to make a dough eaten with soup.

Women preparing gari at a Government Agricultural Development Center, c. 1960s. (Reproduced from the library of the National Archives, Enugu.)

Soon, gari become an important item of trade. By 1938, Igbo traders were sending about 4,000 tons of cassava flour to the north, and by 1942, this had increased to over 6,000 tons of gari per annum. The trade created an opportunity for peasant farmers to increase their income.[32] The Second World War and the high cost of imported food increased the importance of cassava as food for a wider population and the army. By the 1940s, it was spreading in areas such as Abakiliki that had relied heavily on yam production.[33]

Although cassava was seen as less prestigious than yams, and therefore as a woman's crop, its overall importance as a source of income for women increased dramatically over time. Morgan observed that cassava profoundly altered the economic and social relations between husbands and wives and fostered economic opportunities for women as it become an important source of cash income for them.[34] Phoebe Ottenberg, who studied the Afikpo Igbo in the early 1950s, confirms that the introduction of cassava, considered "beneath the dignity of men," was a major source of change in women's economic fortunes.[35] Ottenberg states, "If a woman's husband did not give her food, she was in a sorry plight; [with cassava] now it is possible for her to subsist without the aid of her husband."[36] Though Ottenberg's comments were an uncritical characterization of women's economic position, the crop afforded them opportunities for capital accumulation, self-esteem, and a higher degree of economic independence. "Nowadays women do not care if the husband doesn't give them any food, for they can go to the farm and get cassava," an elderly Afikpo women confirms. "If a woman has any money she buys [rents] land and plants cassava. The year after she does this she can have a crop for cassava meal, which she can sell and have her own money. Then she can say, 'What is man'? I have my own money!"[37] Hence, women strongly resented any attempt to challenge their dominance in cassava production and trade. As early as 1925, for example, they complained about unfair male competition in what was regarded as women's trade.[38]

The high rate of agricultural involution from the end of the Second World War contributed to the apparent decline in yam cultivation and the ascendancy of cassava. By the 1950s, yam was "a rich man's food" and one that required substantial investment in labour and money to produce. "Cassava helps us to feed the family more than any other crop," Robert Ibe said.[39] Linus Anabalam recalls: "Unlike the past when yams and cocoyam were the main crops, it is not unusual to find a farmer with four or five plots under cassava crops alone."[40] Related to the labour question is an aging rural agricultural population. "What can an old man and his wife do as farmers?" Linus Anabalam asked, as he reflected on the labour problem.

Cassava has no specific harvesting age and is, therefore, a convenient crop when alternative income-producing activities "compete for the farmer's time."[41] Given the greater labour involved in yam cultivation, Morgan observed, "more money may be obtained from the growth of cheaper cassava sold in Aba Township or sent in the form of gari to Port Harcourt and Calabar Province."[42] Significant focus shifted to cassava production because it required less labour. The ease of transporting food of increasing popularity among the majority of the employed population, the advantage of a harvest in May and June when no other fresh food is available, and the ability to produce cassava on land with short fallows resulted in the expansion of cassava growing at the expense of yams.[43] While many in the rural areas adapted to the changing agrarian landscape, others, especially the younger population, migrated in search of better opportunities.

MIGRATION AND RURAL LIVELIHOOD

Migration from the Igbo region typically occurred because of demographic and geographical conditions and the pressure on available agricultural land.[44] A disproportionately higher percentage of Igbo migrants came from the densely populated central region than from elsewhere.[45] References to population pressure found in missionary letters and travellers' journals suggest that the size of the population was already an economic problem by the end of the nineteenth century. A missionary from the Owerri region reported in 1866 that "population is so great that if they hear we shall want carriers, they come in great numbers begging to be used, even during the farming season."[46] Thomas Northcote linked the poor quality of the soil, the shortage of land, poverty, and subsistence insecurity in the region in the early twentieth century. Significant labour migration from the barren lands of the Onitsha-Awka axis to more favoured regions already existed by this time.[47] As Kenneth Dike confirms, "The density of population which was and still is a main feature of the I[g]bo country.... Hence the I[g]bos pressing against limited land resources had, of necessity to seek other avenues of livelihood outside."[48]

Like Dike, Simon Ottenberg observed that poor soil incapable of supporting more than subsistence agriculture was a major factor that forced the Igbo to seek sources of livelihood elsewhere.[49] R. K. Udo, writing about Eastern Nigeria, paints a picture of densely populated areas like Mbaise and Awka from which people were forced to migrate as tenant farmers even before the colonial era.[50] British anthropologist Sylvia Leith-Ross described Nguru Mbaise, in central Igboland in 1935, in these words:

> The over-population of this area is well known, with its consequent land and, possibly, food shortages. I saw it at its poorest time, when last season's yams were finished, and only a few of the new season's (women's) yams were ready to be dug.

The population depended on cassava and cocoyam and a small amount of very poor corn for its daily food. The over farmed land produced smaller and smaller crops.[51]

Emmanuel Ude recorded that land scarcity in Mgbowo forced the people to migrate to other parts of Igboland such as Ezioha, Inyi, and Ndeaboh by the beginning of the nineteenth century.[52] Similar observations were made by Ikenna Nzimoro about the Awka, who migrated as squatter farmers to other parts of Igboland because of poor soil. Poor soil in the Nnewi area, he noted, caused a switch from farming to trade.[53] This trend also explains the migration into Owerri (Oratta-Ikwere), and eventually across the Imo River into Aba (Ngwa) and from these across the Aba River to the Aza, where the movement was stopped by the British conquest.[54] The westward migrants settled on the borders of Benin (among the northern and southern Ika). To the northeast, they invaded the Cross River lowlands and established a frontier on the Okpauku River. There the Igbo reproduced the grassland pattern of fortified settlements in which compounds were loosely grouped together. To the south, they reproduced the forest pattern of dispersal.[55]

Linked to the demographic pressure in the Igbo region is the environmental degradation that has become a part of the landscape in several parts of the Igbo country.[56] Continued use of the land and human activity has led to a breakdown in agricultural productivity. Floyd had predicted that this would happen under the traditional farming methods used in the region.[57] For the 1940s, Forde and Jones estimated a population of between 600 and 1,000 persons per square mile over much of Okigwe Division.[58] A population of more than 1,000 persons per square mile has been recorded in northern Ngwa, Owerri, and Orlu.[59] According to the 1963 census, the regional population density had reached more than four times the Nigerian average.[60] Land scarcity, land degradation, and a high level of non-farm activities have been noticeable in areas such as Isu and Mbaise, where population density was over 1,000/sq. km by the middle of the twentieth century.[61] The very high concentration of population in the Igbo region gave rise to extensive modification of the natural environment and exposed the soil in many parts of the region to leaching and erosion.

The pressure on the land is reflected in internal migration rates as well as migration out of the region. Many men from the over-populated parts of Igboland were forced to work as migrant farmers on the lands of others. Accurate figures for these moves are not obtainable, but from oral sources, it can be seen that certain area such as Okigwe, Obigbo, and Etche attracted tenant farmers and labourers from the more crowded areas of Mbaise, Owerri, Mbano, and Etiti, among others. Charles Takes, who carried out rural sociological research in the Okigwe Division of Owerri Province in 1962, noted that there were considerable differences in density. The population in the area north of Okigwe Township still contained virgin land available for cultivation. In the southern part of the division, however, towns such as Mbano and Etiti faced extreme scarcity of land, such that many people no longer found a living in agriculture.[62] Luke Osunwoke of Umuorlu recalls: "Our people went outside the community to look for food. They lacked sufficient land and there were no thick forests. So they often migrated to Elele, Ahaoda, and Ikwerre where they worked as agricultural laborers."[63] The members of the Nguru clan of Mbaise, struggling with the demands of an expanding population on leached, eroding land, supplemented their income by working as migrant labourers for the Etche clan.[64] Some parts of eastern and western Nigeria have served as recipients of Igbo migrants from the less agriculturally favourable areas such as Awka, Owerri, Mbaise, Isu, Mbano, and Obowo, who have relentlessly sought ways to improve their lives through migration. For the Mbaise Igbo in general, high population density and increased intensification became the only way out of their economic problems. Many migrated as tenant and seasonal farmers within and outside the region. According to Isichei, the Ezza, who had ample land, performed herculean feats of industry on their own yam farms and then travelled west to toil on the farms of others.[65]

The land tenure system, which led to progressive fragmentation of farmland and to primogeniture, made life in rural areas difficult.[66] By the 1950s, fallow periods had been significantly reduced in many part of central Igboland such as Mbaise, Mbano, Obowo, and Etiti.[67] The fragmentation of land holdings resulting from the land tenure system made agriculture frustrating and inefficient and prompted many young men to migrate. Stanley Diamond explains the pull and push factors in Igbo migration:

Population pressure and land scarcity remained the most important determinants of migration. Population pressure on deteriorating forestlands (1,000 plus per square mile) in, for example, Owerri Province at the heart of the Eastern Region, had, in conjunction with the social character of the I[g]bo, led to a continuous migration of I[g]bo to all regions of Nigeria; the largest number of migrants, of course, found their way north since the Region so designated represents three-quarters of the country. Moreover, the educational level of I[g]bo was higher than that of the average Northerner, enabling them to get jobs in the civil service, trading, [and] utilities. Nigeria becomes, in effect, an I[g]bo Diaspora.[68]

The population of Igboland has increased progressively. Part of what constituted Owerri Province (in Imo and Abia States) now has an estimated population of over 6.8 million according to the 2006 population census of Nigeria.[69] This population explosion has meant unprecedented pressure on available land, considerable deterioration of the environment, and high levels of poverty. I have seen in my own village that even firewood and water have become commodities that most rural households have to pay for. This was not the case two decades ago.

Colonialism had its inevitable impact on the rate of migration as improved communications and structural changes increased mobility and opportunity. Clearly, the rate of migration was caused by structural changes as well as the motivation of individuals, mainly men, to acquire wealth and improve their lives. During the early parts of the colonial period, large numbers of Igbo people moved out of Igboland following the development of towns and the expansion of the railway and roads. Many also responded to the growth and expansion of the trade in European goods and the public service sector, and the increased opportunities for economic independence.[70]

The discovery of bitumen coal in Udi near Enugu in 1903 was crucial in the shaping of colonial policy toward this section of southeastern Nigeria. After geological assessments in 1903 and further tests in 1908–1909, the government planned a railway to the seaport at Port Harcourt to facilitate the evacuation of coal and other resources from the region. This southern railway, which joined the northern one, reached Makurdi in 1910 and reached its terminus at Oturkpo in the Benue area in the following year. Owerri Province supplied Port Harcourt, the colliery, and the brickfields with a large number of labourers. The railway, which linked up other eastern towns such as Umuahia, Omoba, and Aba, facilitated the evacuation of palm products to the coast. Throughout the first half of the twentieth century, the railway remained the most effective mode of transport, hauling almost all of Nigeria's foreign trade traffic. It was only after the extension of infrastructure by the colonial state (railways, creek clearing) that European firms seriously began building up hinterland stations.[71]

Railways and road construction accelerated the rate of agricultural output and stimulated market opportunities, and administrative reorganization and the introduction of rudimentary technology and research in agriculture motivated and sustained local interest in production. A unique "rail culture," marked by the growth of retail and service sectors dominated by women food vendors, developed along the railroads. Developments in transportation in turn led to the rise of cities and urban areas as trading centres where the raw materials produced by the local people were exchanged for European manufactured goods.[72] These developments created an increasingly mobile Igbo population that swelled the emerging colonial cities and commercial centres. Gradually, inland transport and port facilities were developed, and banking and other financial institutions were organized to facilitate the ever-increasing use of a single modern currency as the means of exchange over all of Nigeria.[73]

The Great Depression of the late 1920s and early 1930s exacerbated the migration from the oil-palm-producing areas, which had been accustomed to cash incomes.[74] Local people were severely affected by the fall in the price of palm oil and kernels. Low returns from export crops and the general economic decline forced many rural dwellers to seek their fortune elsewhere. As Jones observed, "the depression of the palm oil trade and the lack of any alternative exports stimulated the drive towards migrant labour," despite the attempt by the government and commercial agencies to improve palm oil cultivation and local systems of agriculture.[75]

The introduction of taxation in late 1927 created problems for young adult males who could not meet their tax and other needs from the resources available in the rural areas. Taxation forced many young men to respond to opportunities in the expanding cocoa industry in the West and the rubber and timber industry in the mid-west. Migrants' income became the major source of cash for the payment of income tax, for marriage, education, the building of homes, and support for family members, especially the elderly. This trend was already visible by the late 1930s. Sylvia Leith-Ross was told during a visit to [E]Inyiogugu village, Mbaise in 1935, that "about a hundred young men had migrated, during recent times, to Oluko in Umuneke Court area, 'where there is plenty of good land and the people do not mind,'" where they have "settled for good as farmers."[76] Harris's study of the economy of sixteen persons among the Ozuitem Igbo shows that off-farm sources of income including remittance contributed substantially to the total annual income of these individuals in 1939.[77] Although income from outside was changing rural lifestyles and occupations irrevocably, Linus Anabalam remarked:

> The migration of men and young men and women also placed a heavy burden on women and the elderly as the expansion of Western education and opportunities for work in the public service led to a dramatic exodus of young educated people from the rural areas in search of white-collar jobs.[78]

Urban towns drew a large Igbo population, including traders who became very active throughout the country. It is estimated that the Igbo population in Northern Nigeria increased from 3,000 in 1921 to about 12,000 in 1931. By 1953, the Igbo population in the North had reached an estimated 127,000. J. B. Davies, who worked with the United African Company (UAC) for many years in Northern Nigeria, remembers:

> In the early years, they formed a nucleus of the commercial staff of all commercial companies. They filled the clerical jobs, acted as depot clerks and depot buyers. They were very efficient and particularly hard working. During this period, they were also the major transporters in the north and owned most of the commercial vehicles.[79]

Lagos alone had 32,000 of the estimated 57,000 Igbo living in Western Nigerian towns and villages.[80] By the 1950s, the Igbo made up more than half of the non-indigenous population of Lagos, Benin, and the northern towns of Kano and Kaduna.[81] A significant number of the nearly 10,000 easterners in the British mandate territory of Cameroon were Igbo.[82]

Migration intensified after the Second World War. Linus Anabalam, who had migrated to Northern Nigeria in the 1940s with a group of other young men who worked as sawyers, recalled that many young men, some as young as twelve years, migrated to other regions in Nigeria. According to him, "We were all motivated to leave the village when we saw the returning soldiers, the kind of dresses they wore and their new lifestyle. You could not have a life like that from farming in the village."[83] Their high rate of migration meant that the Igbo came to dominate the civil service sector even in predominantly non-Igbo areas of the Eastern Region and Northern Nigeria. The domestic staff of the Calabar Catering Rest House in 1949, for example, was predominantly Igbo, most of them cooks, stewards, and houseboys.[84] Others, including a large number of Igbo ex-servicemen, worked in the rubber estates in the Calabar District. A petition by ex-servicemen to the district officer for Calabar seeking payment of bonuses was signed by a predominantly Igbo group. Over 80 per cent of the 59 petitioners were Igbo, an indication of their overwhelming numbers in other parts of the region.[85]

The report of a commission of inquiry set up to look into the affairs of the formers Eastern Nigeria Development Corporation Plantations found in 1968 that only 27 of the 300 senior staff came from Calabar Province where the plantations were located. The overwhelming majority were Igbo who were viewed as non-indigenes.[86] The report further noted:

> In so far as the functional administration of the former ENDC was concerned, the supreme authority rested with a single executive who was stationed at Enugu with his senior staff, mostly Ibos, spread out to all the plantations. The result was that senior and junior posts including labour were filled by their kith and kin. It therefore became impossible to give effective participation to the indigenes of the areas where the plantations were situated.[87]

The situation left the plantations with a labour crisis when the Igbos left during the civil war – a vacuum that the indigenes were unable to fill. The effects of large-scale migration of the Igbo have thus been a source of conflict with their neighbours as much as an avenue to survival in what has been a difficult economic environment.

Igbo migration escalated beyond the confines of Nigeria in the 1940s. British-administered southern Cameroon attracted a significant Igbo population when Britain inherited the territory as a mandate colony after the First World War.[88] Cameroonian towns such as Kumba, Tiko, and Victoria had a large Igbo population, most of whom were engaged in small-scale distributive trade in foodstuffs and imported goods. The migration of the Igbo into the territory was encouraged by the British to relieve the Igbo region of its very high population.[89]

A large number of Igbo people responded to the opportunities created by dwindling labour in Spanish plantations on the island of Fernando Po.[90] The decline in the indigenous Bubi population by the late nineteenth and early twentieth century forced plantation owners to look elsewhere for labour. The island's palm oil and cocoa plantations become a strong attraction for a large population of Igbo people. Thousands more migrated after the Spanish government of Fernando Po and Rio Muni concluded a labour agreement in 1942 with British Nigeria for the recruitment of Nigerian labourers. The agreement was also meant to check the illegal recruitment of labour from Nigeria as well as regulate the conditions of service for Nigerian migrants in a place that had a reputation for its harsh working conditions. According to the 1944 report of the Nigerian Labor Department, the labour agreement sought to regularize what had become "a large scale traffic in laborers and to endeavor to eliminate the unscrupulous native 'black birder' who earned a lucrative livelihood by kidnapping the ignorant peasants from the Ibo and Ibibio areas."[91]

The propaganda encouraging people to migrate to Fernando Po for work came from the government, churches, friends, and relatives. Many men migrated to "improve their lives," recalled Udochukwu, a former migrant.[92] Although many had migrated to better their lives, the conditions in the Spanish plantations did not create an opportunity for accumulation. Wages were very low and many migrants were afflicted by disease and poverty. Yet, the number of migrants swelled. In 1961, there were around 4,000 Spaniards, 10,000 indigenous people – the Bubis – and over 50,000 Nigerians working on the cocoa plantations.[93] Their number had increased to about 85,000 by the mid-1960s, with Igbo, Ibibio, and Efik comprising two-thirds of the island's population.[94] This agreement remained in force after the independence of both countries in the 1960s.

While the Anglo-Spanish Employment Agency at Calabar recruited only men aged 18 to 45, some women migrated to the islands as wives, while a few more, mostly widows, went on their own. Life for many Igbo migrants to Fernando Po did not improve, as many returned more impoverished than before. The difficulties of plantation life and the very low wages forced some of the women into prostitution. Loise, a former migrant, notes that some men who could not deal with the hardship of plantation life "sent their wives to 'New-Bill,' a public square, for prostitution." Prostitution enabled some families to survive, since the Spanish government "would stop providing food items as was stipulated in the labor agreement, if one left the plantation."[95]

The end of the Nigerian Civil War witnessed further large-scale migration from Igboland. Approximately 90 per cent of young people regularly moved out of the rural areas after the war. They found the challenges of rural areas too great. These movements are connected to the demise of a way of life and are embedded in the individual and collective biographies of many men and women. Alban Eluwa recalls:

> You know in those days, if you did not have the money to attend a secondary school, your next options were to become an apprentice of some sort or move to the city sometimes with no specific aim. You had to leave when your mates had all left the village.[96]

Onyegbule Korieh, a former migrant, recalls, "When I arrived at Obigbo in 1973, I found that there was money to be made from selling gari and part-time farming. I could make enough money out there to feed my family well and send my son to a secondary school in 1973."[97] For many men like Alban and Onyegbule, survival in the rural areas had become a thing of the past due to the devastating impact of the civil war. Alban summed up the dilemma young people faced: "As a young man, if you stay in the village here, people will always suspect you when a neighbor's chicken is missing. We had to seek opportunities to survive by leaving the village."[98]

Overall, the migratory pattern that emerged in Igboland was a response to the economic, ecological, and demographic factors in different parts of the region. Migration has given the Igbo an edge in small retail trade in the urban and rural areas throughout Nigeria and beyond, where they have operated retail shops or worked as artisans. While some elderly folks interpret the absence of young people from farm work as an "unwillingness of young people to cultivate the land," the cash income generated outside Igboland has helped to transform many Igbo societies and provided an alternative income that supported a rapidly disappearing rural agrarian society. Remittance has also increased local purchasing power and the ability of elderly men and women to hire labour for farm work. Onyegbule Korieh explains: "Many of us depend on money remitted by our children to survive today. The rural area is very 'dry.'"[99] Such remittances into Igboland have contributed to the economic advancement of individuals, households, and entire Igbo communities. Indeed the changing rural life, social expectation, support of the elderly, and demands of the extended family system force young men and women to find quicker and more "honourable" ways of earning cash than farming. This pattern has transformed the rural landscape and further reduced the values attached to agriculture.

MIGRATION, HOST COMMUNITIES, AND NEW IDENTITY

The migration trend has continued in recent times and most households have migrants living elsewhere. However, the host communities did not always welcome the Igbo. Indeed, their presence, even in other parts of southeastern Nigeria, often angered locals who resented what they perceived as Igbo aggressive tactics. J. B. Davies, an agent of the United African Company in Northern Nigeria, recalls:

There was always a certain amount of animosity. I think it went right back to the early years. When I first went to Nigeria in the 1930's practically every single commercial employer was other than Hausa, they were mainly I[g]bo's and Yoruba's. Practically every artisan was a westerner or easterner, and in the Public Works Dept of the native authorities, practically all the road labour were Ibo. This worked well for a time, but once the Hausa started to feel his feet and wanted to get on, wanted to start learning, wanted to earn money, wanted to get out of his farm and move into other fields, then he found he was blocked. He found that easterners and westerners were very happy and quite content to block him.... I think this was one of the things that tended to create a big rift between the northerner and the southerner.[100]

In Southern Cameroons, the Igbo dominated public sector employment, trade, and education until the plebiscite of 1961, when the region voted to join Cameroon. There was the notion that the Igbo would unquestionably continue to dominate the local population if the region became part of Nigeria.[101] The Igbo fear-factor was a political and economic reality in the Southern Cameroons from the early 1950s among the indigenous population. Sera Williams, daughter of Manga Williams, who was then king of the Victoria area, campaigned seriously from the early 1950s for "Southern Cameroons" to join the rest of Cameroon. She is reported to have once demanded from a crowd: "Any woman wey Igboman never slapa'am for this market place, make e-put ye hand for up!" No hand was raised in response to her question asking any woman in the crowd who had not been slapped by an Igbo man in the market to raise her hand.[102] W. A. Robinson, British plebiscite administrator, remembers the "obvious distrust of Nigerians and in particular of the I[g]bos who were numerous in the frontier areas."[103] Similar hostility to the Igbo was found in Fernando Po, leading to the expulsion of 40,000 Nigerians, mostly Igbo, from the island in 1975 during Macias Nguema's rule.

New forms of ethnic consciousness and identity often emerged among the Igbo in diaspora communities. But the development of ethnic consciousness or identity is not automatic. Such developments occur in a particular context and are influenced by the receiving community's view of the migrant community. Once outside the homeland, a greater sense of cohesion, cooperation, and identification, at least based on a common language and experience, emerges in response to particular ecological and contextual factors. For the same reason, attitudes toward immigrant populations have often been informed by attempts to protect individual and group interests as host communities perceive their own survival as threatened. Both the perceived threat to the host community and the confrontational attitude toward the immigrant community gives rise to new senses of identity, often conflict-driven.

The Igbo responded to the contestation for resources with the host community by forging a new sense of community in the spirit of *igwe bu ike* (there is strength in numbers). This sense of community led to the formation of what have been described as "home associations" within the Igbo diaspora communities. G. I. Jones notes:

> The greatest advantage possessed by migrants from the I[g]bo area, whether these were traders, craftsmen, labourers or in superior employment, was their segmentary social structure and the attitudes derived from it, and also their traditional trading organisation. Both came to play as soon as I[g]bo moved outside their home neighbourhoods. On their home ground the I[g]bo were an aggregation of independent towns or villages each competing and on guard against its neighbours. In other parts of Nigeria, the I[g]bo felt themselves to be a solid and united group. Anyone speaking the language was a fellow tribesman, a relative with whom one was in duty bound to combine for mutual aid.[104]

Ideally, no Igbo person can have two homes, since "home" is not just a geographical expression but the place where one was born, where the ancestors are buried, or where one can connect with the past. The Igbo society remains one in which kinship plays a crucial and dominant role. The ethnic unions that emerged among Igbo migrants, therefore, acted as a bridge between their temporary location and their original homes. Town unions remained very active in this regard, while ethnic unions remained more effective in protecting Igbo interests in relation to other groups. In this context, *erinma* or solidarity,

... drew inspiration from the awareness that all members of each unit or segment of the Igbo socio-political structure [were] kinsmen or kins-women whose rights and privileges were the concern of all.... Thus, for the Igbo, *erinma* (an abstraction contracted from *eriri omumu nwa* or the umbilical cord) implies familyhood and symbolizes the organic link between people of common ancestry.[105]

As important as *erinma* has been in bridging the divide between individualism and cooperativism, *erinma* seem to have expressed itself among migrant groups in the form of a greater tendency toward cooperation despite a strong desire for individual achievement. This new conception of *erinma* emerges to serve the collective interest of the new community in the diaspora based on common experience and interests rather than on kinship. The expression of *erinma* among host communities becomes a form of collective action expressed by the group to protect its own political, social, and economic interests.

AGRICULTURAL DECLINE AND CHANGING IDENTITIES

We have always been farmers, but today, we depend on the market to survive. – *an Mbaise Elder*

This comment by an elder from Mbaise in 1999 captures the changing rural identity of the Igbo. The agrarian culture and structures of rural populations have been disappearing rapidly. An often-neglected aspect of understanding agricultural crises in rural societies is the impact of values and the constraints they may impose in dealing with contemporary social and economic issues. Values influence how people conceptualize problems and find solutions. While agricultural policies have been set by governments in an attempt to bring about desired ends within a society, they often do not take into account the values that local people attach to agriculture or its link to their identity as individuals or groups. Rural societies, farmers themselves, and the land upon which they farm operate within a structural framework imbued with values and norms, which have consequences for the survival and continuity of rural life. Thus, some of the policies designed to solve the agricultural problem at one time or another have exacerbated the problems by neglecting to consider the way in which rural societies employ local value systems in making economic decisions and structuring people's lives. For instance, the economic crisis in the rural areas of Igboland has affected the quality of life and dietary habits. Much of the local diet is overwhelmingly composed of carbohydrates. Protein-rich food items like beans and meat are not frequently eaten by most families, as was the case in the past, because they cannot now afford them. Dwindling agricultural production and lack of agricultural labour have made rural areas more vulnerable to food shortages for the first time since the end of the civil war. But the attempt to negotiate the changing rural landscape has also entailed fundamental changes in other aspects of rural life including the roles that men and women have historically played.

Yet the most corrosive effects on rural identity have occurred in the context of a national economy that has continued to draw from the rural population. The Igbo have consistently combined farming with other economic activities. The economic returns from trading, for many households, are far greater than what they could ever earn from exhausted lands. Nwanyiafo Obasi maintains, "It is best to combine farming and trade. If your trading capital falls, you can have something to fall back on."[106] Over 80 per cent of household income in central Igboland comes from non-agricultural activities. While most of Igboland remains agrarian in outlook, various forces have acted to modify and transform its agrarian characteristics.

The contested nature of gender ideology, especially contemporary patterns and changes in the self-image of rural men, reveals the most significant change in male and female identity. The challenges to male identity and masculinity have become even more insistent as structural changes connected with the destabilizing effects of agricultural decline transform gender roles and challenge male domination and economic power in rural settings. The change in gender roles and the challenge to the quintessential male authority, identity, and power have often led to conflicts between men and women. Phanuel Egejuru's *The Seed Yam Have Been Eaten* illustrates this change as reflected in the civil war agricultural economy of the Igbo. Jibundu, the protagonist in the novel, expresses perhaps the prevailing view of many Igbo people. "Cassava has displaced yam in our occupation," he laments:

> We clear the bushes as usual and burn them. The women and the children plant cassava. It is less demanding. It leaves us men at home with little or nothing to do. We drink palm-wine from morning till evening when we eat our gari of fermented cassava *foo-foo*, and then resume our drinking till far into the night. Sometimes when we can coax our wives into giving us some extra change, we buy some home brew *akamere* to top off the palm-wine. Have you ever heard a man begging his wife for pocket money? Yes that's what we do now. One must learn to be the vanquished in a war.[107]

Linus Anabalam, like many other men, lamented that "things have changed because cassava is now king."[108] This reflects a fractured identity among many Igbo men since the 1970s as the Nigerian Civil War and the structural changes that followed the development of the petroleum industry have eroded peasant identities, replacing them with multiple sources of livelihood that lay less emphasis on farming.

The changes that have occurred have had fundamental implications for gender relations and the roles men and women historically played in rural society. Women have increasingly dominated the non-agricultural sector through petty trading, food production, and food processing and they shoulder the household food burden. Households have become increasingly dependent on female income, not only in female-headed households but also in many marginal rural households, where female incomes are significant and sometimes constitute higher contributions to total income than those of males.[109] I spoke to men who saw themselves as "good for nothing." They had lost their identity as men. For these men, the decline of a yam-based agrarian culture has engendered a crisis of masculinity and male identity in rural Igboland. "We are like castrated men today," stated a rural dweller, who was ashamed of his dependency on his wife's meagre trade for survival. Jonas Onwukwe, a retired worker at a government rubber estate at Emeabiam in Owerri Province, agreed: "We depend on our wives for subsistence because of their control over cassava production and marketing."[110] The historical trajectory of changes in gender roles is not unique to Igboland, yet its effect on many Igbo men has been quite traumatic. Unlike the situation in many other societies, however, agricultural decline in Igboland, especially the decline in yam production and the palm oil trade, has made men hapless victims of the commercial drift of the 1970s and 1980s.

What has followed is a considerable transgression of gender norms, changes in the nature of family, kinship, human relations, and work, from the 1970s onward. Such transgressions have redefined issues of masculinity, femininity, sexuality, childhood, parenthood, the interaction between gender and sexuality, and household production strategies. All this has brought about a considerable crisis in many households that have to struggle to meet their daily needs. The dramatic changes in the rural economy, traditional norms, and household survival strategies are emblematic of the constantly shifting and renegotiated facets of African domestic and formal economy. These processes are especially complex for societies such as the Igbo because they are intertwined with the national and international economy. As men lose control of the social and economic structures of rural life, many have interpreted these changes as abnormal behaviour, particularly on the part of women.

The most visible change in rural areas has occurred in the control of income from the sale of palm produce, especially in parts of central Igboland, where it remains an important source of rural income. Many informants agree that women have largely taken over the control of income from palm oil – income previously seen as belonging to male household heads. Onyegbule Korieh provides some explanations: "Things have changed. When women process palm produce these days, they take both the oil and kernels. Women now own both the oil and kernels."[111] Linus Anabalam agrees: "Very few men today have control over the palm oil produced in their house. Men do not care much any more.... It is a woman's own today."[112] Eugenia Otuonye's view of the changing nature of resource control is expressed powerfully: "Any man who would demand the money from oil is crazy in the head.... Where should we [women] get the money to feed the family?"[113] While palm oil continues to provide a substantial part of rural household income, households have altered the previous system of allocating oil to the man and kernels to the woman. "Ask my wife," Onyegbule Korieh challenged. "I do not know what happens to the oil sold in my house anymore."[114] For many Igbo households the structural changes in the economy have eroded the long-standing system of provisioning for the household and the control of income from palm produce. The shift highlights the crisis of identity engendered by major structural changes among the Igbo and the ways in which both men and women have tried to negotiate it.

Although the priority given to agriculture has diminished, rural Igboland remains largely agrarian in outlook. Facing increasing difficulties in surviving as farmers because of population growth, poor soil, and the major changes that resulted from the war and the expansion of the oil industry, the Igbo began to adapt. But those who remained in the rural areas have refused to be entirely uprooted from their agrarian roots. The persistent agrarian outlook has influenced the value attached to farming in rural communities and the strong link between farming and rural identity. Rural dwellers still consider themselves primarily farmers, but like people in many other African societies, they have adopted a dual strategy that combines non-agricultural income earning with persistence in subsistence agriculture. Such a strategy has enabled rural African peasants to retain the security that subsistence agriculture offers during periods of economic crisis. The case of central Igboland suggests that agricultural and rural transformation and the ways in which people have responded has been shaped by this psychological dependence on agriculture, but all this has also been mediated by gender and the link with the capitalist world.

Most parts of Igboland have adopted a dual strategy of farming and trade in order to survive rural poverty. This is especially pronounced in central Igboland. While subsistence farming is a way of life that has virtually ceased for the majority of rural dwellers – since many purchase their food from markets – the idea of obtaining part of household subsistence from the farm continues to be highly prized among the elderly, who remain emotionally attached to agriculture. Rural farmers claim, and rightly so, that combining farming with other economic activities is an insurance against insecurity. This perception has remained strong in the psyches of rural Igbo people.

The persistence in agriculture is a rational economic behaviour and a practical expression of the belief among the Igbo that one ought not to depend on the market for basic subsistence. Igbo persistence in farming also has much to do with Igbo identity. Victor Uchendu summarized this notion thus: "To remind an Igbo that he is *ori mgbe ahia loro* [one who eats only when the market holds] is to humiliate him."[115] This observation was quite true when Uchendu wrote it in 1965. However, persistence in farming has followed many trajectories since the end of the civil war. The Igbo have adopted several strategies, developed alternative income-generating activities, and adopted radical changes in their agricultural practices. While there remains a psychological dependence on agriculture, such dependency has mediated new forms of adaptation, especially the increased growth in cassava production.

Yet, the transition from a yam-based subsistence economy and a palm oil cash-based economy to a cassava-based agrarian system reflects the ability of peasants to adapt to a changing social and economic environment. Thus, from their early identification as a "yam people at heart," the Igbo have become a "cassava people at heart."

CONCLUSION

The focus of this book has been to analyze the complexities surrounding the changing world of rural farmers in the context of various historical epochs, highlighting the structural changes that occurred as a result of these changes, and stressing the importance of restoring the voices of rural people in the history of the changes. Drawing on rural farmers' responses to official policies, their memories of events, and the impact of the ecological, environmental, and demographic factors that are endemic to Igbo society, this book has demonstrated the need for a more inclusive framework for explaining the dynamics of agricultural change in an African society. As this case study shows, the trajectory of agricultural change in Igboland has been the result of a complex array of factors – some external in origin, and others the result of factors that derive from internal social, political, demographic, environmental, and economic conditions. The study emphasizes the importance of government policies, resource endowment, demographic factors, changes in national and regional economies, and, not least, the role of social values in the processes of agricultural development and change. It is not enough to show that African societies have experienced significant transformations; it is perhaps more important to explain how different groups, regions, and genders in rural Africa have been affected by state intervention and the other structural changes that have occurred and how they have responded to these challenges. Such an analysis challenges previously held assumptions about African farming systems, in regard to the capacity of peasants to increase productivity, that were based only on the broader picture. The book highlights the need to complement any general theory about African agricultural change with detailed case studies to produce a coherent outline that incorporates local specificities. To fully understand the history of rural Africa in the colonial and post-colonial periods, we must also explore the roles, actions, and responses of the rural population in the context of the changes that occurred. This is what this book has done.

The debate that emerged in the 1980s on the nature and explanations of African agricultural decline exposed complex conceptual and theoretical dilemmas. Clearly, the analysis of agricultural change masks important regional and historical issues about the nature of change in particular settings. The externalist paradigm explains only one aspect of many contributory factors, while the internalist perspective ignores important historical antecedents. Both explanatory models were found wanting due to their overwhelming emphasis on macroeconomic indicators while they ignored peasants as historical actors and the centrality of gender to any meaningful analysis of agricultural change. The role of external and internal factors remains crucial to any meaningful analysis, but only if they are considered together with myriad other variables, including the actions of the rural population. Obviously, empirical evidence challenges general explanatory models, which do not provide important details about the forms and nature of the change, or details of how local variables and social dynamics, including gender, have influenced the nature of change and local responses.

Another conceptual concern in this study centres on the gender question, as it constituted a serious omission in previous analysis. The most obvious limitation of general explanatory models is that they tend to ignore the way that the dynamics of agricultural change were mediated by gender ideology in African farming systems. Assumptions about the sexual division of labour as a given operate at various levels of the discourse on women in agricultural production. For policy-makers, these assumptions, especially the idea that men are the genuine farmers, have informed perceptions and ideas about male and female agricultural roles. Gender as a category of analysis has been taken as self-evident in the study of African agricultural decline. However, this neglect became a significant discursive context for feminists with regard to the roles of women and men in general, and in the debates about the gendered nature of official agricultural and development policies. I have explained how this relates to Igbo agriculture, and the limitations imposed by the neglect of gender analysis. In terms of a feminist-informed political economy, the most obvious limitation of the mainstream analysis of agricultural change was the exclusion of gender as a social category and as an essential framework for the analysis of agricultural change, particularly in Africa. However, this study has shown that ideas about gender as a category of analysis or about the impact of gender ideology in the economy are diverse and vary from one society to another. This book has outlined the source of tension in feminist debates over the notion of gender and the impasse of universalizing gender experiences. Although useful in explaining the realities of colonial and post-colonial African economies, the political economy model, on which the Western conception of gender is based, has marginalized local gender relations and treated male experiences only peripherally. Although conclusions from a particular region and historical context should not be extrapolated to all of Africa, the agricultural and societal transformations in central Igboland have been examined in these broad contexts.

Agriculture and agriculture-related commerce were the central elements of pre-colonial Igbo society and economy, and they defined livelihood and identity. Before the beginning of colonial rule in 1900, the domestic economy was heavily dependent on yam production. Other economic activities and social practices were directly or indirectly linked to this agricultural system. The importance of the yam, the food security it provided, and the social status it conferred on big yam farmers shaped the production pattern of central Igboland in many significant ways. The production of yams was also directly linked to Igbo masculinity and social stratification, gender ideology, and labour practices. Successful production of yams required a large labour force, a considerable amount of time, and significant investment in agricultural inputs. The significance of yams in the life of the Igbo man suggests that there was already an increase in agricultural intensification and ecological change in the nineteenth century. The production of yams for subsistence and prestige purposes encouraged farmers to produce above subsistence levels. This book suggests that intensive yam production explains the high population density of central Igboland and the depleted soil in many parts of the region. Unfortunately, the available data do not permit a refined analysis of rural life for this period. The paucity of demographic and environmental data on Igboland has prevented a detailed examination of the influence of population growth on the rate of environmental transformation and change prior to 1900. However, one can speculate that ecological and demographic factors had already imposed constraints on the ability of farmers to increase agricultural productivity on the eve of colonial rule.

There is no doubt that a complex interaction of internal and external forces shaped the economy of many African societies from the late nineteenth century onward. Africa's encounter with colonialism is particularly significant in this regard. The agrarian system built before European contact had prepared the Igbo to play a central role from the era of the slave trade and throughout the colonial period. The Igbos' agricultural potential, especially in the production of yams and later palm oil, and their population, fed the Atlantic trade. As chapter 1 demonstrates, a high degree of commercialization and commercial relations had developed between the Igbo and other parts of the Atlantic World before the colonial encounter. Both the Igbo and the British built upon this existing network to inaugurate the significant transformation of economic and social life at the beginning of the twentieth century.

When the British first imposed colonial rule in the region in 1900, they were essentially interested in extracting the agricultural products needed to support the colonial economy without changing the fundamental structure of the local production system. The political change that followed in the form of indirect rule was instituted to create an infrastructure that would enable the achievement of these economic objectives. The emphasis on export production, the new regulations encouraging local production, and the ever-increasing requirement for cash pushed the Igbo to expand the production of palm oil and kernels. However, the top-down approach of colonial officials often neglected local production systems while the patriarchal assumption governing African farming systems neglected female farmers. The inability of the colonial state to work within the pre-colonial production system, the channeling of agricultural development programs mostly to male farmers to the exclusion of women, and the neglect of the subsistence sector significantly transformed the agricultural economy of the Igbo. The analysis of gender in this book reveals that the roles of women and men in agriculture became differentiated because of the British notion of the "male farmer." This new gender ideology imposed on the Igbo ran contrary to the complementary nature of men's and women's roles in the production system. Amid the inconsistent agricultural policies of the colonial era, Igbo producers were integrated into the world economy, which fundamentally changed the rural economy and the people.

The impact of agricultural development and the exploitation of the agricultural resource base stimulated increased production and revolutionized production methods in some areas. This led to further agricultural intensification. The colonial economy also created specialist traders and oil palm harvesters, who significantly transformed rural life and employment. Unlike previous works, this study has explored the long-term implications of increased agricultural commercialization and agricultural intensification in a densely populated region of Nigeria. I have argued that the colonial government's development ideology and the transformation of other sectors of the economy encouraged agricultural involution and contributed to the declining importance of agriculture in the region. Like other African societies in the colonial period, the Igbo were part of the making of their own history, but not "necessarily under conditions of their own choosing."[1]

The half-century following colonial rule was marked by significant events, which had a direct impact on Africa and Africans, even though they occurred in faraway Europe. The discussion of global crises such as the Great Depression and the two World Wars suggests that these events had a direct impact on the lives of rural Africans. Chapters 4 and 5 demonstrate that the colonial economy created a local dependency on income from the sale of cash crops and increased the vulnerability of peasants to the slumps that often occurred in the local and international markets. I have shown that the local population was visibly distressed by the declining income from palm oil, the high prices for food, and the insecurity engendered by the depression of the late 1920s and early 1930s. The discussion of the 1929 Women's Revolt in 1929 (in chapter 4) suggests that the revolt was deeply rooted in the agrarian economy of the region. Previous studies have emphasized the feminist origin of the revolt and the introduction of income tax by the British in Eastern Nigeria in 1928. However, the report of the Aba Commission of Inquiry, which examined the immediate and remote causes of the revolt in 1930, and the "Notes of Evidence" recorded during the inquiry demonstrate that the depression, which was rooted in the agricultural economy, was paramount in the minds of the local people. Frequent references were made to the low price of palm produce and the significant rise in the price of foodstuffs. Over-all, the tax problems provided an opportunity for the people to demonstrate their anger regarding the state of the economy and the political policies of the British. The 1929 revolt would become one of several popular protests and locally initiated agitations that would characterize African-British relations in colonial eastern Nigeria into the early 1950s.

The analysis provided of the World War II period in chapter 5 is significant in regard to both the contribution of rural African societies and the impact of the war on their lives. In Nigeria, Britain sought and received the commitment of Nigerians to support the war effort. Across the country, communities mobilized in various ways to support the war effort. This book demonstrates that the local population contributed financial support both directly and indirectly. Igbos supplied soldiers in a variety of capacities and provided resources, including food items for the troops. Igbo farmers were forced to increase the production of palm oil even though prices remained lower than before the war. However, the biggest problem faced by the colonial administration was how to curb the rising cost of living that became prominent due to labour shortages created by the war, low levels of import and export, and shortages of locally produced food items, such as rice, yams, cassava, and salt, and imported products such as sugar.

The demands of the war forced the British to restructure the local economy to ensure that Africans produced the commodities needed to support the British war effort, including food and export products such as palm oil. The government introduced new regulations and laws to effectively control peasant production and the distribution of essential food items through the Nigeria General Defense Regulations (Law No. 75 of 1941). Those directly affected by the new regulations and controls were farmers and traders engaged in the sale of produce including yam and gari. Both local and urban populations were visibly distressed due to the food crisis, especially because of the British management of the local production system and the insecurity that this engendered, as reflected in the petition of a local trader in Aba who petitioned a British district officer to consider "the lives of a family which may perish as a result of the measures ... taken to restrict the garri trade."[2] And Mr. Muoma, who had been prevented from carrying on his normal trade in gari, deemed it "abnormal" and an "injustice," since cutting him out of the gari trade "deprived him of his livelihood."[3] The Aba Community League, an organization representing various community associations and unions, wrote to the local district officer on 12 August 1942 about the negative effects of food restrictions on the residents of the town.[4]

The agrarian roots of the various petitions written by the local population remain evidence of their struggles to survive the depressed economy of the war and their strategies for coping with the crisis engendered by the war. This fills an important gap in the history of World War II in relation to rural African colonial history and challenges the dominant Western-centred narrative of the war that lays less emphasis on the contributions of the African population and the impact of the war on their society as described in chapter 5. Since most of those who petitioned officials during the war were small traders and rural farmers, the exploration of their petitions and the economic, cultural, and social conditions that gave rise to them is a major contribution to the historical analysis of the conditions faced by African societies, especially the lower classes, during the war. It provides scholars with new and groundbreaking materials about World War II and corrects the impression that the effects of the war on the "home front" applied to European societies alone. Yet, the war also created opportunities for Igbo farmers and traders who took advantage of the high prices for *gari* to increase production and sale of the product in Northern Nigerian cities.

It is facile, however, to speak of government policies and external influences as if they represent the overwhelming determinant of agricultural change among the Igbo. Obviously, all these factors must be considered in tandem with the impact of the palm produce trade since the nineteenth-century commercial transition in Igboland. Government policies cannot be isolated from pre-existing ecological and demographic factors that increased the rate of agricultural involution. These factors exacerbated the situation caused by government intervention in the rural economy and were coupled with the problem of the expanding urban sector, which attracted a large portion of the rural population.

The early post-colonial period was revolutionary in many ways. Although the regional government in Eastern Nigeria continued the agricultural programs of the colonial regime, which emphasized production for export, the government also broke away from the earlier rejection of plantation agriculture. The indigenous political elite intervened even more directly in peasant agriculture than the colonial government had done.

From 1962 onward, the regional government tightened its grip on the agricultural economy through the establishment of produce marketing boards, community plantation projects, and new farm settlements. From the government's point of view, agricultural expansion, the participation of the state in agricultural projects, and the integration of rural peasants into these projects were essential to the rapid development of the region. Yet, the condition of peasants deteriorated while the state and marketing boards profited from the control of the peasant surplus. The continued neglect of the perspectives of rural farmers and the gendered ideology of the "male farmer" did not change in the post-colonial period.

By focusing on the Nigerian Civil War and the development of the petroleum industry, the study highlights the impact of these factors on rural agrarian life. The agricultural economy of the Igbo went into a deep crisis because of the civil war. The local food production capacity of Igboland could not meet the needs of the army and the rest of the population, despite the efforts of peasants and the Biafran regime to sustain production during the war. The severe food crisis that emerged after the outbreak of the war in 1967 revealed the inability of the region to feed itself on locally produced food. This had already created a high level of Igbo dependency on other regions before the war broke out. The war, overall, produced a high rate of agricultural involution in terms of a movement toward non-agricultural activity, due in part to people's frustration with rural poverty, especially among the younger members of the population.

The development of the petroleum industry worsened the crisis in the food sector. As oil exports and revenues increased, the overall importance of agriculture declined, while expenditure on food importation increased. The dependency on petroleum revenue had serious negative effects on the rural areas, as it induced a high level of out-migration to the urban areas. The high rate of urban infrastructural development and industrialization in the petroleum era made agriculture unattractive and drained rural labour. The "boom and bust" cycles that followed the development of the petroleum industry affected rural peasants, as the high inflation rate that followed the development of this industry made it more difficult for the rural population to cope. The situation continued into the 1980s and led to a decline in the quality of urban life as well.

Overall, the realities of agricultural decline and food insecurity were influenced by government policies and the general attitude toward agriculture. The agricultural policies of the government often reflected the "boom and bust" cycles of the petroleum economy. The worsening economic situation of the 1980s forced the government to introduce new agricultural programs aimed at increasing the levels of food production. However, government interventions did not result in significant increases in levels of production, due to mismanagement, corruption, and nepotism.

Paralleling the changing nature of rural agriculture there were also significant shifts in the rural economy, in farming practices, and in employment opportunities. Overall, the Igbo have adapted remarkably well. The agricultural crisis in the region compelled the Igbo to increase their non-agricultural forms of employment. These changes, while contributing to income diversification outside agriculture, did not totally disrupt the agricultural base of the rural economy. Some members of the rural population have continued to farm, but under difficult circumstances, including less fertile soil, decreased availability of labour, and scarcity of land. Given the increasingly tight straits in which the Igbo found themselves, it is not surprising that income diversification has become the norm. Younger persons, in particular, have favoured migration, and remittances remain an important source of household sustenance and economic development.

However, the combined effects of a declining agricultural economy and opportunities outside agriculture have undermined aspects of traditional Igbo agricultural ideology. The importance of the yam as the icon of Igbo agriculture and masculinity and its social and ritual importance have declined because of the decreasing importance of agriculture. The increased importance of cassava in Igbo agriculture and food security has increased the role of women in household food security and income. The changes in the local economy and the transformations in the region have worked to empower women in significant ways. Women have assumed greater control over household resources and have ventured into areas traditionally regarded as male spheres of influence. The changing nature of the roles that women and men have played in the economy demonstrates the contested nature of gender ideology and challenges mainstream ideas about gender division of labour and resource control.

The picture of socio-economic change that emerges from this study is a complex one. The combination of state policies, peasant actions, and environmental and demographic factors accounts for the trajectory and pace of change and the rural strategies for survival. For the Igbo, like many other societies in Africa, the internal dynamics of change have been influenced by external factors created by a capitalist world economic system that has depended on developing economies for cheap raw materials. However, the Igbo case points to some of the complexities associated with agricultural transformation and the problems that emerge from a simplistic analysis.

In historical perspective, the tragedy of the Igbo agrarian experience owes much to the policies implemented by the colonial and post-independence authorities. The most common threads in the post-independence era have been the role of the state in the expansion of cash-crop production, the state's direct intervention in the peasant economy, and the lack of any radical change in colonial ideology and attitudes toward the rural farmer. Another common thread has been the neglect of the sociocultural and economic background of local societies by both the colonial and the post-independence states. The neglect of peasant perspectives in the design of government agricultural programs has limited whatever progress the programs might have made in agricultural transformation. To argue that local perspectives matter and that they have much to offer to improve agricultural production is not to suggest that peasants have a monopoly of all of the necessary agricultural knowledge. Rather, agricultural improvement programs would have gained much from incorporating the perspectives of peasants to ensure lasting sustainability. Throughout this study, an attempt has been made to highlight the importance of demography and of the ecology of the Igbo territory in explaining the nature of agricultural change. This book has demonstrated that only through an examination of how both internal and external factors interacted with each other vis-à-vis the local environment, and the inclusion of the actions and voices of rural people as agents of history and change, can the agrarian history and current crisis of the Igbo be understood in its entirety.

In spite of the fact that the state provided the institutional and structural framework within which local farmers operated, the goal of transforming and modernizing agriculture has not always been reached. Rural response to modernization has not always been in congruence with official expectations. The transformations that took place occurred not just as a result of state policies, they happened because African institutions, structures, and initiatives played significant roles. In the political sphere, greater success was achieved despite the constant resistance of the Igbo population to British rule.

In the final analysis, the twentieth century remains a period in which the agricultural past was represented by significant growth as much as by weaknesses and decline. By the end of the twentieth century, the lives of rural dwellers in rural Igboland had been transformed by the effects of public policies and the actions of rural dwellers themselves. The attempt to modernize agriculture and the encouragement given by the colonial state to rural farmers to produce palm oil and kernels transformed people's lives and increased their dependency on the market for survival. In addition, the dual impact of the depression of the late 1920s and early 1930s and the world wars as historical by-products of a globalizing world, coupled with increased structural changes that strengthened in the wake of European contact, facilitated the rapid transformation of rural lives. The result was a profound change in livelihood strategies. The level of non-agricultural income-generating activities in the rural areas increased in the post-independence period, especially from the end of the civil war in 1970, when agriculture ceased to provide a decent standard of living for the rural population.

The high rate of agricultural involution that began at end of the war was accelerated in the late 1970s by the dependency of the state on income from the sale of petroleum. This period has been described as a period of agricultural crisis for many African societies, and this was certainly so for the Igbo. In this economic environment, the importance of agriculture in rural life and livelihood changed significantly with the increased dependency on non-agricultural sources of income for rural dwellers.

NOTES

FOREWORD

1 A. G. Hopkins, *An Economic History of West Africa* (New York: Columbia University Press, 1973), 168–69.

2 See, for example, Thurstan Shaw, *Igbo Ukwu: An Account of Archaeological Discoveries in Eastern Nigeria*, 2 vols. (London, [1970], M. A. Onwuejogwu, "The Dawn of Igbo Civilization," *Odinani* [1971]).

3 Victor Uchendu, *The Igbo of Southeastern Nigeria* (New York: Holt, Rinehart and Winston, 1965), 22.

4 G. I. Jones, *The Trading States of the Oil Rivers* [1963], 13.

5 Robert Stevenson, *Population and Political Systems in Tropical Africa* [1968], 190.

6 Ibid., 192.

7 Korieh, "Introduction," 2.

8 Ibid., 13.

INTRODUCTION: PERSPECTIVES, SETTING, SOURCES

1 Interview with Grace Chidomere, Umuchieze, Mbaise, 13 December 1998.

2 Interview with Chief Francis Enweremadu, Mbutu, Mbaise, 2 January 2000.

3 Interview with Comfort Anabalam, Umuchieze, Mbaise, 13 December 1998.

4 National Archive of Nigeria, Enugu (NAE), ABADIST, 14/1/873, "A. Jamola, to the District Officer, Aba," 21 July 1943.

5 Cited in Elizabeth Isichei, *A History of Nigeria* (London: Longman Group, 1983), 400.

6 For an analysis of the oil palm trade in Nigeria, see, for example, Eno J. Usoro, *The Nigerian Oil Palm Industry: Government Policy and Export Production, 1906–1965* (Ibadan, Nigeria: Ibadan University Press, 1974); and O. N. Njoku, "Trading with the Metropolis: An Unequal Exchange," in *Britain and Nigeria: Exploitation or Development*, ed. Toyin Falola, 124–41 (London: Zed Books, 1987).

7 Igbo culture and ecological areas can be broadly categorized as follows: Western or Delta Igbo (Asaba, Ika, Ndokwa); Northwestern (north and south Niger flood plain: Onitsha, Idemili, Aguata, Nri, Awka [Anambra]; Northern [Awgu, Enugu, Nsukka, Abakaliki: Enugu State and part of Ebonyi State]; Central [Orlu, Owerri, Nkwere, Ideato, and Mbano, Mbaise, Etiti, Okigwe: Imo State]; Southwest [Ohaji, Egbema, Oguta, Ndoni, and Ikwerre: part of Imo and Rivers States]: South [Ngwa, Asa, Etche, Ukwa: Abia State]; and Eastern [Umuahia-Ikwuano, Bende, Ohafia, Afikpo, Aro: part of Abia and Ebonyi States]). Taken from Ogbu U. Kalu, "Osondu: Patterns of Igbo Quest for Jesus Power," unpublished paper.

8 In the settler colonies of southern and eastern Africa, where Africans competed with capitalist agriculture, the labour of African men and the subsistence production of African women also helped to subsidize the state, capitalist agriculture, mining, and industry. See, for example, Colin Bundy, *The Rise and Fall of the South African Peasantry* (Berkeley: University of California Press, 1979).

9 However, the pace of agricultural transformation varied widely from the cash crop producing regions of West Africa to southern and eastern African societies, where farmers faced more direct demands

on their lands and labour from European settlers. The different colonial experiences account for the varied nature of African agricultural transformation, the farmers' responses, and the effects of the decline in the agricultural economy. For an account of the process of agricultural change in Africa, see, for example, H. J. W. Mutsaers, *Peasants, Farmers and Scientists: A Chronicle of Tropical Agricultural Science in the Twentieth Century* New York: Springer, 2007); W. J. Barber, "The Movement into the World Economy," in *Economic Transition in Africa*, ed. M. J. Herskovits and M. Harwitz, 299–29 (Evanston, IL: Northwestern University Press, 1964); Sara Berry, *No Condition is Permanent: The Social Dynamics of Agrarian Change in Sub-Saharan Africa* (Madison: University of Wisconsin Press, 1993); W. R. Duggan, *An Economic Analysis of Southern African Agriculture* (Westport, CT: Praeger, 1986); Anthony G. Hopkins, *An Economic History of West Africa* (New York: Columbia University Press, 1973); Martin A. Klein, ed., *Peasants in Africa: Historical and Contemporary Perspectives* (London: Sage, 1980); David Siddle and Kenneth Swindell, *Rural Change in Tropical Africa: From Colonies to Nation States* (Cambridge, MA: B. Blackwell, 1990); J. Tosh, "The Cash Crop Revolution in Tropical Africa: An Agricultural Reappraisal," *African Affairs* 79 (1980): 79–94.

10   In 1985, for example, an estimated 10 million Africans left their homes and fields because they were unable to support themselves. An additional 20 million were reported to be at risk of debilitating hunger. See Lloyd Timberlake, *Africa in Crisis: The Causes, the Cures of Environmental Bankruptcy* (London: Earthscan, 1985). Numerous World Bank reports since 1981 have indicated an overall pattern of severe economic deterioration and stagnation manifested in food security problems and low levels of growth in the agricultural subsector. See especially World Bank, *Accelerated Development in Sub-Saharan Africa: An Agenda for Action* (Washington, DC: World Bank, 1981); *Towards Sustainable Development in Sub-Saharan Africa: A Joint Program of Action* (Washington,

DC: World Bank, 1984); *Sub-Saharan Africa: From Crisis to Sustainable Development* (Washington, DC: World Bank, 1989). Studies of agricultural sustainability include Abe Goldman, "Threats to Sustainability in African Agriculture: Searching for Appropriate Paradigms," *Human Ecology* 23, no. 3 (1995): 291–334. See also G. K. Douglass, "The Meaning of Agricultural Sustainability," in *Agricultural Sustainability in a Changing World Order*, ed. G. K. Douglas, 3–29 (Boulder, CO: Westview, 1994); George J. S. Dei, "Sustainable Development in the African Context: Revisiting Some Theoretical and Methodological Issues," *African Development* 18, no. 2 (1993): 97–110; and C. K. Eicher, *Sustainable Institutions for African Agricultural Development*, International Service for National Agricultural Research (ISNAR), Working Paper no. 19 (The Hague: ISNAR, 1989).

11   Studies of agricultural change in Nigeria have focused on how state actions transformed rural agricultural economies and threatened agricultural sustainability. See, for example, Jerome C. Wells, *Agricultural Policy and Economic Growth in Nigeria, 1962–1968* (Ibadan, Nigeria: Oxford University Press, 1974); Food and Agricultural Organisation, *Agricultural Development in Nigeria, 1965–1980* (Rome: FAO, 1966).

12   There is also an argument that the economic reforms driven by the IMF and the World Bank in Africa over the last decades have exacerbated the pace of agricultural and economic decline. For the implications of structural adjustment programs (SAPs) on African agriculture, see S. Commander, ed., *SAP and Agriculture: Theory and Practice in Africa and Latin America* (London: Overseas Development Institute, 1989). See also Christina H. Gladwin, ed., *Structural Adjustment and African Women Farmers* (Gainesville: University of Florida Press, 1991); and Commonwealth Secretariat, *Engendering Adjustment for the 1990s: Report of a Commonwealth Expert Group on Structural Adjustment* (London: Commonwealth Secretariat, 1989).

13   There appears to be a consensus on the decline in the level of agricultural production, although there is less agreement

on exactly what are the causes and what should be the remedies. Furthermore, critics seeking general explanatory models of the nature of agricultural crisis have reproduced this error. The current ubiquitous use of the word "crisis" in explaining the decline in African agriculture is not without uses, but it needs the addition of specific local details to avoid over-generalization. The general "crisis" thesis has led to distortions in the description of the nature of the agrarian crisis and sustainability because the discourse has not been adequately grounded in the social structures and everyday life of the studied societies. An in-depth understanding of the varied nature of the African agricultural crisis calls for an exploration of regional variations and experiences. For a general review of the literature on the African agricultural crisis, see Sara Berry, "The Food Crisis and Agrarian Change in Africa: A Review Essay," *African Studies Review* 27, no. 2 (1984): 59.

14   See Berry, "The Food Crisis." Berry extends her argument for a need to reconceptualise African agrarian discourse in *No Condition is Permanent*, especially, 10–16. See also M. F. Lofchie and S. K. Commins, "Food Deficit and Agricultural Policies in Tropical Africa," *Journal of Modern African Studies* 20, no. 1 (1982): 1–25. See also M. F. Lofchie, "The Decline of African Agriculture," in *Drought and Hunger in Africa: Denying Famine a Future*, ed. Michael H. Glantz, 85–110 (Cambridge: Cambridge University Press, 1987); J. Hinderink and J. J. Sterkenburg, "Agricultural Policy and Production in Africa: The Aims, the Methods, and the Means," *Journal of Modern African Studies* 21, no. 1 (1983): 1–23; Michael Watts and Thomas Bassett, "Crisis and Change in African Agriculture: A Comparative Study of the Ivory Coast and Nigeria," *African Studies Review* 28, no. 4 (December, 1986):3–27; R. Baker, "Linking and Sinking: Economic Externalities and the Persistence of Destitution and Famine in Africa," in *Drought and Hunger in Africa*, ed. M. H. Glantz, 149–70 (Cambridge: Cambridge University Press, 1987). See also Ray Bush, "The Politics of Food and Starvation," *Review*

of African Political Economy 68 (1996): 169–95.

15   See, for example, FAO, *Regional Food Plan for Africa* (Rome: FAO, 1980); World Bank, *World Development Report* (Washington, DC: World Bank, 1978); and World Bank, *World Development Report* (Washington, DC: World Bank, 1980).

16   For more on this debate, see Berry, "The Food Crisis," *No Condition Is Permanent*, 10–6; Lofchie and Commins, "Food Deficit and Agricultural Policies in Tropical Africa," 1–25.

17   See, for example, Bade Onimode, *Imperialism and Underdevelopment in Nigeria: The Dialectics of Mass Poverty* (London: Zed Books, 1982). On agriculture and commodity production, see, for example, A. Faloyan, *Agriculture and Economic Development in Nigeria: A Prescription for the Nigerian Green Revolution* (New York: Vantage Press, 1983); J. O. Ahazuem and Toyin Falola, "Production for the Metropolis: Agriculture and Forest Products," in *Britain and Nigeria: Exploitation or Development*, ed. Toyin Falola, 80–90 (London: Zed Books, 1987); Bade Onimode, *Imperialism and Underdevelopment in Nigeria: The Dialectics of Mass Poverty* (London: Zed Books, 1982). See also O. N. Njoku, "Trading with the Metropolis"; Hopkins, *Economic History*; Rodney, *How Europe Underdeveloped Africa*; E. A. Brett, *Colonialism and Underdevelopment in East Africa: The Politics of Economic Change* (London: Heinemann, 1974); G. Kay, *The Political Economy of Colonialism in Ghana* (Cambridge: Cambridge University Press, 1972); I. W. Zartman, ed., *The Political Economy of Nigeria* (New York: Praeger, 1983). For the relationship between peasant agriculture and the government at federal and state levels during the colonial era and the first four years of independence, see Gerald Helleiner, *Peasant Agriculture, Government and Economic Growth in Nigeria* (Homewood, IL: R. D. Irwin, 1966). See also D. Rimmer, "The Economic Imprint of Colonialism and Domestic Food Supplies in British Tropical Africa," in *Imperialism, Colonialism and Hunger: East and Central Africa*, ed. Robert I. Rotberg,

141–65 (Lanham: MD: Lexington Books, 1984).

18 Cited in Huss-Ashmore, "Perspectives in African Food Crisis," 12.

19 Levi and Havinden, *Economics of African Agriculture*, 129–30. See also Ayodeji Olukoju, "The Faulkner 'Blueprint' and the Evolution of Agricultural Policy in Inter-War Colonial Nigeria," in *The Foundations of Nigeria: Essays in Honor of Toyin Falola*, ed. Adebayo Oyebade, 403–22 (Trenton, NJ: Africa World Press, 2003).

20 Michael Watts, *Sillent Violence: Food, Famine and Peasantry in Northern Nigeria* (Berkeley: University of California Press, 1983), xxiii.

21 Sara Berry, "The Food Crisis and Agrarian Change in Africa: A Review Essay," *African Studies Review* 27, no. 2 (1984), 61.

22 Robert E. Clute, "The Role of Agriculture in African Development," *African Studies Review* 25, no. 4 (December 1982): 3. See also Chima J. Korieh, "Food Production and the Food Crisis in Sub-Saharan Africa," in *Africa, Vol. 5: Contemporary Africa*, ed. Toyin Falola, 391–416 (Durham, NC: Carolina Academic Press, 2003); Chima J. Korieh, "Agriculture," in *Africa, Vol. 5: Contemporary Africa*, ed. Toyin Falola, 417–36 (Durham, NC: Carolina Academic Press, 2003).

23 Clute, "The Role of Agriculture," 2–3.

24 Ibid.

25 Ester Boserup, *Women's Role in Economic Development* (London: Allen & Unwin, 1970).

26 The discussion should (arguably) centre on gender because its relational nature would lead to a critical examination of economic, social, and political processes. This is crucial in examining agricultural change, since men and women are defined in terms of one another in the organization of production. For a clear articulation of this position, see Joan Wallach Scott, "Gender: A Useful Category of Historical Analysis," in Joan Wallach Scott, *Gender and the Politics of History* (New York: Columbia University Press, 1988). See also Marnie Hughes-Warrington's fine articulation of

Scott's ideas on gender in *Fifty Key Thinkers on History* (London: Routledge, 2000), 279–80.

27 See A. V. Chayanov, *The Theory of Peasant Economy* (Homewood, IL: R.D. Irwin, 1966). On the evolution of peasant societies in parts of Europe, see, for example, Peter Hoppenbrouwers, Jan Luiten van Zanden, and J. Luiten van Zanden, ed., *Peasants into Farmers? The Transformation of Rural Economy and Society in the Low Countries (Middle Ages–19th Century) in Light of the Brenner Debate* (Turnhout: Brepols, 2001).

28 See Sara Berry, *Cocoa, Custom and Socio-Economic Change in Rural Western Nigeria* (Oxford: Clarendon Press, 1975).

29 See Polly Hill, *The Migrant Cocoa-farmers of Southern Ghana: A Study in Rural Capitalism* (Hamburg: LIT, James Currey with the IAI, 1997).

30 See Johannes Lagemann, *Traditional Farming Systems in Eastern Nigeria* (Munich and New York: Weltforum Verlag Humanities Press, 1977). For other differentiations, see Onigu Christine Okali and C. Otite, ed., *Readings in Nigerian Rural Society and Rural Economy* (Ibadan, Nigeria: Heinemann Educational Books, 1990); and D. W. Norman, *Economic Analysis of Agricultural Production and Labor Utilization among the Hausa in the North of Nigeria*, African Rural Employment Paper no. 4 (East Lansing: Michigan State University, 1973).

31 See Martin, *Palm Oil and Protest*.

32 For a general discussion of the development of agriculture in the post-colonial era, see Jerome C. Wells, *Agricultural Policy and Economic Growth in Nigeria, 1962–1968* (Ibadan, Nigeria: Oxford University Press, 1974); and Tom Forrest, "Agricultural Policies in Nigeria, 1970–78," in *Rural Development in Tropical Africa*, ed. J. Heyer, P. Roberts, and G. Williams, 222–58 (London: Macmillan, 1981).

33 For a social history of the civil war, see Axel Harneit-Sievers, Jones O. Ahazuem, and Sydney Emezue, *A Social History of the Nigeria Civil War: Perspectives from*

Below (Enugu, Nigeria: Jemezie Associates, 1997).

34 By a "food-reserve-deficit" area, I mean an area without the capacity to produce enough for reserve during one farming season. Parts of Igboland, which were food-reserve-deficit areas, depended largely on food produced in other regions.

35 On the development of the palm oil industry, see Eno J. Usoro, *The Nigerian Oil Palm Industry: Government Policy and Export Production, 1906-1965* (Ibadan, Nigeria: University of Ibadan Press, 1974). For developments in the period after the abolition of the slave trade, see Martin Lynn, *Commerce and Economic Change in West Africa: The Palm Oil Trade in the Nineteenth Century* (Cambridge: Cambridge University Press, 2002); Susan Martin, *Palm Oil and Protest: An Economic History of the Ngwa Region, South-Eastern Nigeria, 1800-1980 (Cambridge: Cambridge University Press, 1988);* and Allister E. Hinds, "Government Policy and the Nigerian Palm Oil Export Industry, 1939-49," *Journal of African History* 38 (1997): 459-78.

36 For a good introduction to feminist analyses, see Rosemarie Tong, *Feminist Thought: A More Comprehensive Introduction*, 2nd ed. (Boulder, CO: Westview Press, 1998). For studies of gender in the context of colonialism and imperialism, see, for example, Elizabeth Schmidt, *Peasants, Traders, and Wives: Shona Women in the History of Zimbabwe, 1870-1939* (Portsmouth, NH: Heinemann, 1992); Nancy J. Hafkin and Edna Bay, *Women in Africa: Studies in Social and Economic Change* (Stanford, CA: Stanford University Press, 1976); and Chima J. Korieh, "The Invisible Farmer? Women, Gender, and Colonial Agricultural Policy in the Igbo Region of Nigeria, c. 1913-1954," *African Economic History* 29 (2001): 117-62.

37 See the following works by Jane I. Guyer: *Family and Farm in Southern Cameroon* (Boston: Boston University Africa Studies Centre, 1984); "Naturalism in Models of African Production," *Man* 19 (1984): 355-73; "Multiplication of Labor: Historical Methods in the Study of Gender and Agricultural Change in Modern Africa,"

Current Anthropology 29 (1988): 247-72; and "Female Farming in Anthropology and African History," in *Gender at the Crossroads of Knowledge: Feminist Anthropology in the Postmodern Era*, ed. M. di Leonardo, 257-77 (Berkeley: University of California Press, 1991).

38 Oyeronke Oyewumi, *The Invention of Women: Making an African Sense of Western Gender Discourse* (Minneapolis: University of Minnesota Press, 1997), 121. See also Helen Callaway, *Gender, Culture and Empire: European Women in Colonial Nigeria* (Urbana: University of Illinois Press, 1987); Nupur Chaudhuri and Margaret Strobel, ed., *Western Women and Imperialism: Complicity and Resistance* (Bloomington: Indiana University Press, 1992); and Malia B. Formes, "Beyond Complicity versus Resistance: Recent Work on Gender and European Imperialism," *Journal of Social History* (Spring 1995): 629-41.

39 See, for example, Jean Allman, Susan Geiger, and Nakanyike Musisi, ed., *Women in African Colonial Histories* (Bloomington: Indiana University Press, 2002); Margot Lovett, "Gender Relations, Class Formation and the Colonial State in Africa," in *Women and the State in Africa*, ed. Jane Parpart and Kathleen Staudt, 23-46 (Boulder, CO: Lynne Rienner, 1989); and Mona Etienne, "Women and Men, Cloth and Colonization: The Transformation of Production-Distribution Relations among the Baule (Ivory Coast)," *Cahiers d'Etudes Africaines* 17 (1977): 41-63.

40 See Northcote. W. Thomas, *Anthropological Report on the Ibo Speaking Peoples of Nigeria*, Vol. 1 (London: Harrison and Sons, 1913-1914), 97. See also Elizabeth Isichei, *A History of the Igbo People* (London: Macmillan, 1976), 27, 79; John Iliffe, *The African Poor: A History* (Cambridge: Cambridge University Press, 1987), 92.

41 Barry Floyd, *Eastern Nigeria: A Geographical Review* (London: MacMillan, 1969), 57.

42 For a description of rural poverty among the Igbo by the late nineteenth century, see Iliffe, *The African Poor*, 82-94.

43 See George, *Journal*, 21 January 1866, CMS CA3/O. 18/23; F. M. Denis, *Journal*,

17 November 1908, CMS: UP 4/F2; T. J. Dennis, *Journal*, March 1907, CMS: UP 89/F1, cited in Iliffe, *African Poor*, 82.

44 See Anthony O'Connor, *Poverty in Africa: A Geographical Approach* (London: Pinter, 1991), 4. See also Goldman, "Population Growth."

45 G.E.K. Ofomata, ed., *A Survey of the Igbo Nation* (Onitsha, Nigeria: Africana First Publishers, 2002), especially Part 2.

46 Usoro, *The Nigerian Oil Palm Industry*.

47 Susan M. Martin, "Farming, Cooking, and Palm Processing in the Ngwa Region of Southeastern Nigeria, 1900–1930," *Journal of African History* 25, no. 4 (1984): 411–27.

48 Gloria Chuku, *Igbo Women and Economic Transformation in Southeastern Nigeria, 1900–1960* (New York: Routledge, 2005).

49 Nwando Achebe, *Farmers, Traders, Warriors, and Kings: Female Power and Authority in Northern Igboland, 1900–1960* (Portsmouth, NH: Heinemann, 2005).

50 For dependency theory, see, for example, Walter Rodney, *How Europe Underdeveloped Africa* (Washington DC: Howard University Press, 1984); Giovanni Arrighi and John S. Sane, *Essays on the Political Economy of Africa* (New York: Monthly Review Press, 1973); and Emmanuel Arghiri, *Unequal Exchange: A Study of the Imperialism of Trade*, trans. B. Pearce (New York: Monthly Review Press, 1972). For Nigeria, see Bade Onimode, *Imperialism and Underdevelopment in Nigeria: The Dialectics of Mass Poverty* (London: Zed Books, 1982); A. Faloyan, *Agriculture and Economic Development in Nigeria: A Prescription for the Nigerian Green Revolution* (New York: Vantage Press, 1983); and J. O. Ahazuem and Toyin Falola, "Production for the Metropolis: Agriculture and Forest Products," in *Britain and Nigeria: Exploitation or Development*, ed. Toyin Falola, 80–90 (London: Zed Books, 1987). For similar analysis of other parts of Africa, see E. A. Brett, *Colonialism and Underdevelopment in East Africa: The Politics of Economic Change* (London: Heinemann, 1974); and G. Kay, *The Political Economy of Colonialism in Ghana* (Cambridge: Cambridge University Press, 1972). On

the relationship between peasants and the government, see Gerald Helleiner, *Peasant Agriculture, Government and Economic Growth in Nigeria* (Homewood, IL: R. D. Irwin, 1966). See also D. Rimmer, "The Economic Imprint of Colonialism and Domestic Food Supplies in British Tropical Africa," in *Imperialism, Colonialism and Hunger: East and Central Africa*, ed. Robert I. Rotberg, 141–65 (MA: Lexington Books, 1984).

51 See T. Shanin, *Peasants and Peasant Societies* (Harmondsworth, UK: Penguin, 1976); Bundy, *Rise and Fall*, 4; and Deborah Bryceson, Chrstobal Kay, and Jos Mooji, ed. *Disappearing Peasantries? Rural Labor in Africa, Asia and Latin America* (London: Intermediate Technology Publications, 2000). See also L. A. Fallers, "Are African Cultivators to be called 'Peasants'?" *Current Anthropology* 2, no. 2 (1961): 108–10; and John S. Saul and Roger Woods, "African Peasantries," in *Peasants and Peasant Societies*, ed. Theodor Shanin, 103–14 (Harmondsworth, UK: Penguin, 1971). See also Terence Ranger, "Growing from the Roots: Reflections on Peasant Research in Central and Southern Africa," *Journal of Southern African Studies* 5, no. 1 (1978): 99–133; and Martin A. Klein, ed. *Peasants in Africa: Historical and Contemporary Perspectives* (Beverly Hills, CA: Sage, 1980), 9–14.

52 See Paul Thompson, "Historians and Oral History," in *The Voice of the Past: Oral History* (Oxford: Oxford University Press, 1988). Reproduced in *The Oral History Reader*, ed. Robert Perks and Alistair Thomson, 21–28 (London: Routledge, 1998).

53 For the use of life histories and oral narratives, see Susan Geiger, *Tanu Women: Gender and Culture in the Making of Tanganyika Nationalism, 1955–1965* (Portsmouth, NH: Heinemann; Oxford: James Currey, 1997), 15–19. For the connections between fieldwork experience and the resulting ethnography, see Judith Okely, "Anthropology and Autobiography: Participatory Experience and Embodied Knowledge," in *Anthropology and Autobiography*, ed. Judith Okely and Helen Callaway, ASA Monographs 29, 1–28 (London

and New York: Routledge, 1992). See also Juliana Flinn, Leslie Marshall, and Jocelyn Armstrong, ed., *Fieldwork and Families: Constructing New Models for Ethnographic Research* (Honolulu, HI: University of Hawai'i Press, 1998), 5-6. See George E. Marcus and Michael M. J. Fischer, *Anthropology and Cultural Critique: An Experimental Movement in the Human Sciences* (Chicago: University of Chicago Press, 1986), for a discussion of ethnography's alliance of observation with involvement in the daily life and experiences of local people.

54 For the use of life histories in historical reconstruction and the problems of interpretation and representation, see Geiger, *Tanu Women*, 16. See also Kathleen Barry, "Biography and the Search for Women's Subjectivity," *Women's Studies International Forum* 12, no. 6 (1989): 561-77.

55 Studying one's own society has been an issue elaborately discussed by anthropologists. See, for example, Donald Messerschmidt, *Anthropologist at Home in North America: Methods and Issues in the Study of One's Own Society* (Cambridge: Cambridge University Press, 1981); Akemi Kikumura, "Family Life Histories: A Collaborative Venture," in *The Oral History Reader*, ed. Perks and Thomson, 140-44; and R. Merton, "Insiders and Outsiders: A Chapter in the Sociology of Knowledge," *American Journal of Sociology* 78 (1972): 9-47. For support of insider research, see Victor Uchendu, *The Igbo of Southeastern Nigeria* (New York: Holt, Rinehart and Winston, 1965); and G. K. Nukunya, *Kinship and Marriage among the Anlo Ewe* (New York: Humanities Press, 1969).

56 For a discussion of this, see Enya P. Flores-Meiser, "Field Experience in Three Societies," in *Fieldwork: The Human Experience*, ed. Robert Lawless et al., 49-61 (New York: Gordon and Breach, 1983).

57 Ndaywel E. Nziem, "African Historians and Africanist Historians," in *Profile of a Historiography*, B. Jewsiewicki and D. Newbury, ed., 20-27 (Boulder, CO: Sage, 1986). While this view assumes the unity of African perspective, the fundamental problem of academic literature, Jewsiewicki argues,

however, lies in the question of where and by whom it is produced as well as where and by whom it is read. See Bogumil Jewsiewicki, "African Historical Studies, Academic Knowledge as 'Usable Past': A Radical Scholarship," *African Studies Review* 32, no. 3 (1989): 9.

58 Richard Wright, "Introduction: Blueprint for Negro Writing," in *The Black Aesthetics*, ed. Addison Gayle, Jr., 315-26 (New York: Doubleday, 1972).

59 Obioma Nnaemeka, ed., "Introduction," in *Sisterhood, Feminisms and Power: From Africa to the Diaspora*, ed. Obioma Nnaemeka, 2 (Trenton, NJ: Africa World Press, 1998).

60 Samuel Raphael, "Introduction," in *Village Life and Labour* (London: Routledge and Kegan Paul, 1975).

61 For useful comments on credibility of oral accounts, see Jan M. Vansina, *Oral Tradition: A Study in Historical Methodology*, trans. H. M. Wright (Chicago: Aldine Publishing Company, 1961), first published in 1961 as *De la tradition orale: Essai de méthode historique* and *Oral Tradition as History* (Madison: University of Wisconsin Press, 1985).

62 Hoopes, *Oral History*, 15. For comments on the problematic nature of text and archived materials, see also Ruth Finnegan, *Oral Traditions and the Verbal Arts: A Guide to Research Practice* (London: Routledge, 1992), 82.

63 John Rae, "Commentary," 175. Quoted in Hoopes, *Oral History*, 15.

64 Thomas Spear, *Mountain Farmers: Moral Economies of Land and Agricultural Development in Arusha and Meru* (Berkeley: University of California Press, 1997), 11.

1 "WE HAVE ALWAYS BEEN FARMERS": SOCIETY AND ECONOMY AT THE CLOSE OF THE NINETEENTH CENTURY

1   M. Fortes and E. E. Evans-Pritchard, eds., *African Political Systems* (London: Oxford University Press, 1940), 5–6.

2   M. G. Smith, "On Segmentary Lineage Systems," *Journal of the Royal Anthropological Institute* 86, part 2 (July–Dec., 1956): 39–80. See also Aidan W. Southall, *Alur Society* (Cambridge: Cambridge University Press, 1956); and Paula Brown, "Patterns of Authority in West Africa," *Africa* 21, no. 4 (October 1951): 261–78.

3   See A. E. Afigbo, "The Indigenous Political Systems of the Igbo," *Tarikh* 4, no. 2 (1973): 12–23; M. A. Onwuejeogwu, "Evolutionary Trends in the History of the Development of the Igbo Civilization in the Culture Theater of Igboland in Southern Nigeria" (1987 Ahiajoku Lecture) Owerri, Nigeria: Ministry of Information and Culture, 1987).

4   Afigbo, "The Indigenous," 15. See also Raphael C. Njoku, "Neoliberal Globalism in Microcosm: A Study of the Precolonial Igbo of Eastern Nigeria," *Mbari: The International Journal of Igbo Studies* 1, no. 1 (2008): 46–68; Simon Ottenberg, *Leadership and Authority in an African Society: The Afikpo Village Groups* (Seattle: University of Washington Press, 1971); G. I. Jones, "Ibo Age Organization, with Special Reference to the Cross River and North-Eastern Ibo," *Journal of the Royal Anthropological Institute* 92, part 1 and 2 (1962): 191–21; and Richard N. Henderson, *The King in Every Man: Evolutionary Trends in Onitsha Ibo Society and Culture* (New Haven, CT: Yale University Press, 1972).

5   NAE, ONPROF, 8/1/4702, "Anthropological Report on Onitsha Province," by C. K. Meek, 1931. See also Ikenna Nzimiro, *Studies in Igbo Political Systems: Chieftaincy and Politics in Four Niger States* (London: Frank Cass, 1972).

6   Unlike grains, tubers left no archaeological evidence. For an excellent analysis of plant and crop domestication in Igboland, see Okigbo, *Plant and Food*. See also John E. Njoku, *The Igbos of Nigeria: Ancient Rites, Change and Survival* (Lewiston, NY: Edwin Mellen Press, 1990). For more analysis on the evolution of Igbo agriculture, see L. C. Okere, *The Anthropology of Food in Rural Igboland, Nigeria: Socioeconomic and Cultural Aspects of Food and Food Habit in Rural Igboland* (Lanham, MD: University Press of America, 1983), 29–50.

7   See Bede N. Okigbo, *Plants and Food in Igbo Culture and Civilization: 1980 Ahiajoku Lecture* (Owerri: Ministry of Information and Culture, 1980), 11.

8   Okigbo, *Plants and Food*. See also Echeruo, "Aro and Nri: Lessons," 206.

9   This is confirmed in a 1966 Food and Agriculture Organization of the United Nations (FAO) study. See Food and Agriculture Organization of the United Nations, *Agricultural Development in Nigeria, 1965–1980* (Rome: Food and Agriculture Organization of the United Nations, 1966), 397. For more on Igbo food habits and caloric intakes, see Anita Whitney, *Marketing of Staple Foods in Eastern Nigeria* (East Lansing: Michigan State University, 1968).

10 W. B. Morgan, "The Influence of European Contacts on the Landscape of Southern Nigeria," *Geographical Journal* 125, no. 1 (1959): 53.

11 See Adiele Afigbo, *Ropes of Sand: Studies in Igbo History and Culture* (Ibadan, Nigeria: University Press in Association with Oxford University Press, 1981), 126.

12 Ibid., 126.

13 Ibid., 125–26. See also J. E. Flint, *Nigeria and Ghana* (Englewood Cliffs, NJ: Prentice Hall, 1966), 63.

14 Cocoyam, known as a woman's crop, is ritualized as yam and controlled by similar taboos.

15 Interview with Nwanyiafo Obasi, Umunomo, Mbaise, 25 July 1999.

16 See W. B. Morgan, "Farming Practice, Settlement Pattern and Population Density

in Eastern Nigeria," *Geographical Journal* 121, no. 3 (September 1955): 330.

17   G. I. Jones, *From Slaves to Palm Oil: Slave Trade and Palm Oil Trade in the Bight of Biafra* (Cambridge: African Studies Center, 1989), 1.

18   Ibid., 3.

19   On indigenous trading networks, see David Northrup, *Trade Without Rulers: Precolonial Economic Development in Southeastern Nigeria* (Oxford: Clarendon Press, 1978).

20   Duarte Pacheco Pereira, cited in David Northrup, "The Growth of Trade among the Igbo before 1800," *Journal of African History* 13, no. 2 (1972): 220.

21   For pre-colonial exchange relations, see Northrup, *Trade Without Rulers*. See also Northrup, "The Growth of Trade," 217–36. See also Hermann Koler, *Einige Notizen über Bonny* (Göttingen, 1840), trans. Uche Isichei, in *Igbo Worlds: An Anthology of Oral Histories and Historical Descriptions*, ed. Elizabeth Isichei, 14–17 (Philadelphia: Institute for the Study of Human Issues, 1978).

22   "Mr. John Grazilhier's voyage from Bandy to New Calabar," in John Barbot, *A Description of the Coasts of North and South Guinea* (Vol. V in Churchill's *Voyages and Travels*) (London, 1746), 380–81. Cited in Isichei, *Igbo Worlds*, 10.

23   According to his autobiography, written in 1789, Olaudah Equiano (c. 1745–1797) was born in Igboland. He was kidnapped and sold into slavery when he was eleven years old. His involvement in the movement to abolish the slave trade led him to write and publish *The Interesting Narrative of the Life of Olaudah Equiano, or Gustavus Vassa, the African* (1789). There is heated debate today over Equiano's nativity raised by Vincent Carretta in *Equiano, the African: Biography of a Self-Made Man* (University of Georgia Press, 2005). For a contrary view see, Chima J. Korieh, ed., "Introduction," in *Olaudah Equiano and the Igbo World: History, Society, and Atlantic Diaspora Connections*, 1–20 (Trenton: Africa World Press, 2008).

24   Equiano, *The Interesting Narrative*, 39.

25   Archibald John Monteith's memoir was written by Reverend Joseph Horsfield Kummer in 1853. Kummer served the Moravian Mission in Jamaica and this account was edited by Vernon H. Nelson from the manuscript in the Archives of the Moravian Church, Bethlehem, Pennsylvania. See "Archibald John Monteith: Native Helper and Assistant in the Jamaica Mission at New Carmel," *Transactions of the Monrovian Historical Society* 21, no. 1 (1966): 30. See also Maureen Warner-Lewis, *Archibald Monteath: Igbo, Jamaican, Moravian, Jamaican, Moravian* (Kingston: University of the West Indies Press, 2007).

26   Extracts from Koler, *Einige Notizen über Bonny*, in Isichei, *Igbo Worlds*, 14–17. Yam is the common name applied to about 500 species of the genus *Dioscorea* of the *Dioscoreaceae* family. Tubers vary in size and shape, averaging 3–8 lb., but sometimes reaching more than 60 lb.

27   C.M.S Archives, CA3/010, W. E. Carew, *Journal*, January 1866. Cited in Isichei, *Igbo Worlds*, 210.

28   F.O. 403/233, Harcourt, Report on the Aquette Expedition, 29 February 1896–29 March 1896, cited in Isichei, *Igbo Worlds*, 211.

29   S. R. Smith, "Journey to Nsugbe and Nteje, 1897," *Niger and Yoruba Notes* (1898), 82–83, cited in Isichie, *Igbo Worlds*, 202.

30   Onwuka Njoku, *Economic History of Nigeria: 19th and 20th Centuries* (Enugu, Nigeria: Magnet, 2001), 9.

31   C.O. 520/31, "Political Report on the Eza Patrol," encl. in Egerton to Lyttelton, 16 July 1905, cited in Isichei, *Igbo Worlds*, 242.

32   *Western Equatorial Africa Diocesan Magazine*, 1904, 29ff., cited in Isichei, *Igbo Worlds*, 207–8. Cassava is a perennial woody shrub with an edible root, which grows in tropical and subtropical environment.

33   Morgan, "The Influence," 52.

34   Allison, *The Interesting Narrative*, 39.

35   Victor Uchendu, *The Igbo of Southeastern Nigeria* (New York: Holt, Rinehart and Winston, 1965), 30.

36   Morgan, "The Influence," 49.

37   See Barry Floyd, *Eastern Nigeria: A Geographical Review*. New York: Frederick C. Prager, 1969.

38   Morgan, "The Influence," 52. Population pressure and land scarcity have fundamentally influenced Igbo agriculture where the characteristically poor soil continued to deteriorate rapidly with frequent cultivation. For the impact of soil type on agricultural productivity in Eastern Nigeria, see, for example, G. Lekwa, "The Characteristics and Classification of Genetic Sequences of Soil in the Coastal Plain Sands of Eastern Nigeria" (PhD dissertation, Michigan State University, 1979). See also R. K. Udo, "Pattern of Population Distribution and Settlement in Eastern Nigeria," *Nigerian Geographical Journal* 6, no. 1 (1963): 75.

39   Morgan, "The Influence," 53.

40   The Biafra hinterland was a major source of slaves during the Atlantic trade. For an analysis of Igbo participation in the slave trade see, for example, Ugo Nwokeji, "The Biafran Frontier: Trade, Slaves and Aro Society, c.1750–1905," (PhD thesis, University of Toronto, 1998). On the transition from slave trade to commodity trade, see Robin Law, *From Slavery to 'Legitimate' Commerce: The Commercial Transition in Nineteenth century West Africa* (Cambridge: Cambridge University Press, 1995); and Martin Lynn, *Commerce and Economic Change in West Africa: The Palm Oil Trade in the Nineteenth Century* (Cambridge: Cambridge University Press, 1997). For an analysis of the gender implication of the transitions, see, for example, Martin, "Slaves, Igbo Women."

41   ONPROF, 7/15/135, "World Agricultural Census," Resident, Onitsha to District Officer Awgu, 16 January 1929.

42   See the works of Margaret M. Green, *Land Tenure in an Ibo Village in South-Eastern Nigeria. Monographs on Social Anthropology (London School of Economics Monographs on Social Anthropology no. 6)* (London: Berg, 1941); and J. Harris, "Human Relationships to the Land in Southern Nigeria," *Rural Sociology* 7 (1942): 89–92. See also Abe Goldman, "Population Growth and Agricultural Change in Imo State, South-eastern Nigeria," in *Population Growth and Agricultural Change in Africa*, ed. B. L. Turner II, R. Kates, and G. Hyden, 250–301 (Gainesville, FL: University of Florida Press, 1993).

43   David R. Smock and Audrey C. Smock, *Cultural and Political Aspects of Rural Transformation: A Case Study of Eastern Nigeria* (New York: Praeger, 1972), 21. This high population density is reflected in the 1991 population census. An important demographic characteristic is the high female population ration in the region, which is on the average 10,000 more than the male population in most areas. The demographic composition has gender and development implications including access to resources and contribution to agricultural production. See Federal Office of Statistics "1991 Population of States by Local Government Areas," *Digest of Statistics*, December 1994.

44   Rhodes House Oxford (hereafter RH), Mss. Afr. s. 823 (1), J. R. Mackie Papers on Nigerian Agriculture.

45   W. B. Morgan and J. C. Pugh, *West Africa* (London: Methuen, 1969), 322–23.

46   Ibid.

47   Sylvia Leith-Ross, *African Women: A Study of the Igbo of Nigeria* (London: Faber and Faber, 1939), 48.

48   William Allan distinguishes between obligatory and voluntary shifting cultivation. Voluntary shifting cultivation is found where land is plentiful in relation to population. Here populations could move to new areas without the restrictions imposed by the need to allow cultivated land to regenerate. See William Allan, *The African Husbandman* (New York: Barnes & Noble, 1965), 6–7.

49   See Morgan and Pugh, *West Africa*, 322.

50   Shifting cultivators could also rate the fertility of a piece of land and its suitability for a particular crop by the vegetation that covers it and by the physical characteristics of the soil. For a discussion of the ecological basis of soil and agricultural systems, see Allan, *The African Husbandman*, 3–19.

51 Interview with Mbagwu Korieh, Umuchieze, Mbaise, 18 December 1998.

52 For a study of the impact of fallow on the soil, see B. T. Kang, G.F. Wilson, and T. L. Lawson, *Alley Cropping: A Stable Alternative to Shifting Cultivation* (Ibadan, Nigeria: International Institute of Tropical Agriculture, 1984), 22. See also O. A. Opara-Nadi, "Soil Management Practices and Agricultural Sustainability in Traditional Farming Systems," in *Agriculture and Modernity in Nigeria: A Historical and Contemporary Survey of the Igbo Experience*, ed. Jude C. Aguwa and U.D. Anyanwu, 86–104 (New York: Triantlantic Books, 1998).

53 B. N. Okigbo, "Plant and Agroforestry in Land Use Systems of West Africa," in *Plant Research in Agroforestry*, ed. P. A. Huxley, 25–41 (Nairobi: International Council for Research in Agroforestry, 1993).

54 It is likely that by the end of the eighteenth century most parts of Igboland and neighbouring areas were so well inhabited that founding new communities became nearly impossible. The development of a more permanent agricultural practice, therefore, became inevitable.

55 J. W. Wallace "Agriculture in Abakaliki and Afikpo," *Farm and Forest* 2 (1941): 89–95, cited in Morgan and Pugh, *West Africa*, 69; Forde and Jones, *The Ibo and Ibibio*, 14.

56 Anthony G. Hopkins, *An Economic History of West Africa* (London: Longman, 1973), 38.

57 Ibid., 35.

58 See, for example, C. Meillassoux, "Essai d'interpretation du phénomène économique dans les sociétés tradinonelles d'auto-subsistence," *Cahiers d'études africaines* 4 (1960): 38–67. See also Emmanuel Terray "L'Organisation sociale des Dida de Côte-d'Ivoire," *Annales de l'Université d'Abidjan*, Serie F, vol. I, part 2.

59 On patriarchal mode of production and its relevance in stateless societies in pre-colonial Africa, see Jeanne Koopman Henn, "The Material Basis of Sexism: A Mode of Production Analysis with African Examples" (Boston University Working papers/

African Studies Center, no. 119, Boston, MA: African Studies Center, 1986).

60 For more on African modes of production, see Wim van Binsbergen and Peter Geschiere, eds., *Old Modes of Production and Capitalist Encroachment: Anthropological Explorations in Africa* (London: Routledge and Kegan Paul, 1985).

61 For the classification of land among the Igbo, see S.N.C. Obi, *The Ibo Law of Property* (London: Butterworth, 1963).

62 In reality, the notion of individual ownership is quite alien to Igbo indigenous culture. Land is assumed to belong to the community, lineage, the ancestors and the generations yet unborn. For the models of communal tenure, see, for example, John M. Cohen, "Land Tenure and Rural Development in Africa," in *Agricultural Development in Africa*, ed. R. H. Bates and M. F. Lofchie, 349–99 (New York: Preager, 1980); Horoshi Akabane, "Traditional Patterns of Land Occupancy in Black Africa," *Developing Economies* 8 (1970): 161–79; William Allan, *The African Husbandman* (New York: Barnes & Noble, 1965), 360–74; and Frank Mifsud, *Customary Land Law in Africa* (Rome: Food and Agricultural Organization, 1967).

63 NAE, OWDIST, 9/15/2, file no. 4/29, "Rural Land Policy," District Officer, Owerri to Resident, Owerri Province, Port Harcourt, May 1929.

64 NAE, CALPROF, 14/7/1698, file no. E/2994/12 "Land Tenure in the Aba District" District Commissioner, Aba District to the Provincial Commissioner, Eastern Province, October 1912. See also NAE, CALPROF, 14/7/1698, file no. E/2994/12 "Report on Land Tenure" Acting District Commissioner, Orlu to H.P.C Calabar, November, 1912.

65 W. B. Morgan, "Farming Practice, Settlement Pattern and Population Density in Eastern Nigeria," *Geographical Journal* 121, no. 3 (1955): 326.

66 The importance of age in determining gender and social relations is important in many Nigerian communities. For the case of the Yoruba of southeastern Nigeria, see Oyeronke Oyewumi, "Mothers Not

Women: Making an African Sense of Western Gender Discourse" (PhD dissertation, University of California at Berkeley, 1993), 2. See also Onaiwu Ogbomo, *When Men and Women Mattered: A History of Gender Relations among the Owan of Nigeria* (Rochester, NY: University of Rochester Press, 1997), 2–6; and Karen Sacks, *Sisters and Wives: The Past and Future of Sexual Equality* (Urbana: University of Illinois Press, 1982).

67　The patrilineal systems of many Igbo communities allowed men a high degree of authority in decision-making about land, but not necessarily in agricultural production. For an assessment of the impact of patriarchy on land ownership, see, for example, Uchendu, *The Igbo*, 22; Thomas, *Ibo*, vol. 1, chap. 10, cited in Iliffe, *African Poor*, 92. On gender relations and land in African agriculture, see Jean Davison, "Land and Women and Agricultural Production: The Context," in *Agriculture, Women and Land: The African Experience*, ed. Jean Davison, 1–32 (Boulder, CO: Westview Press, 1988). See also Simi Afonja, "Changing Mode of Production and the Sexual Division of Labour among the Yoruba," in *Women's Work*, ed. Eleanor Leacock and Helen I. Safa, 122–35 (South Hadley, MA: Bergin and Garvey, 1986). For an analysis of gender differences in control over resources and labour at the household level, see Ann Whitehead, "'I'm Hungry, Mum': The Politics of Domestic Budgeting," in *Of Marriage and the Market*, ed. K. Young, C. Wolkowitz and R. McCullagh, 88–111 (London: CSE Books, 1981).

68　Davison, "Land, Women and Agricultural Production," 2. Chubb described land among the Igbo as the *fons et erigo* (fountain and origin) of human morality, productivity, and fertility and therefore, to that extent, the principal legal sanction. See Chubb, *Ibo Land Tenure*, 6–7. See also Uchendu, *The Igbo*, 22.

69　In traditional Igbo society, a variety of factors including initiation into adulthood, age, and marriage determined one's status as an adult, but they also determined when one became economically independent.

70　Interview with Mbagwu Korieh, Umuchieze, Mbaise, 18 December 1998.

71　The Igbo week calendar is made up of eight days. The major market days in the week are *Orie, Afo, Nkwo*, and *Eke* with four minor market days on the same nomenclature.

72　The *Ofo* is the symbol of authority in Igbo society and each lineage head remained the custodian of the *Ofo* until he died. See Anyanwu, "Igbo Family Life," 147–48.

73　Interview with Eleazer Ihediwa, Owerrinta, Isiala Ngwa, 24 July 1998. See also John Oriji, *History of the Ngwa People*, 64–67.

74　Interview with Comfort Anabalam, Umuchieze, Mbaise 13 December 1998. On Isusu in Igbo socio-political economy, see Anthony I. Nwabughuoghu, "The Isusu: An Institution for Capital Formation among the Ngwo Igbo: Its Origins and Development," *Africa* 54 (1984): 46-58.

75　Interview with Ugwuanya Nwosu, Owerri, 20 December 1998.

76　See Don C. Ohadike, *Anioma: A Social History of the Western Igbo People*, (Athens, OH: Ohio University Press, 1994) and "'When Slaves Left, Owners Wept': Entrepreneurs and Emancipation among the Igbo People," in *Slavery and Colonial Rule in Africa: Studies in Slave and Post-Slave Societies and Cultures*, ed. Suzanne Miers and Martin A. Klein, 189–207 (London: Routledge, 1999). See also Carolyn Brown, "Testing the Boundaries of Marginality: Twentieth-Century Slavery and Emancipation Struggles in Nkanu, Northern Igboland 1920–29," *Journal of African History* 37, no. 11 (1996): 51–80.

77　See Afigbo, *Ropes of Sand*, 130.

78　Interview with Ugwuanya Nwosu, Owerri, 20 December 1998.

79　David van Nyendael, "A Description of Rio Formosa, or the River of Benin," cited in G. Ugo Nwokeji, "African Conceptions of Gender and the Slave Traffic," *William and Mary Quarterly* 58, no. 1 (2001): 15.

80　William Bosman, *A New and Accurate Description of the Coast of Guinea* (London, 1705), 344.

81 Denis de Cardi, "A Voyage to Congo," A Collection of Voyages and Travels, 4 vols. (London, 1704), 1–622, 629, 630–31.

82 Walter Rodney, A History of the Upper Guinea Coast 1545–1800 (Oxford: Clarendon Press, 1970), 103.

83 Catherine Coquery-Vidrovitch, African Women: A Modern History, translated by Beth Gillian Raps (Boulder, CO: Westview Press, 1997), 11–12.

84 Hopkins, An Economic History, 21.

85 On African farming systems and division of labour, see Baumann, "The Division of Work," 328.

86 Allison, The Interesting Narrative, 39.

87 Basden, Niger Ibos, 93.

88 Interview with Chief Theophilus Onyema, Umuorlu, Isu, 5 January 2000.

89 Interview with Luke Osunwoke, Umuorlu, Isu, 5 January 2000.

90 Harris, "Paper on Economic," 12, Anyanwu, Igbo Family Life., 137.

91 Chuku, "Igbo Women," 39.

92 Alexander Falconbridge, An Account of the Slave Trade on the Coast of Africa (London, 1788), 21.

93 Cited in Onwuejeogwu, "Evolutionary Trends," 59.

94 Basden, Niger Ibos, 389–90, 394.

95 The Church Missionary Intelligencer, August 1891, 573, cited in Isichei, Igbo Worlds, 256.

96 Interview with Linus Anabalam, Umuchieze, Mbaise, 13 and 14 December 1998.

97 Emmanuel Nlenanya Onwu, "Uzo Ndu an Eziokwu: Towards an Understanding of Igbo Traditional Religious Life and Philosophy," 2002 Ahiajoku Lecture (Owerri, Nigeria: Ministry of Information, 2002).

98 RH, Mss. Afr. s. 1000, Extract from Edward Morris Falk Papers.

99 M. Angulu Onwuejeogwu, Afa Symbolism and Phenomenology in Nri Kingdom and Hegemony: An African Philosophy of Social Action (Benin City: Ethiope Publishing Corporation, 1997), 7–9.

100 Kenneth O. Dike and Felicia Ekejiuba, The Aro of Southeastern Nigeria, 1650–1980: A Study of Socio-economic Formation and Transformation in Nigeria (Ibadan, Nigeria: University Press Limited, 1990), 109.

101 Echeruo, "Aro and Nri: Lessons," 200–1.

102 Ibid.

103 Interview with Agu Elija Ukaeme, Umunomo Mbaise, 3 August 1999.

104 Interview with Nze James Eboh, Obowo, Etiti, 2 January 2000.

105 Telephone interview with Johnston Njoku, Glassboro, New Jersey, 2 February 2007. See also Johnston Njoku, "Transformations in Marriage, Gender, and Class System Resulting from Atlantic Trades and Colonialism in the Bight of Biafra," paper presented at the 4th International Conference of the Igbo Studies Association, Howard University, 31 March to 1 April 2006.

106 See Chinua Achebe, Things Fall Apart (New York: Anchor, 1994).

107 Ibid., 28.

108 Ibid.

109 Phanuel Egejuru, The Seed Yams Have Been Eaten (Ibadan, Nigeria: Heinemann, 1993).

110 Ibid., 74.

111 For more analysis, see Obioma Nnaemeka, "Fighting on All Fronts: Gendered Spaces, Ethnic Boundaries, and the Nigeria Civil War," Dialectical Anthropology 22 (1997): 247–48.

112 Achebe, Things Fall Apart, 16, 33–34.

113 NAE, OWDIST, 4/13/70, file no. 91/27, "Cultivation of Crops, Owerri District," District Officer to Resident Owerri Province, June 1928.

114 Onwuejeogwu, "Evolutionary Trends," 60.

115 Basden, Niger Ibos, 389–94.

116 During my field interviews, people talked about the growing of yams as if it were synonymous with farming.

117 Morgan, "The Influence," 52.

118 Basden, Niger Ibos, 389.

119 Onwuejeogwu, "Evolutionary Trends," 58.

120 Cited in Onwu, Ụzọ Ndu.

121 Onwuejeogwu, "Evolutionary Trends," 59.

122 Interview with Chief Theophilus Onyema, Umuorlu, Isu, 5 January 2000.

123 Isichei, A History of the Igbo, 10.

124 Adiele E. Afigbo, "Trade and Trade Routes in Nineteenth Century Nsukka," Journal of the Historical Society of Nigeria 7, no. 1 (1973): 77–90; Adiele E. Afigbo, "The Nineteenth Century Crisis of the Aro Slaving Oligarchy of Southeastern Nigeria," Nigeria Magazine 110–12 (1974): 66–73; G. I. Jones, "Who are the Aro?", The Nigerian Field 8, no. 3 (1939): 100–3; and G. I. Jones "Native and Trade Currencies in Southern Nigeria during the Eighteenth and Nineteenth Centuries," Africa 28, 1 (1958): 43–54.

125 Ukwu I. Ukwu, "The Development of Trade and Marketing in Iboland," Journal of the Historical Society of Nigeria 3, no. 4 (1967): 650.

126 On Nri civilization, see Thurstan Shaw, Igbo-Ukwu: An Account of Archeological Discoveries in Eastern Nigeria, vol. 1 (Evanston: Northwestern University Press, 1970), 268–84; and Onwuejeogwu, Afa Symbolism. On the Aro, see Dike and Ekejiuba, The Aro .

127 Ukwu, "The Development of Trade," 650.

128 Ibid.

129 Ibid., 651–55. See also Dike and Ekejiuba, The Aro.

130 Ibid.

131 See Eltis and Richardson, "West Africa," 16–35.

132 See David Eltis et al., eds. The Trans-Atlantic Slave Trade: A Database on CD-ROM (Cambridge: Cambridge University Press, 1999).

133 Ibid.

134 David Eltis and David Richardson estimated that about one in seven Africans shipped to the New World during the whole era of the transatlantic slave trade originated from the Bight of Biafra. See David Eltis and David Richardson, "West Africa and the Transatlantic Slave Trade: New Evidence on Long Run Trends," Slavery and Abolition 18, no. 1 (1997): 16–35. See Douglas B. Chambers, "'My Own Nation': Igbo Exiles in the Diaspora," Slavery and Abolition 18, no. 1 (1997): 72–97. For African export figures for 1470s–1699, see Paul E. Lovejoy, "The Volume of the Atlantic Slave Trade: A Synthesis," Journal of African History 23 (1982): 478–81. For 1700–1809, see David Richardson, "Slave Exports from West and West-Central Africa, 1700–1810: New Estimates of Volume and Distribution," Journal of African History 30, no. 1 (1989): 1–22.

135 Chambers, "'My Own Nation," 75–7.

136 Birgit Muller, "Commodities as Currencies: The Integration of Overseas Trade into the Internal Trading Structure of the Igbo of South-East Nigeria," Cahiers d'études africaines 97, 25, no. 1 (1985): 65.

137 Morgan, "The Influence," 53.

138 For the link between the slave trade and the local agrarian economy, see, for example, Martin, "Slaves, Igbo Women and Oil Palm."

139 John Barbot, A Description of the Coasts of North and South Guinea (London, 1746), 465, cited in Jones, From Slaves, 39.

140 Jones, From Slaves, 41.

141 Calculated from Eltis et al., eds. The Trans-Atlantic Slave Trade.

142 Alexander Falconbridge, An Account of the Slave Trade on the Coast of Africa (London, 1788), 21.

143 Barbot, A Description, 379–80.

144 Ibid., 465.

145 Jones, From Slaves, 40.

146 Okigbo, "Towards a Reconstruction," 10.

147 See Robin Law, "The Historiography of the Commercial Transition in Nineteenth Century West Africa," in African Historiography: Essays in Honour of Jacob Ade Ajayi, ed. Toyin Falola, 91–115 (Hawlow, 1993); Martin A. Klein, "Social and Economic Factors in the Muslim Revolution in Senegambia," Journal of African History 13 (1972): 414–41; David Eltis, Economic

Growth and the Ending of the Trans-Atlantic Slave Trade (Oxford: Oxford University Press, 1987); and Ralph A. Austen, "The Abolition of the Overseas Slave Trade: A Distorted Theme in West African History," *Journal of the Historical Society of Nigeria* 5, no. 2 (1970): 257–74. See also Chima J. Korieh, "The Nineteenth Century Commercial Transition in West Africa: The Case of the Biafra Hinterland," *Canadian Journal of African Studies* 34, no. 3 (2000): 588–615.

148 Morgan, "The Influence," 53.

149 See, for example, Anyanwu, "Igbo Family Life," 260; Dike, *Trade and Politics*, 49; and A. J. Latham, *Old Calabar 1600–1891: The Impact of the International Economy upon a Traditional Society* (Oxford: Clarendon Press, 1973).

150 American Memory, "Evidence of Capt, the Hon J. Denman, to the Select Committee on West Coast of Africa, 1942," http://memory.loc.gov/ll/llst/014/0100/01780107.gif, [accessed 19 February 2006].

151 Law, "The Historiography," 91–115.

152 Jones, *From Slaves*, 50.

153 Parliamentary papers 1842 XI pt. 1, appendix and index no. 7, 232, cited in Anyanwu, "Igbo Family Life," 260.

154 Kathleen M. Baker, *Agricultural Change in Nigeria: Case Studies in the Developing World* (London: John Murray, 1989), 3.

155 Dike, *Trade and Politics*, 101.

156 Walter I. Ofonagoro, *Trade and Imperialism in Southern Nigeria 1881–1889* (New York: Nok Publishers, 1979), 319.

157 H. L. Gallwey, "Journeys in the Benin Country, West Africa," *Geographical Journal* 1, no. 2 (1893): 122–30.

158 *New York Times*, 29 March 1863, 8.

159 Isichei, *The Ibo People*, 67; Anyanwu, "Igbo Family Life," 260.

160 Jones, *From Slaves*, 53.

161 Echeruo, "Aro and Nri: Lessons," 207–8.

162 Ibid., 206–7.

163 Dike and Ekejiuba, *The Aro*, 3. See also Michael J. C. Echeruo, "Aro and Nri: Lessons of Nineteenth Century Igbo History," in

The Aftermath of Slavery: Transitions and Transformations in Southeastern Nigeria, ed. Chima J. Korieh and Femi J. Kolapo, 228–47 (Trenton: African World Press, 2007).

164 Waibinte Wariboko, "New Calabar Middlemen, Her Majesty's Consuls, and British Traders in the Niger Delta During the Era of New Imperialism," in *Aftermath of Slavery: Transitions and Transformations in Southeastern Nigeria*, ed. Chima J. Korieh and Femi J. Kolapo, 17–40 (Trenton: Africa World Press, 2007).

165 See Hopkins, *An Economic History*, 216. For more analysis, see Law, *From Slave Trade*, 1995; and Lovejoy and Richardson, "Initial Crisis of Adaptation," in Law, *From Slave Trade*, 1995. See also Martin A. Klein, "The Development of Slavery in West Africa," paper delivered at the Harriet Tubman Seminar, York University, 1996.

166 Martin, "Slaves, Igbo Women," 182.

167 Ibid.

168 Ibid.

169 See Stone, "Women, Work and Marriage," 16.

170 Morgan, "The Influence," 48.

171 Adolphe Burdo, *The Niger and the Benueh: Travels in Central Africa* (London: Richard Bentley and Sons, 1880), 134.

172 Cited in Mba, *Nigeria Women*, 48.

173 Raymond Gore Clough, *Oil Rivers Trader* (London: C. Hurst, 1972), 41–42.

174 Ibid., 41–42.

175 Henderson, *The King*, 230–43. See also Mba, *Nigeria Women*, 49.

176 Mba, *Nigeria Women*, 32.

177 Ibid., 48.

178 NAE, RIPROOF, 8/5/661, file no. OWN 630/17, "Trade Prices at up Country Markets," D. O. Okigwi to Resident Owerri Province, Port Harcourt, 29 November 1917.

2 PAX BRITANNICA AND THE DEVELOPMENT OF AGRICULTURE

1   *New York Times*, 12 April 1903.

2   Ibid.

3   For new perspectives on British imperialism, see P. J. Cain and A.G. Hopkins, *British Imperialism, 1688–2000*, 2nd ed. (Harlow, England: Longman, 2001).

4   Report by The Hon. W.G.A. Ormsby-Gore (Parliamentary Under-Secretary for State for the Colonies) on his Visit to West Africa during the year 1926 (London: HMO, 1926), 77.

5   The works of Robinson and Gallagher have given prominence to the notion of formal and informal empire. See J. Gallagher and R. E. Robinson, "The Imperialism of Free Trade," *Economic History Review* 6, no. 1(1953): 1–15. See also C. R. Fay, *Imperial Economy and its Place in the Foundation of Economic Doctrine, 1600–1932* (Oxford: Clarendon Press, 1934).

6   RH, Mss Afr. s. 1873, Robert B. Broocks Papers.

7   *London Gazette* 5 June 1885, 2581. Cited in J. C. Anene, "The Foundation of British Rule in Nigeria (1885–1891)," *Journal of Historical Society of Nigeria* 1, no. 4 (1959): 253–62. Named after palm oil, the major export product from the region, the protectorate originally also included territories like Benin and Itsekiri, which later became part of Western Nigeria.

8   See U.O.A. Esse, "Introduction," *Catalogue of the Correspondence and Papers of the Niger Coast Protectorate* (CSO 3/1/1–3/5/1, 1894–1999) (Enugu: National Archives of Nigeria, 1988).

9   On the political and economic developments in this period, see J. E. Flint, *Sir George Goldie and the Making of Nigeria* (London: Oxford University Press, 1960); E. J. Alagoa, *The Small Brave City State: A History of Nembe-Brass* (Ibadan, Nigeria: University of Ibadan Press, 1964); S.J.S. Cookey, *King Jaja of the Niger Delta: His Life and Times 1829–1889* (New York: Nok Publishers, 1974); Walter I. Ofonagoro,

Trade and Imperialism in Southern Nigeria 1881–1889 (New York: Nok Publishers, 1979); Obare Ikime, *The Fall of Nigeria: The British Conquest* (Ibadan, Nigeria: Heinemann, 1982); J. C. Anene, *Southern Nigeria in Transition 1885–1906* (Cambridge: Cambridge University Press, 1966); J.U.J. Asiegbu, *Nigeria and its British Invaders 1851–1920* (New York: Nok Publishers, 1984); H. Galway, "The Rising of the Brassmen," *Journal of African Society* 34 (1935): 144–62; C. Gertzel, "Relations between Africans and European Traders in the Niger Delta, 1880–1896," *Journal of African History* 3 (1962): 361–66; J.U.J. Asiegbu, "Some Notes on Afro-European Relations and British Consular Roles in the Niger Delta in the Nineteenth Century," *Journal of Niger Delta Studies* 1, no. 2 (1971): 101–16; and Waibinte Wariboko, "New Calabar Middlemen, Her Majesty's Consuls, and British Traders in the Niger Delta during the Era of New Imperialism," in *The Aftermath of Slavery: Transitions and Transformations in Southeastern Nigeria*, ed. Chima J. Korieh and Femi J. Kolapo, 17–40 (Trenton: Africa World Press, 2006).

10 Wariboko, "New Calabar Middlemen," 28.

11 Ibid., 33.

12 Alagoa, *The Small Brave*, 116.

13 Ukwu, "The Development of Trade," 656.

14 In this period eight British commercial companies, British & Continental African Company Limited, Couper Johnstone & Company, Hatton & Cookson, John Holt & Company, Liverpool African Company, Richard & William King, Smith & Douglas Limited, Taylor Laughland & Company, and Thomas Harrison & Company united in 1889 to form the African Association Limited, but their claim to charter similar to the Royal Niger Company charter to administer the Oil Rivers was refused.

15 Other military expeditions include the Douglas Expedition (*ogu* Douglas), Ahiara Expedition, 1905.

16 On Aro commercial activities, see, for example, Adiele Afigbo, "The Eclipse of the Aro Trading Oligarchy of Southeastern

Nigeria, 1901–1927," *Journal of the Historical Society of Nigeria* 4, no. 1 (1971): 3–24; and Adiele Afigbo, "The Aro Expedition of 1901–1902: An Episode in the British Occupation of Iboland," *ODU: A Journal of West African Studies* 7 (1972): 3–25.

17   Don C. Ohadike's *The Ekumeku Movement: Western Igbo Resistance to the British Conquest of Nigeria, 1883–1914* (Athens, OH: Ohio University Press, 1991). See also Don C. Ohadike, *Sacred Drums of Liberation: Religions and Music of Resistance in Africa and the Diaspora* (Trenton: Africa World Press, 2007) for resistance movements from Africa and the African Diaspora through religion and music.

18   See Ohadike, *The Ekumeku Movement*.

19   See Felix K. Ekechi, "Portrait of a Colonizer: H. M. Douglas in Colonial Nigeria, 1897–1920," *African Studies Review* 26, no. 1 (March 1983): 25–50. See also Felix K. Ekechi, "Episodes of Igbo Resistance to European Imperialism, 1860–1960," in *Olaudah Equiano and the Igbo World: History, Society, and Atlantic Diaspora Connections*, ed. Chima J. Korieh, 229–53 (Trenton: Africa World Press, 2009).

20   The term "indirect rule" is a misnomer. The formulation of policy and its implementation were vested on the few British personnel in the territories. For the application of "indirect rule" system in Eastern Nigeria, see Adiele E. Afigbo, *Warrant Chiefs: Indirect Rule in Southeastern Nigeria, 1891–1929* (New York: Humanities Press, 1972). See also Frederick Lugard, *The Dual Mandate in Tropical Africa* (London: W. Blackwood, 1926), for the articulation of the indirect rule policy. See also Michael Crowder, *West Africa under Colonial Rule* (Evanston: Northwestern University Press, 1968); and Michael Crowder, *Colonial West Africa: Collected Essays* (London: Frank Cass, 1978).

21   Lugard, *Dual Mandate*, 193.

22   RH, Mss Afr. s. 1073, "Extract from a circular from the secretary, Northern Provinces Kaduna, to all Residents, Northern Provinces," 23 November 1928.

23   Ibid.

24   Ibid.

25   Chima J. Korieh, "Islam and Politics in Nigeria: Historical Perspectives," in *Religion, History and Politics in Nigeria: Essays in Honour of Ogbu U. Kalu*, ed. Chima J. Korieh and G. Ugo Nwokeji, 109–24 (Lanham, MD: University Press of America, 2005).

26   Although some Muslim intellectuals like Aminu Kano served as a voice of dissent and dissatisfaction with the Anglo-Fulani hegemonic collaboration, the ruling class collaborated with the British largely to preserve their own authority and privileges.

27   RH, Mss Afr. s. 1073, "Extract from a circular from the secretary."

28   Ibid.

29   For an analysis of indirect rule in Western Nigeria, see, J. A. Atanda, *The New Oyo Empire: Indirect Rule and Change in Western Nigeria 1894–1934* (New York: Humanities Press, 1973).

30   RH, Mss Afr. s. 1073, "Extract from a circular from the secretary."

31   RH, Mss Afr. s. 1551, J.G.C. Allen Papers, "Nigerian Panorama, 1926–1966."

32   Lugard, *Dual Mandate*, 193.

33   Lugard was also committed to containing the expansion of Islam to non-Muslim groups. Lugard did not want non-Muslim groups to be forcibly placed "under Moslem rule (which in practice means their conversion to the Moslem faith) even though that rule may be more advanced and intelligent than anything they are as yet capable of evolving themselves." See Muhammad S. Umar, *Islam and Colonialism: Intellectual Responses of Muslims of Northern Nigeria to British Colonial Rule* (Leiden: Brill, 2006), 34.

34   Walter Elliot, "The Parliamentary Visit to Nigeria," *Journal of the Royal African Society* 27, no. 107 (1928): 215–16. Based on an address delivered by Major Walter Eliot, M.C., M.P. (Under-Secretary of State for Scotland; Chairman of the Research Grants Committee of the Empire Marketing Board and Chairman of the Delegation of the Empire Parliamentary Association which visited Nigeria 1927–

28), at a Dinner of the African Society on 13 March 1928.

35 R H, Mss Afr. s. 1551, J.G.C. Allen Papers.

36 On the policy of indirect rule in Eastern Nigeria, see Adiele Afigbo, *The Warrant Chiefs: Indirect Rule in Eastern Nigeria 1891-1929* (Ibadan, Nigeria: Longman, 1972).

37 RH, Mss Afr. s. 2288/2, Alex J. Braham papers.

38 Report by The Hon. W. G. A. Ormsby-Gore, 115.

39 Ibid.

40 RH, Mss Afr. s. 1000 (1), Edward Morris Falk papers.

41 RH, Mss Afr. s. 1873, Robert B. Broocks, papers.

42 Ibid.

43 Ibid.

44 RH, Mss Afr. s. 1551, J.G.C. Allen Papers.

45 RH, Mss Afr. s. 1881. . A.F. B. Bridges papers.

46 RH, Mss Afr. s. 1551, J.G.C. Allen Papers.

47 See also Enyeribe Onuoha, *The Land and People of Umuchieze* (Owerri, Nigeria: Augustus Publishers, 2003), 17–21.

48 RH, Mss Afr. s. 1551, J.G.C. Allen Papers.

49 Ibid.

50 Ibid.

51 Interview with Onyegbule Korieh, Ihitteafoukwu, Mbaise, 17 December 1998.

52 RH, Mss Afr. s. 546, "Reminiscences of Sir F. Bernard Carr—Administrative Officer, Nigeria 1919-1949," Carr Frederick Bernard papers.

53 RH, Mss Afr. s. 1924, A.E. Cooks papers.

54 RH, Mss Afr. s. 1000 (1), Edward Morris Falk papers.

55 RH, MSS Afr. s. 1551, J.G.C. Allen Papers.

56 RH, Mss Afr. s. 1873, Robert B. Broocks papers.

57 RH, Mss Afr. s. 1551, J.G.C. Allen Papers.

58 RH, Mss Afr. s. 1520, Sylvia Leith-Ross papers.

59 RH, Mss Afr. s. 1873, Robert B. Broocks papers

60 Ibid.

61 Ibid.

62 C.O. 520/26, Egerton to C.O. (Confidential), November 5, 1904.

63 C.O. 520/14, Moor to C.O. No. 183, April 17, 1902.

64 Kannan K. Nair, *Politics and Society in South Eastern Nigeria, 1841-1906: A Study of Power, Diplomacy, and Commerce in Old Calabar* (London: Frank Cass, 1972), 39.

65 See Kenneth O. Dike, *Trade and Politics in the Niger Delta, 1830-1885* (Oxford: Clarendon Press, 1956), 4.

66 On agricultural change in Africa, see, for example, W. J. Barber, "The Movement into the World Economy," in *Economic Transition in Africa*, ed. M. J. Herskovits and M. Harwitz, 299–329 (Evanston, IL: Northwestern University Press, 1961); Sara Berry, *No Condition is Permanent: The Social Dynamics of Agrarian Change in Sub-Sahara Africa* (Madison: University of Wisconsin Press, 1993); Hopkins, *An Economic History*; Martin A. Klein, ed., *Peasants in Africa: Historical and Contemporary Perspectives* (London: Sage, 1980); J. Tosh, "The Cash Crop Revolution in Tropical Africa: An Agricultural Appraisal," *African Affairs* 79, no. 314 (1980): 79–94.

67 Several attempts were made to introduce the production of cotton in the region. Such attempts were frustrated by a combination of factors, including the lack of interest on the part of farmers, and the unsatisfactory soil conditions in many parts of Igboland.

68 W. B. Morgan, "Farming Practice, Settlement Pattern and Population Density in Eastern Nigeria," *Geographical Journal* 121, no. 3 (1955): 331.

69 R. E. Dennett, "Agricultural Progress in Nigeria," *Journal of the Royal African Society* 18, no. 72 (1919): 266.

70 Dennett, "Agricultural Progress," 267.

71 Fredrick. D. Lugard, *The Rise of our East African Empire, Vol. 1.1, The Years of Adventure 1858-1898* (London: Collins, 1893), 381.

72 Ekechi, "Portrait of a Colonizer," 40.

73 CO 520/31, W. Egerton, "Overland journey – Lagos to Calabar, via Ibadan, March 9–April 18, 1905," 29.

74 NAE, OWDIST 6/2/6, "Report on the Export Products of Owerri Province, Nigeria," 3/11/14.

75 *Address by Governor Hugh Clifford, to the Nigerian Council* (Lagos, December, 1920), cited in Usoro, *The Nigeria Oil Palm Industry*, 186-7. See also Fredrick Lugard, "British Policy in Nigeria," *Africa* 10, no. 4 (1937): 385. Lugard discussed the difficulty of acquiring land for plantation development because of the Igbo land tenure system. See also Usoro, *The Nigeria Oil Palm Industry*, 390.

76 For the attempts by Lever to establish plantations in West Africa, see David Meredith, "Government and the Decline of the Nigerian Oil-Palm Export Industry, 1919-1939," *Journal of African History* 25 (1984): 311-29. See also Nworah K. Dike, "The Politics of Lever's West African Concession, 1907-13," *International Journal of African Historical Studies* 5 (1972): 248-64.

77 *West Africa*, 26 July 1924.

78 Ibid.

79 Ibid.

80 See Fredrick Lugard, "British Policy in Nigeria," *Africa* 11, no. 4 (1937): 384; K. M. Buchanan and J. C. Pugh, *Land and People in Nigeria* (London: University of London Press, 1958), 103. See also NAE, CSE 1/861102. Smith Buchannan, Commissioner of Lands, Lagos. Memorandum on the existing methods of Government control of Crown and other lands in Southern Nigeria for the W & ALC. 1892: 11; NAE, CSE 6607/Cv. Lt. Col. Roweg, "Confidential Report on Land Questions in Southern Nigeria: NAE, CSO 583/146. File no. 6990 and 6991: "West African Lands Committee Question." See also: Alan Pim, *Colonial Agricultural Production: The Contribution Made by Native Peasants and by Foreign Enterprise* (London: Oxford University Press, 1946), 133; and O. N. Njoku, "Oil Palm Syndrome in Nigeria: Government Policy and Indigenous Response, 1918–1939," *Calabar Historical Journal* 2, no. 1 (1978): 84.

81 Nigeria, *Annual Report on the Agricultural Department, 1932*, 29.

82 Report by The Hon. W.G.A. Ormsby-Gore, 107.

83 See Meredith, "Government and the Decline," 311. See also Hopkins, *An Economic History*, 210-16; Usoro, *The Nigeria Oil Palm Industry*, 36-40; D. K. Fliedhouse, *Unilever Overseas* (London: Croom Helm, 1978), ch. 9; and Berry, *Fathers Work*, 23.

84 F. M. Dyke was struck by the high degree of skill and knowledge shown by the native farmers in the care of their palm. See *Report on the Oil Palm Industry in British West Africa*, cited in Usoro, *The Nigeria Oil Palm Industry*, 37.

85 For example, in 1910, W. H. Johnson, Director, Department of Agriculture, Southern Nigeria described the methods employed by the native agriculturalists as "extremely crude," employing only "a short-handled hoe and the cutlass." For the analysis of the production systems in Eastern Nigeria, see Usoro, *The Nigeria Oil Palm Industry*; Susan Martin, "Igbo Women and Palm Oil"; and John Oriji, "A Study of the Slave Trade and Palm Produce among the Ngwa Igbo of Southeastern Nigeria," *Cahiers d'études africaines* 91(1983): 311-28.

86 RH, Mss. Afr. s. 823 (4), J. R. Mackie Papers.

87 Elliot, "The Parliamentary Visit," 205-18.

88 Ibid., 216-17.

89 *The Nigeria Handbook* (Lagos: Government Printer, 1927), 255-56.

90 Nigeria, *The Nigeria Handbook, 1929*, 246.

91 See J. E. Gray, "Native Methods of Preparing Palm Oil," in *Second Annual Bulletin of the Agricultural Department for 1923/1924* (Lagos: Government Printer, 1925), 29. See also Helleiner, *Peasant Agriculture*, 5.

92 Ibid.

93 I.E.S. Amdii, "Revenue Generating Capacity of the Nigerian Customs and Excise: 1875–1960," in *100 Years of the Nigerian Customs and Excise: 1891–1991*, ed. I.E.S. Amdii, 12–47 (Zaria, Nigeria: Ahmadu Bello University Press, 1991).

94 See Uchendu, *The Igbo of Southeast Nigeria*, 26.

95 Ibid.

96 See, for example, J. E. Gray, "Native Methods of Preparing Palm Oil," *First Annual Bulletin of the Agricultural Department* (Lagos, 1922); and O. T. Faulker and C. J. Lewis, "Native Methods of Preparing Palm Oil – II," *Second Annual Bulletin of the Agricultural Department* (Lagos, 1923), 6–10.

97 Usoro, *The Nigeria Oil Palm Industry*.

98 Interview with Christina Marizu, Nguru, 25 December 1999.

99 Clough, *Oil Rivers Trader*, 38.

100 RH, Mss Afr. s. 1000, Edward Morris Falk papers.

101 Clough, *Oil Rivers Trader*, 51–52.

102 Echeruo, "Aro and Nri," 240.

103 RH, Mss Afr. s. 1924, A. E. Cooks papers.

104 Ibid.

105 Interview with Eugenia Otuonye, Umuchieze, 23 December 1998.

106 RH, Mss Afr. s. 1924, A. E. Cooks papers.

107 Isichei, *A History of the Ibo People*, 67.

108 Gerald D. Hursh et al., ed., *Innovation in Eastern Nigeria: Success and Failure of Agricultural Programs in 71 Villages of Eastern Nigeria* (East Lansing: Michigan State University, 1968), 193–94.

109 CO 583/193/8, 'Palm Oil Industry.'

110 Nigeria, *Annual Report on the Agricultural Department, 1932*, 22.

111 RH, Mss Afr. s. 546, Frederick Bernard Carr paper.

112 Interview with F. Enweremadu, Mbutu Mbaise, 2 January 2000.

113 Nigeria, *Report on the Agricultural Department, 1912*, 23.

114 RH, Mss Afr. s. 546, Frederick Bernard Carr paper.

115 Ibid.

116 W. B. Morgan, "Farming Practice, Settlement Pattern and Population Density in Eastern Nigeria," *Geographical Journal* 121, no. 3 (1955): 332.

117 CO 583/193/8, "Palm oil Industry."

118 Ibid.

119 Spear, *Mountain Farmers*, 1997.

120 Interview with Onyegbule Korieh, Mbaise, 17 December 1998.

121 RH, Mss Afr. s. 1873, Robert Bernard Broocks Papers.

122 RH, Mss Afr. s. 1000, Edward Morris Falk papers.

123 Ibid.

124 Ikem Stanley Okoye, "'Biafran' Historicity: Ife, Okrika, and Architectural Representation," in *The Aftermath of Slavery*, ed. Chima J. Korieh and F. J. Kolapo, 158–96 (Trenton: Africa World Press, 2007).

125 Nigeria, *Report in the Agricultural Department, 1912*, 23–24.

126 RH, Mss Afr. s. 1000, Edward Morris Falk papers.

127 Nigeria, *First Annual Bulletin of the Agricultural Department* (Lagos: Government Printer, 1922), 11.

128 Report by The Hon. W.G.A. Ormsby-Gore, 77.

129 *The Nigeria Handbook* (Lagos: Government Printer, 1927), 140.

130 Nigeria, *Annual Report in the Agriculture Department: Southern Provinces for the Year, 1918* (1919), 20.

131 Ibid., 20.

132 Nigeria, *Annual Report in the Agriculture Department: Southern Provinces for the Year, 1919*, 18.

133 See NAE, ABADIST, 14/1/397, file nos. 851; NAE, AHODIST, 14/1/465, relating to pawning of persons as security for debts.

134 Interview with Linus Anabalam, Mbaise, 13 December 1998.

135 Iliffe, *The African Poor*, 92.

136 See J. S. Harris, "Some Aspects of the Economics of Sixteen Ibo Individuals," *Africa* 14 (1943–4): 302–35.

137 S. N. Nwabara, *Iboland: A Century of Contact with Britain, 1860–1960* (Atlantic Highlands, NJ: Humanities Press, 1978), 15. See also C. K. Meek, *Land and Authority in a Nigeria Tribe: A Study in Indirect Rule* (New York: Barnes and Noble, 1970), 1.

138 Meek, *Land and Authority*, 15–16.

139 Green, *Ibo Village Affairs*, 43.

140 Takes, "Socio-Economic Factors," 6.

141 Udo, *Geographical Regions*, 83. See also W. B. Morgan and J. C. Pugh, *West Africa* (London: Methuen, 1969), 9.

142 Udo, *Geographical Regions*, 83.

143 Philip Raikes, "Modernisation and Adjustment in African Peasant Agriculture," in *Disappearing Peasantries? Rural Labour in Africa, Asia and Latin America*, ed. Deborah Bryceson, Cristobal Kay, and Jos Mooij, 79 (London: Intermediate Technology Publications, 2000).

144 Jones, *From Slaves*, 1.

145 Freund, *The Making of Modern Africa*, 98.

146 Interview with Eleazer Ihediwa. See also J. Tosh, "The Cash Crop Revolution in Tropical Africa: An Agricultural Reappraisal," *African Affairs* 79, no. 314 (1980): 79–94; and J. Heyer, P. Roberts, and G. Williams, ed., *Rural Development in Tropical Africa* (New York: St. Martin's Press, 1981), 168–92.

147 Some scholars see the capitalist transformation of African colonial economies as clearly determined by the colonizing power rather than "natural" developments within indigenous communities. See Clive Y. Thomas, *The Rise of the Authoritarian State in Peripheral Societies* (New York: Monthly Review Press, 1984), 10–19.

148 Geoffrey B. Kay, *The Political Economy of Colonialism in Ghana: A Collection of Documents and Statistics 1900–1960* (Cambridge: Cambridge University Press, 1972), 331. See also Berry, *No Condition is Permanent*, 1993.

149 Walter Rodney, *How Europe Underdeveloped Africa* (Washington: Howard University Press, 1974).

150 Barron S. Hal, *Mixed Harvest: The Second Great Transformation in the Rural North, 1870–1930* (Chapel Hill, NC: University of North Carolina Press, 1997), 15.

3 GENDER AND COLONIAL AGRICULTURAL POLICY

1 See Adiele E. Afigbo, *The Warrant Chiefs: Indirect Rule in Southeastern Nigeria, 1891–1929* (London: Longman, 1972). See also C. K. Meek, *Law and Authority in a Nigerian Tribe: A Study in Indirect Rule* (London: Oxford University Press, 1937). On the decline in the status of women in Africa under colonialism, see, for example, Jane Parpart, "Women and the State in Africa," in *The Precarious Balance: State and Society in Africa*, ed. Donald Rothchild and Naomi Chazan, 208–15 (Boulder, CO: Westview Press, 1988); Jane Parpart and Kathleen A. Staudt, eds., *Women and the State in Africa* (Boulder, CO: Lynn Rienner Publishers, 1989); Hafkin and Bay, eds., *Women in Africa*; and Claire Robertson and Iris Berger, *Women and Class in Africa* (New York: Africana Publishing, 1986). For the case of Igboland, see Nina Mba's *Nigerian Women Mobilized*.

2 The recognition of difference and diversity is a common trend running through the writings of African feminist scholars. See, for example, Obioma Nnaemeka, ed., *Sisterhood, Feminism and Power: From Africa to the Diaspora* (Trenton, NJ: Africa World Press, 1998). For the work of a culturally sensitive writer, see, for example, Chilla Bulbeck, *Re-orienting Western Feminism: Women's Diversity in a Postcolonial World* (Cambridge: Cambridge University Press, 1998).

3 See Igor Kopytoff, "Women's Roles and Existential Identities," in *African Gender Studies: A Reader*, ed. Oyeronke Oyewumi, 127–44 (New York: Palgave Macmillan, 2005).

4 See, for example, NAE, OWDIST, 4/13/70, file no. 91/27, "Cultivation of Crops, Owerri District," District Officer to Resident Owerri Province, June 1928. The District Officer acknowledged that statistics were not available for women's crops.

5 Cited in Dennett, "Agricultural Progress," 266–67.

6 See NAE, CALPROF, 14/8/712, file no. E/1019/13, "Report on travelling and agricultural instructional work," Superintendent of Agriculture, Eastern Province, Calabar, 1913.

7 NAE, CALPROF, 14/8/711, file no. 1018/13, *The Quarterly Report of the Agricultural Department, 1918.* See also Superintendent of Agriculture, Eastern Province, "Report on the Progress of Pupils attached to the Agricultural Department."

8 Nigeria, *First Annual Bulletin of the Agricultural Department* (Lagos: Government Printer, 1922), 17.

9 Chuku, "The Changing Role," 150.

10 NAE, ONDIST, 12/1/578, "Instruction for Farmers' Sons," H. G. Poynter to superintendents of agriculture, 14 December 1933.

11 Nigeria, *Report on the Agricultural Department, 1934* (Lagos: Government Printer, 1935), 26.

12 NAE, ONDIST, 12/1/578, "Instruction for Farmers' Sons." District officer, Nsukka Division to Resident, Onitsha Province, 3 February 1934.

13 NAE, ONDIST, 12/1/578, "Instruction for Farmers' Sons," B. C. Stone, District officer, Onitsha to Superintendent of Agriculture, Onitsha, 13 February 1934.

14 NAE, ONDIST, 12/1/578, "Instruction for Farmers' Sons," B. W. Walter, District officer, Udi Division to Superintendent of Agriculture, Onitsha, 6 February 1934.

15 RH, Mss Afr. s. 1779. Norman Herington papers.

16 Ibid.

17 Ibid.

18 Ibid.

19 NAE, EKETDIST, 1/2/50, file no. 499, "Co-operative Agricultural Settlements for Nigeria, Registrar of Co-operative Societies to the Chief Secretary to the Government," Lagos, April 1940.

20 NAE, EKETDIST, 1/2/50, "Cooperative Agricultural Settlements."

21 Interview with Eneremadu, Mbutu, 2 January 2000.

22 NAE, EKETDIST, 1/2/50, "Cooperative Agricultural Settlements."

23 NAE, RIVPROF, 8/5/661, "Cooperative Agricultural Settlements for Nigeria," Registrar of Cooperatives Societies to Chief Secretary to the Government, Lagos, 1940.

24 NAE, RIVPROF, 8/5/661, "Cooperative Agricultural Settlements."

25 Ibid.

26 Ibid.

27 Ibid.

28 Ibid.

29 Ibid.

30 RH, Mss Afr. s. 1975, Michael Mann, "Community Development in Okigwi Division," n.d., c. 1950–53.

31 Ibid.

32 Ibid.

33 Ibid.

34 Ibid.

35 Eastern Region, *Annual Report on the Department of Agriculture (Eastern Region), 1952/53,* 2.

36 Ibid.

37 Ibid.

38 Ibid.

39 RH, Mss Afr. s. 1779, Norman Herington papers.

40 Ibid.

41 This was the situation in most of the colonial territories. See, for example, Janice Jiggins, *Gender-Related Impacts and the Work of the International Agricultural Centres, CGIAR Study Paper* 17 (1986), 1–2.

42 RH, Mss Afr. s. 862 Swaisland H. Eastern Nigeria papers.

43 Ibid.

44 See Richard Goodridge, "Women and Plantation in Western Cameroon Since 1900," in *Engendering History: Caribbean Women in Historical Perspective*, ed. Verene Shepherd, Bridget Brereton, and Barbara Bailey, 394 (New York: Palgrave Macmillan, 1995).

45 Eastern Region, *Annual Report of the Agricultural Department, 1952–53*, 1.

46 Eastern Region, *Annual Report of the Agricultural Department, 1956–57*, 3.

47 See Eastern Region, *Annual Report, Agriculture Division, 1958/59*.

48 NAE, ARODIV, 19/1/18, "Quarterly Reports, Aro District," Agricultural Station Arochukwu to Agricultural Officer, Abak, 19 December 1952.

49 For report from various agricultural divisions, see NAE, LD 51-ESIALA, 27/1/53, "Matter relating to agricultural loans."

50 Ibid.

51 See C. K. Laurent, *Investment in Nigerian Tree Crops: Smallholder Production* (Ibadan: NISER, University of Ibadan, 1968), 2 and 11, Rigobert Oladiran Ladipo, "Nigeria and Ivory Coast: Commercial and Export Crops since 1960," in *African Agriculture: The Critical Choice: Studies in African Political Economy*, ed. Hamid Ait Amara and Bernard Founou-Tchuigoua, 101–20 (London: Zed Books, 1990).

52 Allan McPhee, *The Economic Revolution in British West Africa* (London: Frank Cass), 8.

53 Ibid. See also NAE, ONPROF, 1-1-111144, "Report on Oil Palm Survey, Ibo, Ibibio and Cross River Areas," by A.F.B. Bridges, 1938; Imperial Institute Handbook, *The Agricultural and Forest Products of British West Africa* (London, 1922).

54 Nigeria, *Second Annual Bulletin of the Agricultural Department*, 1923. See also Gray, "Native Methods of Preparing Palm Oil," 29.

55 See Lugard, *The Dual Mandate*, 268–69.

56 Kathleen M. Baker, *Agricultural Change in Nigeria: Case Studies in the Developing World* (London: John Murray, 1989), 49.

57 Baker, *Agricultural Change*, 22.

58 See George Dei, "Sustainable Development in the African Context: Revisiting Some Theoretical and Methodological Issues," *African Development* 18, no. 2 (1993), 97–110.

59 Nigeria, *Annual Report on the Agricultural Department 1938*, 27–30.

60 See Anyanwu, *The Igbo Family Life*, 200.

61 A.F.B. Bridges, "Reports on Oil Palm Survey, Ibo, Ibibio and Cross River Areas," 1938.

62 Usoro, *The Nigerian Oil Palm Industry*, 93.

63 Morgan, "Farming Practice," 332.

64 Ibid.

65 Ibid., 330.

66 Ibid., 331.

67 Eastern Nigeria, *Agricultural Division Annual Report, 1959/1960* (Enugu: Government Printer, 1961), 41–42.

68 Mba, *Nigerian Women Mobilized*, 75.

69 Martin, *Palm Oil and Protest*, 1988. See also Mba, *Nigerian Women Mobilized*.

70 See Ukaegbu, "Production," 233. See also Mba, *Nigerian Women Mobilized*, 106.

71 NAE, CALPROF 7/1/2339, file no. 4577, "Restlessness among the Annang Women." District Officer Opobo to The Senior Resident, Calabar Province, 27 February 1952. For the threat on the traditional rights of women with the introduction of new technology, see Margery Perham, ed., *Native Economies of Nigeria* (London: Faber and Faber, 1946), 229.

72 NAE, CALPROF, 7/1/2339, file no. 4577, "Agenda from Annag Women Association to be discussed with ADO, Opobo, 2 September, 1952," E. S. James, Assistant District Officer to the District Officer, 10 February 1952.

73 Margaret M. Green, *Igbo Village Affairs: Chiefly with Reference to the Village of Umueke Agbaja* (London: Frank Cass, 1964).

74 Interview with Christopher Chidomere, Mbaise, 13 December 1998.

75 Susan Martin, "Slaves, Igbo Women and Oil Palm," in *From Slavery to 'Legitimate' Commerce: The Commercial Transition in Nineteenth Century West Africa*, ed. Robin Law, 182 (Cambridge: Cambridge University Press, 1995).

76 For a useful critique of the broad and generalized framework on understanding the role of gender in economic change, see Margaret P. Stone, "Women, Work and Marriage: A Restudy of the Nigerian Kofyar" (PhD dissertation, Department of Anthropology, University of Arizona, 1988), 16.

77 See, for example, Ukaegbu, "Production in the Nigerian," 31–36. Susan Martin extended Ukaegbu's argument in the following articles, "Gender and Innovation," 411–22, and "Igbo Women and Palm oil," 180.

78 Ukaegbu, "Production in the Nigerian," 31–36.

79 Sara Berry, *Cocoa, Custom, and Socioeconomic Change in Rural Western Nigeria* (Oxford: Clarendon Press, 1975). Polly Hill's study of the development of rural capitalism in Ghana may also be noted. These changes in the direction of capitalist societies were largely driven by the market rather than ideology, although they ultimately transformed social structures including gender. See also Polly Hill, *The Migrant Cocoa-farmers of Southern Ghana: A Study in Rural Capitalism* (Hamburg: LIT, James Currey with the IAI, 1997).

80 NAE, MINLOC, 6/1/175-EP 8840a, "Intelligence Report on Ekwereazu and Ahiara Clans," 24.

81 A. M. Iheaturu, interview with Andrew Anyanwu, aged 80, Ogbe Ahiara, 30 August 1972 and 16 December 1972, transcribed in Isichei, *Igbo Worlds*, 81.

82 NAE, RIVPROF, 8/5/661, file no. OW 630/17, "Trade Prices at Up Country Markets, 1917," D. O. Okigwe to Resident Owerri Province, Port Harcourt.

83 See Felicia Ekejuba's biographical sketch of Omu Okwei, "Omu Okwei – The Merchant Queen of Ossomari: A Biographical Sketch," *Journal of the Historical Society of Nigeria* 3, no. 4 (1967): 633–46. See Leith Ross, *African Woman*, wherein Ruth is described as "the veritable Amazon among traders," 343.

84 Mba, *Nigerian Women Mobilized*, 47.

85 Ibid.

86 Interview with Eliazer Ihediwa, Owerrenta, 24 July 1999.

87 Interview with Linus Anabalam, Mbaise, 13 December 1998.

88 Interview with Nwanyiafo Obasi, Umunomo, Mbaise, 25 July 1999.

89 Interview with E. Ihediwa, Owerrenta, 24 July 1999.

90 Interview with Serah Emenike, Owerri, 22 December 1999.

91 Interview with Francis Eneremadu, 31 December 1999.

92 Interview with Christiana Marizu, Nguru Mbaise, 25 December 1999.

93 Women in various parts of the Eastern Region protested men's participation in what they regarded as women's spheres and often asked colonial officials to intervene on their behalf.

94 Klein, *Peasants in Africa*, 20.

95 Hart, *The Political Economy*, 97–98.

96 W. B. Morgan, "Farming Practice, Settlement Pattern and Population Density in Eastern Nigeria," *Geographical Journal* 121, no. 3 (Sept. 1955): 330.

97 See Mba, *Nigerian Women Mobilized*, 112–14.

98 See *Eastern Region: Annual Report for the Department of Agriculture for 1953/54*.

99 Federation of Nigeria, *Annual Abstract of Statistics, 1960* (Lagos: Federal Government Printer, 1960), 4. See also H. I. Ajaegbu, *Urban and Rural Development in Nigeria* (London: Heinemann, 1976), 32.

4 PEASANTS, DEPRESSION, AND RURAL REVOLTS

1   James C. Scott, *Weapons of the Weak: Everyday Forms of Peasant Resistance* (New Haven, CT: Yale University Press, 1987), xvii.

2   Ibid.

3   James C. Scott, *The Moral Economy of Peasant: Rebellion and Subsistence in Southeast Asia* (New Haven, CT: Yale University Press, 1977).

4   Goran Hyden, *No Shortcut to Progress: African Development Management in Perspective* (Berkeley: University of California Press, 1983), 128.

5   Paul Richards, "To Fight or to Farm? Agrarian Dimensions of the Mano River Conflicts (Liberia and Sierra Leone)," *African Affairs* 104, no. 417 (2005): 571–90.

6   Osumaka Likaka, "Rural Protest: The Mbole against the Belgian Rule, 1894–1959," *International Journal of African Historical Studies* 27, no. 3 (1994): 589–617.

7   On the emergence of "peasant intellectuals," to use Feierman's phrase, see Steven Feierman, *Peasant Intellectuals: Anthropology and History in Tanzania* (Madison: University of Wisconsin Press, 1990), 18.

8   On social movements and the link to economic crisis in rural Africa, see for, example, David M. Rosen, "The Peasant Context of Feminist Revolt in West Africa," *Anthropological Quarterly* 56, no. 1 (Jan. 1983): 35–43.

9   NAE, UMPROF, 1/5/2, file no. C.53/929, vol. I, part 2, Resident Owerri to Secretary Southern Provinces, March 1930.

10 Ibid.

11 This was often carried out on a man who abuses his wife or commits other serious offences against the women of a village of the community. See Judith Van Allen, "'Aba Riots' or 'Igbo Women's War'? Ideology, Stratification, and Invisibility of Women," in *The Black Woman Cross-Culturally*, ed. F.C. Steady, 60 (Cambridge, MA: Schenkman, 1981); and "Sitting on a Man: Colonialism and the Lost Political Institutions of the Igbo," *Canadian Journal of African Studies* 6, no. 11 (1972): 178.

12 *Commission of Inquiry*, 96.

13 Leith-Ross, *African Women*.

14 See, for example, Margery Perham, *Native Administration in Nigeria* (London: Oxford University Press, 1937), 214; Harry A. Gailey, *The Road to Aba: A Study of British Administrative Policy in Eastern Nigeria* (London: University of London Press, 1970); James S. Coleman, *Nigeria: Background to Nationalism* (Berkeley: University of California Press, 1960); and U. C. Onwuteaka, "The Aba Riot of 1929 and its Relation to the System of 'Indirect Rule,'" *Nigerian Journal of Economic and Social Studies* 7 (1965): 273–82. For a general anthropological overview of Igbo Women, see Sylvia Leith-Ross, *African Women* (London: Routledge & Kegan Paul, 1965).

15 For feminist perspectives, see, for example, Susan Rogers, "Anti-Colonial Protest in Africa: A Female Strategy Reconsidered," *Heresies* 9, no. 3 (1980): 22–25. See also Hanna Judith Lynne, "Dance, Protest, and Women's Wars: Cases from Nigeria and the United States," in *Women and Social Protest*, ed. Guida West and Rhoda Lois Blumberg, 333–45 (New York: Oxford University Press, 1990). For more feminist perspectives and analysis of the 1929 women's protest, see Caroline Ifeka-Moller, "Female Militancy and Colonial Revolt," 127–57; Shirley Ardener, "Sexual Insult and Female Militancy," *Man* 8 (1973): 422–40; and Sylvia Tamale, "Taking the Beast by its Horns: Formal Resistance to Women's Oppression in Africa," *African Development* 21, no. 4 (1996): 5–21.

16 See Van Allen, "Sitting on a Man," for further explanation of the act of "sitting on a man." See also Judith Van Allen, "'Aba Riots' or Igbo 'Women's War'? Ideology, Stratification, and Invisibility of Women," in *Women in Africa: Studies in Social and Economic Change*, ed. Nancy Hafkin and Edna Bay, 59–86 (Stanford: Stanford University Press, 1976); Caroline Ifeka-Moller, "Female Militancy and Colonial Revolt: The Women's War of 1929, Eastern Nigeria" in *Perceiving Women*, ed. Shirley Ardener, 128–32 (New York: John Wiley

& Sons, 1975); and Nina Mba, "Heroines of the Women's War," in *Nigerian Women in Historical Perspective*, ed. B. Awe, 75–88 (Ibadan: Sankore/Bookcraft 1992).

17 Nkiru Nzegwu, "Confronting Racism: Toward the Formation of a Female-Identified Consciousness," *Canadian Journal for Women and the Law* 7, no. 1 (1994): 30.

18 Ibid., 20. See also Ifeka-Moller, "Female Militancy and Colonial Revolt," 128–32.

19 Aba Commission of Inquiry, 19.

20 NAE, AWDIST, 2/1/57, file no. 62/1925, "Anti-Government Propaganda in Abakiliki," District Officer Awgu to Senior Resident, Onithsa, 12 March 1926. See also NAE, ONPROF, 7/12/92, file no. 391/1925, J. C. Iwenofu to District Officer, Awgu, 3 November 1925.

21 See NAE, AWDIST, 2/1/57, file no. 62/1925 and NAE, OMPROF, 7/12/92, file no. 391/1925, "Reports on Women's Disturbances." See also Report of the Aba Commission of Inquiry, "Memorandum as to the Origins and Causes of the Recent Disturbances in the Owerri and Calabar Provinces," Appendix III (1), 11–12.

22 Aba Commission of Inquiry, 19.

23 NAE, ONPROF, 7/12/92, file no. 391/1925, "Bands of Women Dancers Preaching Ideas of Desirable Reforms," Senior Resident, Onitsha Province to District Officer, Awka, 9 November 1925.

24 See NAE, CSE, 3.17.15, file no. B 1544, (1925–1926), AW 80 Q, AW 2/1/57., Anti-Government Propaganda Women Dancers (1925). See also AWDIST, 2/1/56, AW 80, (1919–1920).

25 NAE, OWDIST, 4/13/70, file no. 91/27, "Assessment Report, 1927." "Assessment Report," District Officer Owerri to the Resident Owerri Province June 1928.

26 NAE, UMUPROF, 1/5/2, file no. C53/929, vol. 1, part 2, "Women's Movements – Aba Patrol."

27 See *Commission of Inquiry Notes of Evidence*, and Mba, *Nigerian Women Mobilized*, 74.

28 *Commission of Inquiry*, 33.

29 Ibid., 54.

30 Interview with Eleazer Ihediwa, Owerrenta, 24 July 1999.

31 See, for example, I. D. Talbott, *Agricultural Innovation in Colonial Africa: Kenya and the Great Depression* (Lewiston, NY: Edwin Mellen, 1990).

32 Raymond Gore Clough, *Oil Rivers Trader* (London: C. Hurst, 1972), 97.

33 Ibid.

34 This is usually a piece of paper issued to oil sellers, which they used to exchange their oil for various goods in European trading factories. Clough, *Oil Rivers*, 99.

35 Ibid.

36 For example, the United States Department of Agriculture created the Farm Security Administration (FSA) in 1937. The FSA and its predecessor, the Resettlement Administration (RA) were New Deal programs designed to assist poor farmers who suffered from the Dust Bowl and the Great Depression.

37 For more details, see Morgan, "Farming Practice," 332.

38 Basden, *Niger Ibos*, 337.

39 Ibid.

40 Interview with Linus Anabalam, Mbaise, 13 December 1998.

41 See Commission of Inquiry, Appendix III (1), 37. Figures obtained from the Supervising agent of the United African Company, Limited (Opobo).

42 Clough, *Oil Rivers Trader*, 98.

43 NAE, UMUPROF, 1/5/5, file no. C53/1929, "Women Movement Aba, Bende," E. S. Wright to District Officer, Umuahia, 2 January 1930. NB: Mbawsi price 1d per measure more than Umuahia; Uzuakoli price – 1 pence per measure less than Umuahia.

44 See *Notes of Evidence*.

45 NAE, UMUPROF, 1/5/3, cited in Mba, *Nigerian Women Mobilized*, 75.

46 NAE, MINLOC, 6/1/175-EP 8840A, "Intelligence Report on Ekwerazu and Ahiara Clans, Owerri Division," by G. I. Stockley, Assistant District Officer.

47 NAE, MINLOC, 6/1/215, file no. EP 10595A, "Intelligence Report on Obowo and Ihitte Clans."

48 NAE, MINLOC, 6/1/175-EP 8840A, "Intelligence Report."

49 NAE, MINLOC, 6/1/215-EP 10595A, "Intelligence Report on the Obowo and Ihitte Clans," by N.A.P.G. MacKenzie, 1933.

50 Claude Ake, *A Political Economy of Africa* (Essex, England: Longman Press, 1981), 333–34; Rodney, *How Europe*, 165. On the French experience, see Catherine Coquery-Vidrovitch, "French Colonization in Africa to 1920: Administration and Economic Development," in *Colonialism in Africa, 1870–1914, vol. 1: The History and Politics of Colonialism, 1870–1914,* ed. L. H. Gann and P. Duignan, 170–71 (Cambridge: Cambridge University Press, 1969).

51 Alice L. Conklin, *A Mission to Civilize: The Republican Idea of Empire in France and West Africa, 1895–1930* (Stanford, CA: Stanford University Press, 1997), 144.

52 RH, Mss Afr. s. 16, S. M. Jacob, "Report on the Taxation and Economics of Nigeria 1934."

53 Taxation was often seen as the only method of compelling Africans to enter the cash economy through employment or producing for the market. See Leslie Raymond Buell, *The Native Problem in Africa,* vol. 1 (New York: Macmillan, 1928), 331. Cited in Mathew Forstater, "Taxation: A Secret of Colonial Capitalist (So-Called) Primitive Accumulation" (Center for Full Employment and Price Stability Working Paper No. 25, May 2003), 8–9.

54 F. D. Lugard, "Lugards Political Memoranda: Taxation, Memo No. 5" [1906, 1918], in *The Principles of Native Administration in Nigeria: Selected Documents, 1900–1947,* ed. A. H. M. Kirk-Greene, 118, 132 (London: Oxford University Press, 1965).

55 F. D. Lugard, "Lugard's Political Testimony," [1922], in *The Principles of Native Administration in Nigeria: Selected Documents, 1900–1947,* ed. A. H. M. Kirk-Greene, 129–30 (London: Oxford University Press, 1965b).

56 Ibid.

57 Ibid., 173.

58 Report by The Hon. W.G.A. Ormsby-Gore, 115.

59 See *Commission of Inquiry,* 193.

60 PRO, CO, 583/159/12, "Introduction of Direct Taxation in Southern Provinces – Petition Regarding."

61 Ibid.

62 Ibid.

63 PRO, CO, 589/159/12, "Deputy Governor Baddeley to CMS Amery, Secretary of State for Colonies," 16 April 1928.

64 PRO, CO, 589/159/12, "Native Revenue Amendment Ordinance," W. Buchanan Smith, Resident Onitsha Province to Secretary of Southern Provinces, 20 January 1928.

65 Ibid.

66 PRO, CO, 589/159/12, W. Buchanan Smith, Resident Onitsha Province to Secretary of Southern Provinces, 20 January 1928.

67 PRO, CO, 589/159/12, Resident, Ogoja Province to Secretary Southern Provinces.

68 PRO, CO, 589/159/12, "Petition by Ezzi Chiefs Against taxation," R.H.J. Sasse to Secretary, Southern Provinces, Lagos, 11 March 1928.

69 Ibid.

70 PRO, CO, 583/159/12, "Introduction of Direct Taxation."

71 RH, Mss Afr. S. 1924, "District Officer: Memoir," A. E. Cooks papers.

72 PRO, CO, 589/159/12, "Native Revenue Amendment Ordinance," W. Buchanan Smith, Resident Onitsha Province to Secretary of Southern Provinces, 20 January 1928.

73 See *Commission of Inquiry,* 4.

74 Ibid., 8.

75 During the 1926 tax assessment, the people of Oloko and Ayabe had been told that the counting of persons was simply part of the census. Then taxation was introduced in 1928. On this occasion, the women felt

that the authorities could not be trusted. See Mba, *Nigerian Women Mobilized*, 76.

76 The Commission of Inquiry acknowledged this view. See *Commission of Inquiry*, 96.

77 Morgan, "Farming Practice," 331.

78 NAE, OWDIST, 4/13/70, file no. 91/27, "Cultivation of Crops, Owerri District," District Officer to Resident Owerri Province, June 1928.

79 *Commission of Inquiry*, 93.

80 Mba, *Nigerian Women Mobilized*, 76.

81 *Commission of Inquiry*, 12.

82 Interview with Nwanyiafo Obasi, Umunomo, Mbaise, 30 July 1999.

83 *Notes of Evidence*, 13.

84 Adiele Afigbo, "Revolution and Reaction in Eastern Nigeria, 1900–1929," *Journal of the Historical Society of Nigeria* 3, no. 3 (1966): 553.

85 Ibid.

86 PRO, CO, 583/176, "Native Unrest."

87 RH, Mss Afr. s. 16, S. M. Jacob, "Report on the Taxation and Economics of Nigeria 1934."

88 Ibid.

89 *West Africa*, "The Disturbances in S. E. Nigeria," 11 October 1930.

90 On Ghana, see, for example, Stanley Shaloff, "The Income Tax, Indirect Rule, and the Depression: The Gold Coast Riots of 1931," *Cahiers d'études africaines* 14, no. 54 (1974): 359–75. See also B. Jewsiewicki, "The Great Depression and the Making of the Colonial Economic System in the Belgian Congo," *African Economic History* 4 (Autumn, 1977): 153–76.

91 See *Commission of Inquiry*, 103.

92 Ibid., Appendix III (1), 32.

93 Nwanyeruwa of Oloko, *Notes of Evidence*, 24–30.

94 Ibid.

95 Onuoha, *The Land and People*, 18.

96 *Notes of Evidence*, 114.

97 NAE, UMUPROF, 1/5/11, file no. C. 53/1929/vol. X, "Assault on Customary heads and Court members during recent disturbances," the Resident, Owerri Province to the Honourable, the Secretary, Southern Provinces, 12 February 1930.

98 Ibid.

99 *Notes of Evidence*, 98.

100 Feierman, *Peasant Intellectuals*, 103.

101 *Notes of Evidence*, 57. The evidence by administrative officers, police officers, missionaries, and others of many years' experience in the area indicates that local officials were very corrupt. This was not an issue for women alone. Many male witnesses strongly raised the issue of corruption by the native courts and warrant chiefs.

102 PRO, CO, 583/176, "Native Unrest," Correspondence arising out of the Report of the Aba Commission.

103 RH, Mss Afr. s. 1000, Edward Morris Falk Papers,

104 NAE, UMPROF, 1/5/1, "Women's Movement, Aba."

105 NAE, UMUPROF, 1/5/2, file no. C. 53/929, vol. 1, part 2, "Women's Movement – Aba Patrol." J. Cook. A.D.O., Bende to the Resident Owerri Province, 28 November 1929.

106 Clough, *Oil Rivers*, 110–11.

107 Van Allen, "Sitting on a Man," 174.

108 NAE UMPROF, 1/5/21, file no. C.53/929, vol. XXI. "Women's Movement."

109 Ibid.

110 PRO, CO, 583/176/9 "Native Unrest in Calabar and Owerri Provinces: Correspondence Arising from, 1930."

111 Feierman, *Peasant Intellectuals*, 3.

112 *Notes of Evidence*, 57.

113 NAE, UMUPROF, 1/5/11, file no. C. 53/1929, vol. X, "Disturbances – South Eastern Province," The District Officer, Owerri to the Resident, Owerri Province, 21 February 1930.

114 *Commission of Inquiry*, 260.

115 PRO, CO, 583/176, "Native Unrest."

116 *Commission of Inquiry*, 19.

117 Interview with Onyegbule Korieh. Ihitteafoukwu, Mbaise, 17 December 1998.

118 Clifton E. Marsh, "A Socio-Historical Analysis of the Labor Revolt of 1878 in the Danish West Indies," *Phylon* 42, no. 4 (1981): 335–45.

119 See *Report of the Commission of enquiry appointed to inquire into the disturbance in the Calabar and Owerri Provinces, December 1929* (Sessional Paper No. 28), hereafter (*Commission of Inquiry*). See also *Notes of Evidence Taken in the Calabar and Owerri Provinces on the Disturbance 1930*, 8 vols. *Minutes of Evidence Taken at Owerri 1930*, *Minutes of Evidence Taken at Opobo 1930*.

120 NAE, UMUPROF, 1/5/11, file no. C.53/1929/vol. X, "List of Chiefs suspended during the recent Disturbances," District Officer Owerri to The Resident, Owerri Province, 24 April 1930.

121 Ibid.

122 Mba, *Nigerian Women Mobilized*, 96.

123 Oriji, "Igbo Women."

124 RH, Mss Afr. s. 546, F.B Carr papers.

125 Ibid.

126 Ibid.

127 NAE, UMPROF, 1/5/24, file no. C. 53/1929, vol. 26, "Situation: Owerri Province," C. H. Ward, A.C.P, Okpala, 22 January 1931.

128 RH, Mss Afr. s. 546, F. B. Carr papers.

129 Ibid.

130 Ibid.

131 See NAE, UMUPROF, 1/5/4, file no. C 53/1929/vol. 26, Women Movement, Aba Patrol Report to SSP Part II 1/3/30, Resident Owerri to Secretary, Southern Provinces, 14 November 1930. The residents report were based on confidential memos from reports on the recent unrest in the Orlu District of the Okigwi Division by the District Officer, Mr. Homfray; by the acting Resident, Mr. Cochrane; by the Deputy Superintending Inspector of Produce Mr. Sabiston.

132 Ibid.

133 Ibid.

134 Ibid.

135 PRO, CO, 583/193/8, "Palm Oil Industry," E. Beddington to Governor of Nigeria, 17 May 1933.

136 PRO, CO, 583/242/22, "Disturbance in Okigwi," Colonial Governor to Secretary of State for the Colonies, 26 January 1939.

137 PRO, CO, 583/242/22, "Disturbance in Okigwi."

138 Mba, *Nigerian Women Mobilized*, 98.

139 Ibid., 99.

140 Ibid., 107.

141 Interview with Chief Francis Eneremadu.

142 NAE, UMUPROF, 1/5/4, file no. C 53/1929/vol. 26.

5 THE SECOND WORLD WAR, THE RURAL ECONOMY, AND AFRICANS

1 RH, Mss Afr. s. 546, F. B. Carr papers.

2 Ibid.

3 Ibid.

4 See NAE, OP 122II-ONDIST, 12/1/90, Matters relating to the effects and implications of war conditions on Nigeria. See also Nigeria, *Annual Report of the Agricultural Department*, especially 1939/1940.

5 Nigeria, *Annual Report on the Agricultural Department, 1938*.

6 RH, Afr. s. 546, F. B Carr papers.

7 See NAE, ABADIST, 1/2/908, file no. 1642/11, "Memos by D.O. Aba to Ngwa Native Authority, 1940–44," Sir F. Stockdale, *Report of the Mission Appointed to Enquire into the Production and Transport of Vegetable Oils and Oil Seeds Produced in the West African Colonies* (London: H.M.S.O., 1947), Appendix X and XI. See also Michael Crowder, *West Africa Under Colonial Rule* (London: Hutchinson, 1973), 490–98.

8 NAE, EKETDIST, 1/2/50, file no. 499, "Food Production in Nigeria, 1935–1951," 7.

9 Nigeria, *Annual Report of the Agricultural Department, 1938*.

10 RH, Afr. s. 546, F. B. Carr papers.

11 Nwabughuogu, "British War-Time Demands," 8–9.

12 NAE, ABADIST, 1/26/907, file no. 1642, "Palm Produce Production," Chief Secretary to the Government.

13 Ibid.

14 Ibid.

15 NAE, CSE, 1/85/9929, file no. 19947, "Conference of Production Drive Team," L. T. Chubb to Deputy Controller of Palm Produce, Umuahia, 17 October 1943.

16 RH, Afr. s. 546, F. B. Carr papers.

17 Nigeria, *Annual Report of the Agricultural Department Report, 1938*, 2.

18 Nigeria, *Annual Report of the Agricultural Department Report, 1941*, 1.

19 FAO, *Year Book of Agriculture Statistics, Trade*, vol. VII, part 2, 1954 (Rome, 1955), 170–71.

20 See K. Brandt, *The German Fats Plan and the Economic Setting* (Food Research Institute: Stanford University, 1938), 221–71. See also Nzeribe, *Economic Development*, 44.

21 Nzeribe, *Economic Development*, 44–54.

22 Nigeria, *Annual Report of the Agricultural Department Report, 1940*, 2.

23 Ibid.

24 Nigeria, *Annual Report of the Agricultural Department Report, 1941*, 2.

25 RH, Mss. Afr. s. 823 (4), J. R. Mackie Papers.

26 NAE, CSE, 1/85/8621, file no. 18038/70, vol. II, "Production, Onitsha Province," Kernels Production Officer to Deputy Controller of Kernels, Eastern Zone, 14 June 1943.

27 Interview with Eleazer Ihediwa Owerrenta, 24 July 1999, aged c. 71.

28 NAE, ONDIST, 12.1/104, file no. OP 130, "Palm oil Production," Resident Onitsha Province to The Secretary, Eastern Provinces, Enugu, 25 November 1939.

29 Ibid.

30 David Anderson and David Throup, "Africans and Agricultural Production in Colonial Kenya: The Myth of the War as a Watershed," *Journal of African History* 26 (1985): 327. I have addressed aspects of the impact of the war on Nigerian urban population. See Chima J. Korieh, "Urban Food Supply and Vulnerability in Nigeria during the Second World War," in *Nigeria Cities*, ed. Toyin Falola and Steven J. Salm, 127–52 (Trenton, NJ: Africa World Press, 2003).

31 NAE, ABADIST, 1/26/907, file no. 1642, "Palm Produce Production," and P. L. Allpress to Resident, Owerri Province, 4 October 1945.

32 Legislative Council Debates, 13 March 1944, 89, cited in Esse, "The Second World War," 176.

33 See NAE, OBUDIST, 4/1/309, file no. OB 699/vol. II, "Produce Drive: Kernel and Rubber Return Prosecutions."

34 NAE, OBUDIST, 4/1/309, file no. OB 699/vol. II, "Prosecution under palm kernels and rubber Regulation." The cases here were from December 1943 to July 1944.

35 *West African Pilot*, "Our War Production," 23 February 1942.

36 Deborah Bryceson, "Household, Hoe and Nation: Development Politics of the Nyerere Era," in *Tanzania after Nyerere*, ed. Michael Hodd, 39 (London and New York: Pinter, 1989).

37 Nigeria, *Annual Report on the Agricultural Department, 1939–40*, 1.

38 NAE, ONDIST, 12/1/90, file no. OP 122 II, and Nigeria, *Report of the Agricultural Department, 1939–1940*, 1. See also Bernard Bourdillon, *Legislative Council Debate*, 4 December 1939 (Lagos: Government Printer, 1940), 4–13. See also Legislative Council Debates, 17 March 1941, 11.

39 NAE, EKETDIST, 1/2/50, "Food Production in Nigeria, 1935–1951."

40 NAE, RIVPROF, 8/5/430, "Policy of the Agricultural Department," Circular Memo D. A. 14/252, J.R. Mackie to the Chief Secretary to the Government, 26 October 1939.

41 Nigeria, *Annual Report of the Agricultural Department, 1938*. See also Bernard Bourdillon, "Legislative Council Debate, 4 Dec. 1939" (Lagos, 1940), 4–13, "Legislative Council Debates, 17 March 1941," (Lagos, 1942), 11. See also Ayodeji Olukoju, "The Faulkner 'Blueprint' and the Evolution of Agricultural Policy in Inter-War Colonial Nigeria," in *The Foundations of Nigeria: Essays in Honor of Toyin Falola*, ed. Adebayo Oyebade, 403–22 (Trenton, NJ: Africa World Press, 2003).

42 NAE, ABADIST, 1/26/907, file no. 1642, "Palm Produce Production," J.A.G. McCall, Controller of Oil Palm Production, Owerri Province to District Officers, Owerri Province, 3/2/44; Ogoja Province: *Annual Report, 1943*, 8.

43 Esse, "The Second World War," 157–58.

44 Nigeria, *Annual Reports on the Agricultural Department, 1940*, 1, 46–47.

45 RH, Mss Afr. s. 1779, Norman Herington papers.

46 Esse, "The Second World War," 175.

47 Nigeria, *Annual Report of the Agricultural Department, 1942*, 2.

48 NAE, EKETDIST, 1/2/50, file no. 499, "Food Production," Memo: Agricultural Officer Abak to D.O. Calabar, 12 July 1940.

49 NAE, ONDIST, 42/1/1264, file no. OP. 1865/vol. VI, "Annual Report, Onitsha Province," 1943, 65.

50 J. S. Harris, "Some Aspects of the Economics of Sixteen Ibo Individuals," *Africa* 14 (1944): 303.

51 NAE, ABADIST, 14/1/872 file no. 1647.

52 NAE, ONDIST, 12//92, file OP IV, "Food Control," S.A.S. Leslie, Nigerian Secretariat, Lagos, 22 April 1941.

53 Ibid.

54 NAE, ABADIST, 1/26/958, J.V. Dewhurst to the Resident, Owerri Province, Port Harcourt, 12 August 1943.

55 NAE, AIDIST, 2/1/433, file no. IK: 401/18, "Food Control," The Acting District Officer, Ikom to the District Officer, Abakiliki, 18 June 1945.

56 NAE, AIDIST, 2/1/433, file no. AB: 1373/11, "Memo" to District Officer, Abakiliki Division to C. N. C. Igbeagu, 1945.

57 NAE AIDIST, 2/1/433, file no. OG: 2920/140, "Food Control," P. M. Riley, Resident Ogoja Province to District Officers.

58 NAE, AIDIST, 2/1/433, "Nigerian General Defence regulation Order: Gari and Yams," Resident Onitsha Province to District Officer and other Competent Authorities, 16 May 1945.

59 NAE, AIDIST, 2/1/433, "Nigerian General Defence." For cassava restrictions, see also NAE, AID 2/1/433, OG: 2513/1265, "Nigeria General Defence Regulations: Order," P. M. Riley, Resident Ogoja Province to District Officers, 26 June 1945.

60 *West African Pilot*, 23 February 1944.

61 NAE, ABADIST, 14/1/875, file no. 1646, vol. IV, "Gari Control," The Nigerian Police, Aba to District Officer, Aba, 13 May 1944.

62 National Archives Ibadan (hereafter NAI) CSO, 26, file no. 36378. Cited in Wale Oyemakinde, "The Pullen Marketing Scheme: A Trial in Food Price Control in Nigeria, 1941–1947," *Journal of the Historical Society of Nigeria* 6, no. 4 (1973): 413.

63 For more on food control during the Second World War, see Akanmu Adebayo, "The Fulani, Dairy Resources, and Colonial Animal Products Development Programs," in *Nigeria in the Twentieth Century*, ed. Toyin Falola, 201–24 (Durham, NC: Carolina Academic Press, 2002); and Toyin Falola, "'Salt is Gold': The Management of Salt Scarcity in Nigeria During World War II," *Canadian Journal of African Studies* 26, no. 3 (1992): 417.

64 Nigeria, *Annual Report of the Agricultural Department Report, 1940*, 2.

65 NAE, ABADIST, 14/1/875, file no. 1646, vol. IV, Clerk of Omuma Native Court to District Officer Aba Division, 11 November 1944.

66 Calabar Provincial Office: 7/1/1329, "Unrest among the Women of Ikot Ekpene," cited in Mba, *Nigeria Women Mobilized*, 103.

67 NAE, ONDIST, 13/1/2, file no. EP OPC 122, vol. VII, "Food Control," Onitsha Rice and Gari Traders to Resident Onitsha Province, 1 July 1942.

68 NAE, ONDIST, 13/1/2, file no. EP OPC 122, vol. VII, "Doninic Ezenwa and 20 others representing garri market traders to Resident Onitsha Province," 17 July 1942.

69 See, for example, Michael Crowder, "World War II and Africa: Introduction," *Journal of African History* 26 (1985): 287–88; Killingray and Rathbone, ed., *Africa and the Second World War*. For Nigeria, see Toyin Falola, "Cassava Starch for Export in Nigeria during the Second World War," *Journal of African Economic History* 18 (1989): 73–98; Falola, "Salt is Gold," 417; and A. Olorunfemi, "Effects of War-Time Trade Control on Nigerian Cocoa Traders and Producers, 1939–1945: A Case Study of the Hazards of a Dependent Economy," *International Journal of African Historical Studies* 13, no. 4 (1980): 672–87.

70 Many aspects of the Second World War as a potent force for economic, social, and political change in Africa have been the subject of detailed monographs and articles. See, for example, Michael Crowder, "World War II and Africa: Introduction," *Journal of African History* 26 (1985): 287–88; David Killingray and Richard Rathbone, ed., *Africa and the Second World War* (London: Macmillan, 1986). Most of the early literature, however, focused on the impact of the war on nationalism, as well as local contribution of personnel to the war effort. See, for example, F.A.S. Clark, "The Development of the West African Forces in the Second World War," *Army Quarterly* (1947): 58–72; Michael Crowder, "The Second World War: Prelude to Decolonization in Africa," in *History of West Africa, II*, ed. J. F. Ade Ajayi and M. Crowder (London: Longman, 1974); Trevor R. Kerslake, *Time and the Hour: Nigeria, East Africa, and the Second World War* (London: Radcliffe Press, 1997); and G. O. Olusanya, *The Second World War and Politics in Nigeria, 1939–1945* (Lagos: University of Lagos Press, 1973). Exceptions include Toyin Falola's detailed account of the salt crisis during World War

II and the response of the colonial authorities in Nigeria to what they saw a potential source of discontent and anti-colonial sentiment. See Toyin Falola, "Cassava Starch for Export in Nigeria during the Second World War," *Journal of African Economic History* 18 (1989): 73–98; and A. Olorunfemi, "Effects of War-Time Trade Control on Nigerian Cocoa Traders and Producers, 1939–1945: A Case Study of the Hazards of a Dependent Economy," *International Journal of African Historical Studies* 13, no. 4 (1980): 672–87.

71 NAE, ABADIST, 14/1/872, file no. 1646, "Gari Control," O. O. Muoma to District Officer, Aba, 1 July 1943.

72 Ibid., J. O. Okorocha to District Officer, Aba, 7 July 1943.

73 NAE, AIDIST, 2/1/433, 450, "G. I. Udeh to District Officer, Abakiliki," 19 May 1945.

74 Ibid.

75 NAE, ABADIST, 1/26/958, file no. 668, J. E. Akajiofo to District Officer, Aba, 1 September 1943.

76 NAE, ABADIST, 1/26/958, file no. 668, "Application for Grant Export of Yams under Permit," E. M. Eze, Trader to District Officer Aba, 13 August 1943.

77 Ibid.

78 NAE, ABADIST, 14/1/873, file no 1646, "Gari Control," A. Jamola to D.O., Aba District, 21 July 1943.

79 NAE, ABADIST, 14/1/872, file no. 1646, "Gari: Control" David H. Kubiri to D.O., Aba, 5 July 1943.

80 The League represented the majority of unions, ethnic groups and communities in Aba Township. See NAE, ABADIST, 1/26/958, file no. 668, "Foodstuffs: Yams, Plantains, Cocoyams, etc, Requested Prohibition of Railment or Exportation of in Future," Honorary Secretary, Aba Community League to the District Officer, Aba, 2 August 1943.

81 On employment and average rates of wages for various districts in Eastern Nigeria, see, for example, NAE, RIVPROF 9/1/135, "Annual Report on the Social and Economic Progress of the People of Nigeria," for various years.

82   NAE, RIVPROF, 9/1/135, file no. Ow: 1636, "Annual Report on the Social and Economic Progress of the People of Nigeria for the Year 1933," District Officer, Okigwi to Resident, Owerri Province, 5 December 1933.

83   NAE, ABADIST, 14/1/873, file no. 1646, "A Resolution," Gari Traders association, Aba to the Resident, Owerri Province, 29 July 1943.

84   Ibid.

85   Ibid.

86   NAE, ABADIST, 1/26/958, file no. 668, "Unauthorised Markets outside the Township," Secretary Aba Community League to District Officer, Aba, 2 August 1943.

87   NAE, ABADIST, 14/1/875, file no. 1646, vol. IV, "Export of gari to the North by gari traders will stop on 1 September," Nigerian Eastern Mail Press Representative to the District Officer, Aba, 29 August 1944.

88   Harris, "Some Aspects of the Economics," 303.

89   Ibid.

90   NAE, ABADIST, 14/1/875, file no. 1646, vol. IV, "Gari Control," District Traffic Superintendent, Port Harcourt to the Local Authority, Aba, 15 July 1944.

91   Ibid.

92   See NAE, AIDIST, 2/1/433, "Food Control."

93   NAE, ABADIST, 14/1/875, file no. 1646 vol. IV, "Gari Control," H.L.M. Butcher to The Councils and Court Clerks, Aba Division, 2 September, 1944.

94   Ibid.

95   NAE, ABADIST, 14/1/874, file no. 1646 vol. IV, "Gari Control."

96   NAE, ABADIST, 14/1/875, file no. 1646 vol. IV, "Gari Control."

97   NAE, ABADIST, 14/1/872, file no. 1646, "Gari: Control of," S.O. Enyiomah and others to District Officer, Aba, 28 June 1943.

98   Ibid.

99   NAE, ABADIST, 14/1/872, file no. 1646, "Gari: Control of," Agnes Garuba to District Officer Aba, 22 July 1943.

100   NAE, ABADIST, 14/1/872, file no. 1646, "Gari: Control of," C. O. Muoma to District Officer, Aba, 11 July 1943.

101   See, for instance, British Central Office of Information, *Constitutional Development in the Commonwealth: United Kingdom Dependencies* (London: British Central Office of Information, 1955), Part 1.

102   See Shenton, "Nigerian Agriculture," 48.

103   On British policy in the post-war period, see, for example, S.A.H. Haqqi, *The Colonial Policy of the Labor Government, 1945-51* (Aligarh: Muslim University, 1960), 128; and David Goldsworthy, *Colonial Issues in British Politics 1945–1961* (London: Oxford University Press, 1971), 12.

104   NAE, ONDIST, 12/1/1737, file no. OP 2611, "Credit Facilities for farmers, peasant industries and demobilized soldiers," L. T. Chubb, to the Chief Secretary to the Government, Lagos, 12 October 1945.

105   See NAE, C.S.E., 1/85/8584, file no. 18038/38, vol. VII, "Secretary, Eastern Provinces to Deputy Controller of Motor Transport, Aba," October, 1946.

106   There was the 1929 Women's War in Eastern Nigeria, the West African cocoa hold-ups and riots in the West Indies. See Rod Alence, "Colonial Government, Social Conflict and State Involvement in Africa's Open Economies: The Origins of the Ghana Cocoa Marketing Boards, 1939–46," *Journal of African History* 42 (2001): 398.

107   Ibid. See also J. M. Lee and Martin Petter, *The Colonial Office, War and Development Planning: Organisation and the Planning of a Metropolitan Initiative, 1939–45* (London: Maurice Temple Smith, 1982); and D. J. Morgan, *The Official History of Colonial Development, vol. 1: The Origins of British Aid Policy, 1924–45* (Atlantic Highlands, NJ: Humanities Press, 1980).

108   Alence, "Colonial Government," 398. See also David Meredith, "The Colonial Office, British Business Interests and the Reform of Cocoa Marketing in West Africa, 1937–

45," *Journal of African History* 29 (1988): 285–300; David Fieldhouse, "War and the Origin of the Gold Coast Cocoa Marketing Board, 1939–40," in *Imperialism, the State and the Third World*, ed. Michael Twaddle, 153–82 (London: British Academic Press, 1992); and David Fieldhouse, *Merchant Capital and Economic Decolonisation: The United African Company, 1929–87* (Oxford: Oxford University Press, 1994).

109 For studies of the Marketing Boards, see, for example, H.M.A. Onitiri and D. Olatunbosun, ed., *The Marketing Board System: Proceedings of an International Conference* (Ibadan, Nigeria: Nigerian Institute of Social and Economic Research, 1974); P. T. Bauer, "Statistics of Statutory Marketing in West Africa, 1939–51," *Journal of the Royal Statistical Society* 117 (1954): 1–30; and G. K. Helleiner, "The Fiscal Role of the Marketing Boards in Nigerian Economic Development, 1947–61," *Economic Journal* 74 (1964): 582–610.

110 See *Laws of Nigeria*, 1948–9 (Supplement), 237–86, cited in H. A. Oluwasanmi, *Agriculture and Nigerian Economic Development* (Ibadan and Oxford: Oxford University Press, 1966), 160.

111 On the role of the Marketing Boards, see Nigeria, *Handbook of Commerce and Industry in Nigeria* (Lagos, n.d.), 112.

112 For the conditions for approval as a buying agent, see, for example, Nigeria Cocoa Marketing Board, *Annual Report, 1947–48* (Lagos, 1949) 30; Eastern Regional Marketing Board, *Annual Report, 1958*, 93–5; Nigeria Groundnut Marketing Board, *Annual Report*, 1949–50, 29; and Northern Regional Marketing Board, *Annual Report, 1954–55*, 86–87.

113 NAE, ABADIST, 1/26/908, file no. 1642/vol. II, "Secretary, Eastern Province to Resident, Owerri Province," 17 October 1945.

114 Oluwasanmi, *Agriculture and Nigerian Economic Development*, 163.

115 Ibid., 164.

116 See Eastern Nigeria, *Eastern Nigeria Marketing Board Annual Report*, 1960, 18. See also Oluwasanmi, *Agriculture and Nigerian Economic Development*, 166.

117 Wells, *Agricultural Policy*, 40, Note 5. See also *Statement of Future Marketing of West African Cocoa* (London, 1946), 8. See also Oil Palm Produce Marketing Board, "Statement of the Policy Proposed for the Future Marketing of Nigerian Oils, Oil seeds and Cotton," *Sessional Paper*, No. 18 of 1948 and Nigeria Oil Palm Produce Marketing Board, *First Annual Report 1949* (Lagos, 1950), 5.

118 See P. T. Bauer, "Origins of the Statutory Export Monopolies of British West Africa," *The Business History Review* (September, 1954): 197-213. See also Usoro, *The Nigerian Oil Palm Industry*, 111.

119 See Usoro, *The Nigerian Oil Palm Industry*, 74.

120 See, for example, C.E.F. Beer, *The Politics of Peasant Groups in Western Nigeria* (Ibadan: Ibadan University Press, 1976), chap. 2; Nigeria, *Annual Report of the Registrar of Co-operative Societies*, 1950/51 (Lagos: Government Printer, 1952), 3–4.

121 Interview with Gilbert Uzor, Umunomo, Mbaise, 22 July 2000.

122 Interview with Linus Anabalam, umuchieze. Mbaise, 14 December 1998.

123 See also NAE, MINLOC, 6/1/215, file no. EP 10595A. N.A.P.G. MacKenzie ADO, "Intelligence Report of Obowo and Ihitte Clans, Okigwe Division, Owerri Province."

124 Interview with Anex Ibeh, Umunomo, Mbaise, 17 December 1998.

125 Interview with Michael Iheaguta, umuchieze, Mbaise, 2 August 1998.

126 Throughout African, the populations of major cities increased substantially. Freund, *The Making of Contemporary Africa*, 169–170. See also Marvin Miracle, *Maize in Tropical Africa* (Madison: University of Wisconsin Press, 1966), 34–35.

127 NAE, C.S.E., 1/85/8584, fine no. 18038/38, vol. VII, Resident, Onitsha Province to The Secretary, Eastern Provinces, 18 November 1945.

128 Interview with Chief Francis Eneremadu, Mbutu, Mbaise, 2 January 2000.

129 For the study of rate of migration in Igbo-land, see, for example, Ikenna Nzimiro, "A Study of Mobility among the Ibos of Southern Nigeria," *International Journal of Comparative Sociology* 6, no. 1 (1965): 117–30.

130 Interview with Serah Emenike, Owerri, 22 December 1998.

131 Interview with Christina Marizu, Nguru, Mbaise, 25 December 1999.

132 Interview with Chief Francis Eneremadu, Mbutu, Mbaise, 2 January 2000.

6 THE AFRICAN ELITE, AGRARIAN REVOLUTION, AND SOCIO-POLITICAL CHANGE, 1954–80

1 The 1954 Lyttelton Constitution introduced a federal system of government in Nigeria. On the role of the Eastern Region government in agriculture from this period, see Nigeria, *Annual Report on the Department of Agriculture (Central), 1953–1954* (Lagos: Government Printer), 1. See also Eastern Region, *Annual Report: Department of Agriculture, 1954–55* (Enugu: Government Printer, 1955), 1.

2 Food and Agricultural Organization, *Agricultural Development in Nigeria*, 3.

3 For the implementation of this policy in the Eastern Region, see *Annual Report of Agriculture, 1954–55*. See also Eastern Nigeria, *Development Programme, 1958–1962, Official Document No. 2 of 1959* (Enugu: Government Printer, 1959); and *Report of the Economic Mission, 1961: Led by Dr. the Hon. M. I. Okpara, Premier, Eastern Nigeria, Official Document No. 5 of 1962* (Enugu: Government Printer, 1962), 3–8.

4 The ERDB replaced the Eastern Regional Production Development Board (ERPDB), which had coordinated the production and marketing of palm produce since 1949. For the operation of the ERPDB, see Nigeria Oil Palm Produce Marketing Or-dinance (No. 12 of 1949), revised in 1954. See also Toyin Falola, *Economic Reforms and Modernization in Nigeria* (Kent, OH: Kent State University Press, 2004), 122.

5 Eastern Region, *Annual Report on the Department of Agriculture, 1952–53*, 2.

6 Ibid., 1.

7 Eastern Region, *Annual Report for the Department of Agriculture, 1954–55*, 8.

8 Usoro, *The Nigerian Oil Palm Industry*, 88, 90.

9 Uchendu, *The Igbo*, and Mba, *Nigeria Women*, 75.

10 Eastern Nigeria, *Agricultural Division Annual Report, 1959/1960*, 41–42.

11 Usoro, *The Nigeria Oil Palm*, 88.

12 Eastern Region, *Fifth Annual Report of the Eastern Regional Production Board 1953/54* (Enugu, 1954), Appendix III.

13 *The Eastern Outlook and Cameroon Star*, "Oil Mill Production Reaches All-time High," 18 February 1954.

14 Eastern Region, *Annual Report, Agriculture Division, 1956–59*, 11–12.

15 Eastern Region, *Annual Report of the Eastern Regional Production Development Boards*, 1949/50, 4; Eastern Region, *Annual Report of the Eastern Regional Production Development Boards, 1951/52*, Part 1, 13–15; Eastern Region, *Annual Report of the Eastern Regional Production Development Boards, 1952/53*, 4; and Eastern Region, *Annual Report of the Eastern Regional Production Development Boards*, 1953/54, 13.

16 See, for example, A. I. Nwabughuogu, "Oil Mills Riots in Eastern Nigeria, 148–51: A Study in Indigenous Reaction to Technological Innovation," *African Development: A Quarterly Journal of the Council for the Development of Economic and Social Research in Africa* 7, no. 4 (1982): 66–84.

17 See Martin, "Gender, Oil palm and Protest."

18 Mba, *Nigeria Women Mobilized*, 106.

19 Interview with Chief Eneremadu.

20   A. Martin, *The Oil Palm Economy of the Ibibio Farmer* (Ibadan: Oxford University Press, 1956), 12.

21   S. I. Orewa, "Designing Agricultural Development Projects for the Small Scale Farmers: Some Lessons from the World Bank Assistance Small Holder Oil Palm Development Scheme in Nigeria," *Journal of Applied Sciences* 8, no. 2 (2008): 295.

22   Nigeria, *National Development Plan: Progress Report 1964* (Lagos, 1964), 63. See also Malcolm J. Purvis, *Report on a Survey of the Oil Palm Rehabilitation Scheme in Eastern Nigeria - 1967* (CSNRD, Report No. 10, 1968), 2, for a review of the agricultural programs in the region.

23   See Eastern Nigeria, *Eastern Nigeria Development Plan, 1962–1968: Official Document No. 8* (Enugu: Government Printer, 1962).

24   Purvis, *Report on a Survey*, 2.

25   Hursh et al., *Innovation in Eastern Nigeria*, 22.

26   Interview with Philip S. Njoku, Nguru, Mbaise, 12 January 2000.

27   See Eastern Nigeria, *Eastern Nigeria Development Plan, 1962–1968* (Enugu: Government Printer, 1962), 2. See also Eastern Nigeria, *Development Programme, 1958–1962: Eastern Region Official Document No. 2 of 1959* (Enugu: Government Printer, 1959), 12. See also Purvis, *Report on a Survey*, 16.

28   The Eastern Nigeria Development Corporation (ENDC) managed many commercial enterprises, such as cold stores, a soft drink factory, the Obudu cattle ranch, the state owned Progress Hotels, in addition to the government's large agricultural projects. See Eastern Nigeria, *The E.N.D.C. in the First Decade, 1955–1964* (Enugu: ENPC), n.d.

29   Floyd, *Eastern Nigeria*, 219.

30   Ibid.

31   For the location of different plantations, their sizes and labourers, see ibid., 219–20.

32   See, for example, Gyasi "State Expropriation."

33   Interview with former employee of the Emeabiam Rubber Estate.

34   Floyd, *Eastern Nigeria*, 221–222.

35   Ibid.

36   See M. K. Mba, *The First Three Years: A Report of the Eastern Nigeria Six-Year Development Plan* (Enugu: Government Printer 1965), 9. Cited in Floyd, *Eastern Nigeria*, 223.

37   Floyd, *Eastern Nigeria*, 219.

38   NAE, RIVPROF, 8/5/661, Registrar of Co-operatives Societies to Chief Secretary.

39   D. Elson, ed., *Male Bias in the Development Process* (Manchester and New York: Manchester University Press, 1995), 9.

40   Interview with Zebulon Ofurum, aged 68 years, Emeabiam, 28 May 2001.

41   Floyd, *Eastern Nigeria*, 216.

42   Ibid.

43   Ibid.

44   Floyd, *Eastern Nigeria*, 216–17.

45   Ibid., 213. For a historical analysis of the development of plantation in the region, see R. K. Udo, "Sixty Years of Plantation Agriculture in Southern Nigeria, 1902–1962," *Economic Geography* 12 (1965): 356–68.

46   See "The Agricultural Sample Survey for 1963–1964," cited in Hursh et al., *Innovation*, 19.

47   Purvis, *Report on a Survey*, 24.

48   Ibid.

49   Interview with Philip Njoku, Nguru, 12 January 2000.

50   Eastern Nigeria, *Report of the Economic Mission*, 1.

51   Government Press Conference, *ENIS Bulletin* No. E2,200 (Enugu, January, 1961). Cited in Barry Floyd and Monica Adinde, "Farm Settlements in Eastern Nigeria: A Geographical Appraisal," *Economic Geography* 43, no. 3 (1967): 189–230.

52   See, *Nigerian Spokesman*, 10 April 1964.

53   Eastern Nigeria, *Agricultural Extension Newsletter* (Enugu: Government Printer, 1963).

54 Eastern Nigeria, *Report of the Economic Mission*, 6.

55 Ibid.

56 For the Israeli model, see, for example, T. C. Yusev, *The Economics of Farm Settlements in Israel* (New York: Express Printers, 1963); and F. C. Gorman, *Social Relations in Israeli Farm Settlements* (Tel-Aviv: Zester and Rox, 1957).

57 Similar schemes had been established in the Western Region in 1959 to demonstrate that by careful planning, farms could be operated by young people to provide a comfortable standard of living comparable with or even higher than that gained by persons of the same status in other forms of employment. For a review of the Western Nigeria experiment, see O. Okediji, "Some Socio-cultural Problems in the Western Nigeria land Settlement Scheme: A Case Study," *Nigerian Journal of Economic and Social Studies* (1966): 301–10; W. Roider, *Farm Settlements for Socio-Economic Development: The Western Nigeria Case* (Munich: Weltforum, 1971); and O. F. Ayadi and C. O. Falusi, "The Social and Financial Implications of Farm Settlements in Nigeria," *Journal of Asian and African Studies* 31, nos. 3–4 (1994): 191–206.

58 For some of the foreign interests and international bodies, including the Food and Agricultural Organization, interested in the agricultural development of the Eastern Region, see Eastern Nigeria, *Annual Report of the Agricultural Division, 1960–1962*, 3.

59 The USAID provided a heavy equipment advisor; the U.K. also provided technical assistance in the form of soil surveyor and an analytical chemist. Japan also provided such technical support. See Eastern Nigeria, *Annual Report, Agricultural Division, 1963–1964, Official Document No. 2 of 1966* (Enugu: Government Printer, 1966), 1–2. See also Eastern Nigeria, *The First Three Years: A Report of the Eastern Nigeria Six-Year Development Plan* (Enugu: Government Printer, 1965); and *First Progress Report, Eastern Nigeria Development Plan, Ministry of Planning Official Document No. 15 of 1964* (Enugu, Nigeria: Government Printer, 1964).

60 On Israeli technical assistance, see Moshe Schwartz and A. Paul Hare, *Foreign Experts and Unsustainable Development: Transferring Israeli Technology to Zambia, Nigeria and Nepal* (Aldershot: Ashgate, 2000).

61 Floyd, *Eastern Nigeria*, 219. For details of acreage planted and locations, see NAE, ESIALA, 64/1/1, "Eastern Nigeria Development Corporation Agricultural and Plantations Division Situation Report," 1963.

62 Eastern Nigeria, *Development of Agriculture in Eastern Nigeria* (Enugu: Eastern Nigeria Printing Corporation, 1965), 5. For details of the crops grown in each settlement and the number of settlers, see H. I. Ajaegbu, *Urban and Rural Development in Nigeria* (London: Heinemann, 1976), 65.

63 Floyd, *Eastern Nigeria*, 219.

64 See Eastern Nigeria, *The First Three Years: A Report of the Eastern Nigeria Six-Year Development Plan* (Enugu: Government Printer, 1965),

65 See, for example, T. C. Yusev, *The Economics of Farm Settlements in Israel* (New York: Express Printers Inc., 1963); and F. C. Gorman, *Social Relations in Israeli Farm Settlements* (Tel-Aviv: Zester and Rox, 1957). See also Schwartz and Hare, *Foreign Experts and Unsustainable Development*, 6.

66 NAE, ESIALA, 64/1/1, "Eastern Nigeria Development Corporation," 196. Major critics of the settlement scheme included the opposition party in the Eastern Nigeria House of Assembly (The Action Group). See *Daily Express*, 10 November 1962.

67 The change of environment imposed a lot of psychological strain on the settlers. J.C.U. Eme, "Sociological Problems Connected with Farm Settlement Schemes", *Technical Bulletin*, No. 4 (1963): 61. See also Njaka Imelda, "Socio-Psychological Problems in the Farm Settlements" (paper presented at the Conference of Agricultural Officers, Abakiliki, 8 September 1964), 6–9, See Floyd and Adinde, "Farm Settlements," 223.

68 Caleb O. Okoro, "The Uzouwani Farm Set-tlement and Socio-Economic Development in the Anambra Basin, 1961–1971,"(M.A. thesis, Department of History, University of Nigeria, 1986), 44.

69 Floyd and Adinde, "Farm Settlements," 223.

70 Okoro, "The Uzouwani Farm Settlement," 45.

71 Ibid., 4.

72 Eugene Nwana, "Ohaji Farm Settlement: A Flash in the Pan," in Schwartz and Hare, Foreign Experts, 115.

73 Floyd, Eastern Nigeria, 233.

74 Ibid.

75 Floyd and Adinde, "Farm Settlements," 193.

76 Okoro, "The Uzouwani Farm Settlement," 29.

77 See "Eastern Nigeria Farm Settlement Scheme," Agricultural Bulletin No. 2 (n.d.), 10.

78 Ibid., 3.

79 Nwana, "Ohaji Farm Settlement," 110.

80 Okoro, "The Uzouwani Farm Settlement," 43.

81 H. I. Ajaegbu, Urban and Rural Development in Nigeria (London: Heinemann, 1976), 65.

82 For estimates of percentages of population and area experiencing pressure on land, see ibid., 15.

83 Floyd, Eastern Nigeria, 163.

84 Okoro, "The Uzouwani Farm Settlement," 11.

85 See NAE, ESIALA, 64/1/1 for details of these amenities completed by December 1963.

86 Floyd and Adinde, "Farm Settlements," 193.

87 Floyd, Eastern Nigeria, 232.

88 For discussion on mode of production in the African context, see, for example, Bowlig Simon, Peasant Production and Market Relations: A Case Study of Western

Ghana (Copenhagen: Third World Observer, 1993), 7.

89 Eastern Nigeria, Annual Report, Agricultural Division, 1963–1964, 2.

90 A. Kolawole, "Agricultural Stagnation, Food Crisis and Rural Poverty in Nigeria" (paper presented at the Ahmadu Bello University seminar series, 1984).

91 On the role of the marketing board, see Nigeria, Handbook of Commerce and Industry in Nigeria (Lagos, [not dated]), 122; Gerald K. Helleiner, "The Fiscal Role of the Marketing Boards in Nigerian Economic Development, 1947–61," Economic Journal 74, no. 295 (1964): 582–610; and H.M.A. Onitiri and D. Olatunbosun, ed., The Marketing Board System: Proceedings of an International Conference (Ibadan, Nigeria: Nigerian Institute of Social and Economic Research, 1974). See also Martin, Palm oil and Protest, 124.

92 See Eastern Nigeria, Eastern Nigeria Development Plan, 1962–1968: Official Document No. 8 (Enugu: Government Printer, 1962).

93 Purvis, Report on a Survey, 34.

94 Eastern Nigeria, Annual Report, Agriculture Division, 1963–64. Official Document No. 2, 1966 (Enugu: Government Printer 1966), 10.

95 Eastern Nigeria, Annual Report, Agriculture Division, 1963–64. See Table 3, 11.

96 Hursh, Innovation, 213.

97 Eastern Nigeria, Annual Report, Agriculture Division, 1963–64, 13.

98 Ibid.

99 For the quantity of gari export to Northern Nigeria between 1959 and 1963, see Eastern Nigeria, Annual Report, Agriculture Division, 1963–64, Table 5, 13.

100 Interview with Sarah Emenike, Owerri, 22 December 1998.

101 Eastern Nigeria, Programme of Work, 59.

102 Eastern Nigeria, "Programme of Work," Technical Bulletin 12 (Enugu: Agriculture Division, Ministry of Agriculture, 1966), 59.

103 Interview with Onyegbule Korieh, Mbaise, 17 December 1998.

104 Purvis, *Report on a Survey*, 16.

105 Interview with Sybilia Nwosu, c. 85 years, Nguru, Mbaise, 12 December 1998.

106 On agricultural policy, see Jerome C. Wells, *Agricultural Policy and Growth in Nigeria, 1962-1968* (Ibadan: Oxford University Press, 1974). On economic planning, see Peter B. Clark, "Economic Planning for a Country in Transition: Nigeria," in *Planning Economic Development*, ed. Everett E. Hagen, 252-93 (Homewood, IL: Richard D. Irwin, 1963).

107 Interview with Amarahiaugwu Korieh, Umuchieze, Mbaise, 13 December 1998.

108 Interview with Onyegbule Korieh, Umuchieze, Mbaise, 17 December 1998.

109 Interview with Alpelda Korie, Umuchieze, Mbaise, 23 December 1998.

110 Interview with Grace Chidomere, Umuchieze, Mbaise, 13 December 1998.

111 Interview with Susan Iwuagwu, Umunomo, Mbaise, 31 July 1999.

112 For historical studies of the Biafra-Nigeria Civil War, see Mok Chiu Yu and Lynn Arnold, ed., *Nigeria-Biafra: A Reading into the Problems and Peculiarities of the Conflict* (Adelaide: Adelaide University Quaker Society, 1968); Frederick Forsyth, *The Making of an African Legend: The Biafra Story* (New York: Penguin Books, 1977); and Herbert Gold, *Biafra Goodbye* (San Francisco: Twowindows Press 1970).

113 W. T. Morrill, "Immigrants and Associations: The Ibo in the Twentieth Century Calabar," *Comparative Studies in Society and History* 5, no. 4 (July 1963): 425.

114 NAE, OWDIST, 4/13/70, file no. 91/27, "Cultivation of Crops, Owerri District," District Officer to Resident Owerri Province, June 1928.

115 *Globe and Mail*, 2 October 1968.

116 PRO, FCO, 65/384, "Intelligence Memorandum," CIA Food Crisis in Eastern Nigeria.

117 Ibid.

118 See Reuben N. Ogbudinkpa, *The Economics of the Nigeria Civil War and its Prospects for National Development* (Enugu, Nigeria: Fourth Dimension, 2002), 58. See also Zdenek Cervenka, *The Nigerian Civil War 1969-70* (Frankfurt am Main: Bernard and Graefe Verlag für Wehrwesen, 1971), 73.

119 *Time*, 23 August, 1968.

120 Ibid.

121 The Committee for Peace in Nigeria (CPN) was established because of the Nigerian Civil War (1967-70). The Committee, which acted as an independent body and was headed by Lord Fenner Brockway, was active from 1968. Members included leading political figures in Britain, representatives from the missionary societies working in Nigeria, former members of the Colonial Service in Nigeria, business representatives, and Africans from both the federal and the Biafran sides of the conflict.

122 PRO, FCO, 65/384, "Intelligence Memorandum."

123 Ibid.

124 Ibid.

125 Ibid.

126 *Globe and Mail*, 2 October 1968.

127 PRO, FCO, 65/384, "Memorandum."

128 Martin A. Klein, correspondence, 28 December 2006.

129 American Jewish Committee Archive, "Biafran Issues and Background for July 25 Meeting," Marc H. Tanenbaum to Morris B. Abram, 22 July 1968, http://www.ajcarchives.org/AJCArchive/DigitalArchive.aspx, accessed 28 June 2006.

130 American Jewish Committee, "Biafran Issues."

131 PRO, FCO, 65/384, "Memorandum."

132 Interview with Mbagwu Korieh, 18 December 1998.

133 Interview with Jonah Okere, Umuekwune, Ngor Okpala, 12 December 1999.

134 Interview with Francis Ihuoma, aged 78, Mbaise, 17 December 1998.

135 E. Wayne Nafziger, "The Economic Impact of the Nigerian Civil War," *Journal of Modern African Studies* 10, no. 2 (1972): 241.

136 Ibid., 240.

137 Ibid., 241. See also Nigeria, *Annual Abstract of Statistics 1967*, 3.

138 Hursh et al., *Innovation in Eastern Nigeria*, 213.

139 PRO, FCO, 65/384, "Intelligence Memorandum."

140 Ibid.

141 Ibid.

142 Interview with Iwuagwu, Chilaka, Umunomo, Mbaise, 31 July 1999.

143 PRO, FCO, 65/384.

144 For government priority in agriculture before the war, see *Eastern Nigeria Economic Development Plan 1962-1968, Official Document No. 8* (Enugu: Government Printer, 1962), 8.

145 For the activities of the BDC, see NAE, ESIALA, 63/1/70-SEC/217, vol. 1, "Emergency Food Production," Director, Food Production Directorate to the Chairman, BDC, 5 February 1968.

146 NAE, ESIALA, 63/1/70, file no. SEC/217, vol. 1, "Emergency Food Production."

147 Interview with Maria Gold Egbunike, 65 years, transcribed in Azuka Nzegwu, "These Women are Brave: Biafra War/Nigeria Civil War, 1967-1970." http://www.westafricareview.com/war/vol2.2/biafra/student.htm, accessed 31 January 2003.

148 Interview with Ezenwanyi Anichebe, Eziowelle town Anambra State, 22 October 2007, transcribed in Chimee N. Ihediwa, "The Role of Women in Post War Economic Transformation of Igboland, 1970-1985," (unpublished paper).

149 PRO, FCO, 65/384, "Memorandum."

150 Ibid.

151 Interview with Amarahiaugwu Korieh, 65 year, Mbaise, 13 December 1998.

152 Oruene Taiwo Olaleye, *Nation Builders: Women of Nigeria* (London: International Report, Women and Society, 1985), 5.

153 Interview with Comfort Anabalam, Umuchieze, Mbaise, 13 December 1998.

154 Interview with Onyegbule Korieh, Umuchieze, Mbaise, 17 December 1998.

155 Interview with Nwadinma Agwu at Ishiagu, 29 October 2007, transcribed in Ihediwa, "The Role of Women."

156 Interview with Edna Okoye, Akanogu Umudunu Village Abagana, 26 June 2006, transcribed in Ihediwa, "The Role of Women."

157 Interview with Margaret Nwanevu, Amumara Mbaise, 30 October 2007, transcribed in Ihediwa, "The Role of Women."

158 Interview with Chinyere Iroha, Uvuru, Mbaise, 30 October 2007, transcribed in Ihediwa, "The Role of Women."

159 Iyegha, *Agricultural Crisis*, 35. See also Federal Office of Statistics, *Review of External Trade* (Lagos, 1979), 83.

160 Peter Kilby, "What Oil Wealth Did to Nigeria," *Wall Street Journal*, 25 November 1981.

161 Central Intelligence Agency, "Intelligence Memorandum," and "Nigeria: The War's Economic Legacy," 10 May 1971.

162 Central Intelligence Agency, "Intelligence Memorandum," 10.

163 Andrew Okolie, "Oil Revenues, International Credits and Food in Nigeria, 1970–1992," (PhD thesis, Sociology Department, University of Toronto, 1995), 98. See also Watts and Lubeck, "The Popular Classes," 107-8.

164 Joyce Kolko, *Restructuring the World Economy* (New York: Pantheon, 1988), 37.

165 The estimates of oil revenue have not been consistent from different sources. This is a problem associated with much of the official data from Nigeria. On petroleum production and revenue from 1973 to 1979, see *International Financial Statistics* 33, no. 12 (1980): 288. See also Myer, "This Is Not Your Land," note 4, 136.

166 Peter Kilby, "What Oil Wealth Did to Nigeria," *Wall Street Journal*, 25 November 1981.

167 Watts and Lubeck, "The Popular Classes," 108.

168 Ibid.

169 Myers, "This Is Not Your Land," 94.

170 For further discussion, see, for example, Siyanbola, *Structure of the Nigeria Economy* (London, 1979), 27. See also Angaye Gesiye, "Petroleum and the Political Economy of Nigeria," in *The Nigerian Economy: A Political Economy Approach*, Nigerian Economic Society (London: Longman, 1986), 50–69.

171 Berry, *No Condition*, 77.

172 Julius O. Ihonvbere and Timothy M. Shaw, *Towards a Political Economy of Nigeria: Petroleum Policy at the (Semi)-Periphery* (Aldershot: Avebury, 1982), 7.

173 On the paradox of poverty amidst plenty, see Nicholas Shaxson, *Poisoned Wells: The Dirty Politics of African Oil* (New York: Palgrave Macmillan, 2007). See also Hans-Otto Sano, *The Political Economy of Food in Nigeria, 1960–1982: A Discussion on Peasants, State, and World Economy* (Uppsala: Scandinavian Institute of African Studies, 1983), 2.

174 See Andrew C. Okolie, "Oil Revenues, International Credits and Food in Nigeria, 1970–1992" (PhD thesis, Department of Sociology, University of Toronto, 1995). On food out and imports between 1970 and 1977, see various issues of *Nigeria Quarterly Economic Review* published by the Federal Office of Statistics. On food output from 1971–79, see Federal Office of Statistics, *Nigeria Quarterly Economic Review* (Lagos: FOS, 1980). See also T. Forrest, "Agricultural Policies in Nigeria 1900–1978," in *Rural Development in Tropical Africa*, ed. G. Williams et al., 222–58 (London: Macmillan, 1981); and Siyanbola, *Structure*, 27.

175 Iyegha, *Agricultural Crisis*, 37; O. Awoyemi, "Character of Nigerian Agriculture," *Central Bank Bullion* 6, no. 4 (1983): 2, cited in T. S. B. Aribisala, "Nigeria's Green Revolution: Achievements, Problem and Prospects," NISER Distinguished Lecture Series, 8.

176 Nigeria, *The Shagari Administration*, 7.

177 See Morgan and Solarz, "Agricultural Crisis," 59.

178 See Government of East Central State of Nigeria, *Report of Rural Economic Survey of the East Central State of Nigeria* (Enugu: Government Printer, 1977), 41.

179 Watts, *Silent Violence*, 467.

180 Federal Office of Statistics, *Annual Abstract of Statistics* (Lagos, 1986), 15.

181 Discussion with Alban Onyesoh, aged c. 39, Umuchieze, Mbaise, December 1998.

182 Interview with Onyegbule Korieh, Umuchieze, Mbaise, 17 December 1998.

183 Watts, *Silent Violence*, 467, 486.

184 There emerged absentee farmers, who hired labour to work on rented farms.

185 Federal Office of Statistics, *Annual Abstract of Statistics* (Lagos 1961), 141.

186 Ibid., 20.

187 Interview with Uzodinma Ibekwe, Nguru, Mbaise, 25 December 1998.

188 Interview with Onyegbule Korieh, Umuchieze, Mbaise, 17 December 1998.

189 For some treatment of the boom and bust period in Nigeria and the impact on farmers, see, for example, Michael Watts, ed., *State, Oil and Agriculture in Nigeria* (Berkeley: Institute of International Studies, University of California), 19.

190 Watts, *Silent Violence*, 483.

191 See Nigeria, *Third National Development Plan*.

192 This argument is borne out in my research and oral interviews with rural farmers. Many rural farmers consistently argued that political officials who were not farmers often controlled the sale and distribution of fertilizers.

193 See Oguntoyinbo, "The Changing Trends," 9.

194 On the Land Use Act, see, for example, Myers, "This Is Not Your Land," 29; Ega L. Alegwu, "Implications of the 1978 Land Use Act for Agricultural Development in Nigeria," *Issues in Development* 1, no. 2 (1985): 40–50; and Fabiyi Yakubu, "Land Tenure Reform in Nigeria: Implications of the Land Use Decree (Act) for Agricultural Development," *Ife Journal of Agriculture* 1, no. 2 (1979): 235–57.

195 The Land Use Decree gave the federal and state governments power to take over any land within their jurisdiction without compensation. The decree has been extensively used to alienate land in areas close to the urban centres.

196 See Myers, "This Is Not Your Land," 29. In this period, there was interest in agricultural investment by foreign capital. In 1978, the federal commissioner for agriculture informed a Dutch delegation that the Land Use Decree would help the government to acquire land for foreign investors. See Watts and Lubeck, "The Popular Classes," 123.

197 Imo State, *Briefs on Imo State* (Owerri, Nigeria: Government Printer, 1984), 13.

198 Imo State, *Government White paper on the Report of the Judicial Commission of Inquiry into the ADC, Owerri* (Owerri: Government Printer, 1980), 17.

199 Gavin Williams, "The World Bank and the Peasant Problems," in *Rural Development in Tropical Africa*, ed. Judith Heyer, Pepe Roberts, and Gavin Williams, 25 (London, 1981).

200 Imo State of Nigeria, *Government White Paper*, 23.

201 Ibid., 24.

202 Ibid.

203 On the Green Revolution, see Adekunle Folayan, *Agriculture and Economic Development in Nigeria: A Prescription for the Nigerian Green Revolution* (New York: Vantage Press, 1983); and Peter Lawrence, "The Political Economy of the 'Green Revolution' in Africa," *Review of African Political Economy* 15, no. 42 (1988): 59–75.

204 See Federal Republic of Nigeria, *Fourth National Development Plan, 1981–1985*, 88.

205 E. A. Bamisaiye, "Solving the Food Crisis in Africa: The Role of Higher Education," *Journal of African Studies* 11, no. 4 (1985): 182–88.

206 See Berry, *No Condition is Permanent*, 182.

207 Gains were limited. For a review of irrigated rice farming schemes in five West African countries, see, for example, Robin Palmer and Neil Parsons, ed., *The Root of Rural Poverty in Central and Southern Africa* (London, 1981), 412–13.

208 See Nigeria, *Third National Development Plan, 1975–80.*

209 Imo State of Nigeria, *ISADAP after Three Years. Information Bulletin No. 4* (n.d.), 2.

7 ON THE BRINK: AGRICULTURAL CRISIS AND RURAL SURVIVAL

1 Some studies on agricultural sustainability include, Abe Goldman, "Threats to Sustainability in African Agriculture: Searching for Appropriate Paradigms," *Human Ecology* 23, 3 (1995): 291–334. See also G. K. Douglass, "The Meaning of Agricultural Sustainability," in *Agricultural Sustainability in a Changing World Order*, ed. G. K. Douglass, 3–29 (Boulder, CO: Westview, 1994); George J. S. Dei, "Sustainable Development in the African Context: Revisiting Some Theoretical and Methodological Issues," *African Development* 18, no. 2 (1993): 97–110, and C. K. Eicher, *Sustainable Institutions for African Agricultural Development*, International Service for National Agricultural Research (ISNAR), Working Paper No. 19 (The Hague, 1989).

2 The term *agricultural crisis* is used here to describe the manifestation of low agricultural productivity, food insecurity and a decreasing farming population as experiences in central Igboland. The term is useful in explaining trends in the agricultural history of the region in terms of productivity and farming population. For a review of the literature seeking to establish the causes and potential remedies for agricultural decline, see Sara Berry, "The Food Crisis and Agrarian Change in Africa: A Review Essay," *African Studies Review* 27, no. 2 (1984): 59–112. On internal-external factors of African agrarian problems, see M. F. Lofchie and S. K. Commins, "Food Deficit and Agricultural Policies in Tropical

Africa," *Journal of Modern African Studies* 20, no. 1 (1982): 1–25. See also M. F. Lofchie, "The Decline of African Agriculture," in *Drought and Hunger in Africa: Denying Famine a Future,* ed. Michael H. Glantz, 85–110 (Cambridge: Cambridge University Press, 1987). For a Nigerian case study, see Iyegha, *Agricultural Crisis in Africa,* 1988. See also Ray Bush, "The Politics of Food and Starvation," *Review of African Political Economy* 68 (1996): 169–95.

3 See Brad David Jokisch, "Landscape of Remittance: Migration and Agricultural Change in the Highlands of south-central Ecuador" (PhD dissertation, Clark University, 1998), 100.

4 Federal Office of Statistics, *Poverty and Agricultural Sector in Nigeria* (Abuja, Nigeria: Federal Office of Statistics, 1999).

5 Ibid., 7.

6 Raluca Iorgulescu Polimeni and John M. Polimeni, "Structural Adjustment and the Igbo Extended Family," *International Journal of the Humanities* 5, no. 5 (n.d.): 77–82.

7 Interview with Eugenia Otuonye, Umuchieze, Mbaise, 23 December 1998.

8 Interview with Isidore Korieh, Umuchieze, Mbaise, 5 May 2008.

9 Imo State of Nigeria, *ISADAP After Three Years. Information Bulletin No. 4* (n.d.), 1.

10 Ibid., 2–3.

11 Ibid.

12 Ibid., 2.

13 Ibid., 3–5.

14 See, Ngobiri C. Chioma, "The Problems and Prospects of Women in Agriculture: A Case Study of Women farmers in Orlu Local Government Area," (Research Project PGD, Agricultural Economics and Extension, Imo State University, 1998).

15 See Government of East Central State of Nigeria, "Report of Rural Economic Survey of the East Central State of Nigeria," (Enugu, Nigeria: Government Printer, 1977), 41.

16 For further discussion, see, for example, F. A. Olaloku, *Structure of the Nigeria Economy* (London: Macmillan, 1979), 27.

See also Angaye Gesiye, "Petroleum and the Political Economy of Nigeria," in *The Nigerian Economy: A Political Economy Approach,* ed. J. A. Ajayi et al., 50–69 (London: Longman, 1986).

17 Bongo Adi, "Determinants of Agricultural and non-Agricultural Livelihood Strategies in Rural Communities: Evidence from Eastern Nigeria," *Project Muse* 93, 93–109, accessed 30 January 2008.

18 Interview with Nwanyiafo Obasi, Umunomo, Mbaise.

19 Adi, "Determinants," 95.

20 Interview with Eugenia Otuonye, Umuchieze, Mbaise, 23 December 1998.

21 For more on the spread and impact of cassava, see, for example, H. Rosling, *Cassava Toxicity and Food Security: A Review of Health Effects of Cyanide Exposure from Cassava and of Ways to Prevent the Effects* (Ibadan, Nigeria: IITA-UNICEF Program on Household Food Security and Nutrition, 1987), 29. See also: E. O. Onabanjo, *A Selected Bibliography on Cassava* (Lagos: Institute of Industrial Research, 1979); S. A. Agboola, "Introduction and Spread of Cassava in Western Nigeria," *Nigeria Journal of Economic and Social Studies* 10, no. 3 (1968): 369–85; and E. C. Anumihe, "The Role of Cassava in Ensuring Small Farm Household Food Security in Imo State." (PhD thesis, Imo State University, 1998), 7.

22 Ottenberg, *Farmers and Townspeople,* 162–63.

23 Martin, *Palm Oil and Protest.*

24 Njoku, *Economic History of Nigeria,* 13.

25 George, *Journal,* 11 March 1863, CMS C.A3/O.18/20, cited in Iliffe, *The African Poor,* 92.

26 Interview with Ugwuanya Nwosu, Owerri, 22 December 2000.

27 Don C. Ohadike, "The Influenza Pandemic of 1918–19 and "The Spread of Cassava Cultivation on the Lower Niger: A Case Study of Historical Linkages," *Journal of African History* 22 (1981): 379–91. See also Ohadike, *Anioma: A Social History of the Western Igbo People* (Athens, OH: Ohio University Press, 1994), 198–207.

28 W.G.A. Ormsby-Gore, "Some Contrasts in Nigeria," *Geographical Journal* 69, no. 6 (1927): 504.

29 NAE, OWDIST, 4/13/70, file no. 91/27, "Cultivation of Crops, Owerri District," District Officer to Resident Owerri Province, June 1928.

30 Interview with Luke Osunwoke, Umuorlu, Isu, 7 January 2000.

31 Morgan, "Farming Practice," 331.

32 For quantity of gari export to northern Nigeria, see NAE, ABADIST, 14/1/875, file no. 1646, vol. IV, "Gari Control," District Officer Aba, to the Senior Resident, Port Harcourt, 20 July 1942.

33 Simon Ottenberg, *Farmers and Townspeople in a Changing Nigeria: Abakiliki during Colonial Times (1905–1960)* (Ibadan: Spectrum Books, 2005), 162–63.

34 Morgan, "Farming Practice," 331.

35 Phoebe V. Ottenberg, "The Changing Economic Position of Women Among the Afikpo Ibo," in *Continuity and Change in African Cultures*, ed. William Russell Bascom and Melvin J. Herskovits (Chicago: University of Chicago Press, 1956), 214.

36 Ibid., 215.

37 Ibid.

38 NAE, CSE, 3.17.15., B 1544, (1925–1926); NAE, AW 80 Q, AW 2/1/57, Anti-Government Propaganda Women Dancers (1925).

39 Interview with Robert Ibeh, Umucieze Mbaise, 28 July 1999.

40 Interview with Linus Anabalam, Umuchieze, Mbaise, 13 December 1998.

41 W. P. Falcon, W. O. Jones, and S. R. Pearson, *The Cassava Economy of Java* (Stanford, CA: Stanford University Press, 1984), 20.

42 Morgan, "Farming Practice," 331.

43 Ibid.

44 See particularly the works of M. M. Green, *Land Tenure in an Ibo Village in Southeastern Nigeria*. Monographs on Social Anthropology 6 (London: LSE Monographs on Social Anthropology, 1941); and J. Harris, "Human Relationships to the Land in Southern Nigeria," *Rural Sociology* 7 (1942): 89–92.

45 See Abe Goldman, "Population Growth and Agricultural Change in Imo State, Southeastern Nigeria," in *Population Growth and Agricultural Change in Africa*, ed. B. L. Turner II, Goran Hyden, and Robert W. Kates, 250–301 (Gainesville, FL: University of Florida Press, 1993). For contemporary population estimates, see Government of Nigeria, "1991 Population Census" (Lagos: National Population Commission, 1991).

46 See George, *Journal*, 21 January 1866, CMS CA3/O, 18/23; F. M. Denis, *Journal*, 17 November 1908, CMS: UP 4/F2; T. J. Dennis, *Journal*, March 1907, CMS: UP 89/F1, cited in Iliffe, *African Poor*, 82.

47 See N. W. Thomas, *Anthropological Report on the Ibo Speaking Peoples of Nigeria*, vol. 1 (London: Harrison and Sons, 1913–14), 97; Elizabeth Isichei, *A History of the Igbo People* (London: Macmillan, 1976), 27, 79; and Iliffe, *African Poor*, 92.

48 Dike, *Trade and Politics*, 28.

49 Simon Ottenberg, "Ibo Receptivity to Change," in *Continuity and Change in African Cultures*, ed. William R. Bascom and Melville J. Herskovitz, 140 (Chicago: University of Chicago Press, 1959). See also Dmitri Van Den Barsselaar, "Imagining Home: Migration and the Igbo Village in Colonial Nigeria," *Journal of African History* 46 (2005): 57.

50 R. K. Udo, "The Migrant Tenant Farmer of Eastern Nigeria," *Africa: Journal of the International African Institute* 34, no. 4 (Oct. 1964): 326–39.

51 Mss. Afr. s. 1520, Sylvia Leith-Ross papers.

52 Interviews with Oboko Ude of Amata, aged c. 80 years, 20 August 1977; Ezenta Okpo of Ezioha, aged c. 90 years, 26 August 1977, transcribed in Ude Emmanuel N. "Agriculture and Trade in Mgbowo, 1850–1950" (B.A. History project, University of Nigeria, 1978), 3–4.

53 Nzimiro, "A Study of Mobility," 124.

54 For the early kinds of migration in response to population pressures, see John

N. Oriji, *Traditions of Igbo Origin: A Study of Pre-Colonial Population Movements in Africa* (New York: Peter Lang, 1990).

55 See, Davidson, *West Africa*, 125–26; and Rodney, *A History of the Upper Guinea Coast*, 253.

56 Huss-Ashmore, *Perspectives*, 11.

57 Floyd, *Eastern Nigeria*, 57.

58 Official provincial estimates per square mile recorded by Forde and Jones were as follows: Onitsha (1921) 306, (1931) 224; Owerri (1921) 268, (1931) 154. It is obvious that these official figures are unreliable. The population dynamics would suggest that the population density for 1931 would have been relatively higher than the density for 1921. Despite this negative growth, which cannot be explained, contemporary estimates suggest that most of the Igbo territory experienced very high densities. See Daryll Forde and G. I. Jones, *The Ibo and Ibibio Speaking Peoples of Southeastern Nigeria* (London: International African Institute, 1950), 10–13.

59 See Onwuejeogwu, *Evolutionary Trends*, 21.

60 David R. Smock and Audrey C. Smock, *Cultural and Political Aspects of Rural Transformation: A Case Study of Eastern Nigeria* (New York: Praeger, 1972), 21.

61 For current population estimates, see *Nigeria: 1991 Population of States by Local Government Areas* (Lagos: Federal Office of Statistics, 1994).

62 See Charles A. P. Takes, "Socio-Economic Factors Affecting the Productivity of Agriculture in Okigwi Division (Eastern Nigeria): Preliminary Report" (Ibadan, Nigeria: Nigerian Institute of Social and Economic Research, 1963), 6.

63 Interviews with Theophilus Onyema, Umuorlu, Isu, 5 January 2000, and Luke Osunwoke, Umuorlu, Isu, 7 January 2000.

64 R. K. Udo, "Influence of Migrations on the Changing Cultural Map of West Africa," *Ikenga: A Journal of African Studies* 1, no. 29 (1972): 50.

65 Elizabeth Isichei, *A History of Nigeria* (London: Longman, 1983).

66 Azuka A. Dike, "Urban Migrants and Rural Development," *African Studies Review* 25, no. 4 (1982): 86.

67 See Goldman, "Population Growth."

68 Stanley Diamond, *Nigeria: Model of a Colonial Failure* (New York, 1967), 44.

69 See Nigeria, *Official Gazette* (FGP 71/52007/2,500(OL24): 2006 National Population census.

70 Ikenna Nzimiro, "A Study of Mobility among the Ibos of Southern Nigeria," *International Journal of Comparative Sociology* 6, no. 1 (1965): 123–24.

71 Walter Ofonagoro, *Trade and Imperialism in Southern Nigeria, 1881–1960* (New York: Nok, 1979), 109.

72 See, for example, Eicher and Liedholm, *Growth and Development*, 78.

73 Hogendorn, *Nigerian Groundnut Export*, 18–21.

74 G. I. Jones, *From Slaves to Palm Oil: Slave Trade and Palm Oil Trade in the Bight of Biafra* (Cambridge: African Studies Center, 1989), 104–5.

75 Jones, *From Slaves to Palm Oil*, 104.

76 RH, Mss Afr s 1520, Sylvia Leith-Ross papers.

77 Harris, "Some Aspects of the Economics of 16 Ibo individuals," *Africa* 14 (1944): 302–35.

78 Interview with Linus Anabalam, Umuchieze, Mbaise, 12 December 1999.

79 RH, Mss. Afr. S. 1428, 16. Transcript of interview with J. B. Davies, CMB OBE of the United African Company, on 9 March 1971, interviewed by Dr. Collin Newbury, University Lecturer in Commonwealth Studies at Oxford University.

80 Van Den Barsselaar, "Imagining Home," 58. See also G.E.K. Ofomata, *Nigeria in Maps: Eastern States* (Benin City: Ethiope, 1975), 64–6.

81 See Isichei, *A History of the Igbo People*, 210.

82 The region was a former German colony until World War I. On 4 February 1916, the English and the French shared the

Cameroon territory that they had just forcefully gained from the Germans.

83 Interview with Linus Anabalam, 12 December 1999.

84 NAE, CADIST, 3/3/840, Denis Ugwu and others to the Senior District Officer, Calabar, 4 June 1949.

85 NAE, CADIST, 3/3/686, I. Uchendu and others to District Officer, Calabar, 17 February 1949.

86 Report of the Commission of Enquiry into the Affairs (Appointment and Promotion) of the Formers ENDC Plantations from May 1967 to 5 November 1968 (Calabar: Ministry of Home Affairs and Information, 1969), 6.

87 Ibid.

88 Van Den Barsselaar, "Imagining Home," 58. See also Ofomata, Nigeria in Maps, 64–66.

89 Gerald K. Kleis, "Confrontation and Incorporation: Igbo Ethnicity in Cameroon," African Studies Review 23, no. 3 (1980): 91.

90 On the labour migration to Fernando Po, see, for example, PRO, CO 554/127/5, "Labour for Fernado Po."

91 See Nigeria, Annual Report of the Department of Labor for the Year 1944.

92 See Anthony Oham, "Labor Migration from Southeastern Nigeria to Spanish Fernando Po, 1900–1968," (MA thesis, Department of History, Central Michigan University, 2007).

93 RH, Mss. Afr. s. 2033 (1) Robinson, W. A. Lt.-Col, Memoir, 20.

94 See Oham, "Labor Migration," 33.

95 Loise, interviewed by Anthony Ohams, 30 December 2005.

96 Interview with Alban Onyesoh, Umuchieze, Mbaise, 18 December 1998.

97 Interview with Onyegbule Korieh, Umuchieze, Mbaise, 17 December 1998.

98 Interview with Alban Eluwa, Umuchieze, Mbaise, 18 December 1998.

99 The expression 'dry' means lack of resources. Interview with Onyegbule Korieh, Mbaise, 17 December 1998.

100 RH, Mss. Afr. s. 1428, 17, transcript of interview with J. B. Davies, CMB OBE of the United African Company on 9 March 1971; interviewed by Dr Collin Newbury, University Lecturer in Commonwealth Studies at Oxford University.

101 Victor Bong Amaazee, "The 'Igbo Scare' in British Cameroons," Journal of African History 30 (1990): 281.

102 "Inside Contemporary Cameroon Politics," available from avadocs.indymedia. org/pub/Main/JusticeMbuh/ICCPC1.doc. [accessed 16 February 2008].

103 RH, Mss. Afr. s. 2033, W. A. Robinson papers.

104 Jones, From Slaves to Palm Oil, 107.

105 U. D. Anyanwu, "The Aftermath of the Atlantic Slave Trade: Two Settlement Patterns in Southeastern Nigeria," in The Aftermath of Slavery: Transitions and Transformation in Southeastern Nigeria, ed. Chima J. Korieh and Femi J. Kolapo, 59–74 (Trenton: Africa World Press, 2007).

106 Interview with Jane Elizabeth Obasi, Umunomo, Mbaise, 30 July 1999.

107 Egejuru, The Seed Yams, 219.

108 Interview with Linus Anabalam, Umuchieze, Mbaise, 13 December 1998.

109 For the role of female income and wage work in household sustainability, see Janice Jiggins, Gender-Related Impacts and the Work of the International Agricultural Research Centers. CGIAR Study Paper No. 17 (Washington, DC: World Bank, 1986), 51; and FAO, "Special Problems of Female Heads of Households in Agriculture and Rural Development in Asia and the Pacific," (Bangkok: FAO, 1985).

110 Interview with Jonas Onwukwe, aged c. 70, Umunkwo-Emeabiam, May 2001.

111 Interview with Onyegbule Korieh, Umuchieze, Mbaise, 17 December 1998.

112 Interview with Linus Anabalam, Umuchieze, Mbaise, 13 December 1998.

113 Interview with Eugenia Otuonye, Umuchieze, Mbaise, 23 December 1998.

114 Interview with Onyegbule Korieh, Umuchieze, Mbaise, 17 December 1998.

115 Uchendu, *The Igbo*, 30.

CONCLUSION

1   Allen Isaacman and Richard Roberts made this argument in the case of cotton producing areas. See Allen Isaacman and Richard Roberts, *Cotton, Colonialism, and Social History in Sub-Saharan Africa* (Portsmouth, NH: Heinemann, 1995), 2.

2   NAE, ABADIST, 14/1/872.

3   Ibid.

4   NAE, ABADIST, 1/26/958.

BIBLIOGRAPHY

I. MANUSCRIPTS AND OTHER UNPUBLISHED PUBLIC DOCUMENTS

Nigeria. Department of Agriculture (federal)

Annual Report upon the Agricultural Department, 1910–1913.

Annual Report on the Agricultural Department, 1914–1915.

Annual Report on the Agricultural Department, 1922.

Annual Report on the Agricultural Department, 1931–1932.

Annual Report on the Agricultural Department, 1937–1942.

Annual Report of the Agricultural Department, 1952–1953.

Annual Report, Agriculture Division, 1956–1959.

Digest of Statistics. Lagos: Federal Office of Statistics, 1994.

Estimates of the Federal Government of Nigeria, 1964–1965.

First Bulletin of the Agricultural Department, 1922.

Fourth National Development Plan, 1981–1985.

Legislative Council Debate, 4 December 1939.

Legislative Council Debates, 17 March 1941.

Legislative Council Debates, 13 March 1944.

National Agricultural Sample Census, 1993/1994 – Imo State. Lagos: Federal Office of Statistics, 1996.

Notes of Evidence Taken in the Calabar and Owerri Provinces on the Disturbance 1930, 8 vols.

Population Census. Lagos: Federal Office of Statistics, 1991.

Poverty and Agricultural Sector in Nigeria. Abuja: Federal Office of Statistics, 1999.

Programme of Work, Technical Bulletin No. 12. Agriculture Division, Ministry of Agriculture, 1966.

Report of a Commission of Inquiry appointed to Inquire into certain incidents at Opobo, Abak and Utu-Etim-Ekpo in December 1929 (Sessional Paper No. 12) 1929–1930, No. 12.

Report of the Commission of enquiry appointed to inquire into the disturbance in the Calabar and Owerri Provinces, December 1929 (Sessional Paper No. 28) 1930, No. 28.

Report of the Aba Commission of Inquiry, "Memorandum as to the Origins and Causes of the Recent Disturbances in the Owerri and Calabar Provinces," Appendix III (1), 11–12.

Report of the Mission Appointed to Enquire into the Production and Transport of Vegetable Oils and Oil Seeds Produced in the West African Colonies (London: H.M.S.O., 1947).

Report on a Survey of the Oil Palm Rehabilitation Scheme in Eastern Nigeria – 1967 (CSNRD, Report No. 10) (1968).

Second Annual Bulletin of the Agricultural Department, 1923.

The Nigeria Handbook. Lagos: Government Printer, 1929.

Third Annual Report of the Agricultural Department, 1924.

Nigeria. National Archives Enugu (NAE)

Aba District

Annual Report on Aba District, 1943–1945.

ABADIST, 1/26/907, File no. 1642, "Palm Produce Production," 1945.

ABADIST, 1/26/958, File no. 668, "Foodstuffs: Yams, Plantains, Cocoyams, etc, Requested Prohibition of Railment or Exportation of in Future," 1943.

Provinces at Government Lodge, Enugu, 13 December 1948.

ABADIST, 14/1/872 File no. 1646, "Garri: Control," 1943.

ABADIST, 14/1/875, "Gari Control," 1944.

ABADIST, 1/26/907, "Palm Produce Production," 1944.

ABADIST, 14/1/873, File no. 1646 "A Resolution," Gari Traders association, Aba, 1943.

ABADIST, 1/2/908, "Memos by D.O. Aba to Ngwa Native Authority, 1940–44."

Abakaliki District

Annual Report, 1945.

AIDIST, 2/1/433 OG: 2513/1265 "Nigeria General Defence Regulations: Order," 1945.

AIDIST, 2/1/433, File no. AB: 1373/11 "Memo," District Officer Abakiliki Division to C.N.C. Igbeagu, 1945.

Aro Division

Annual Report on Aro District, 1952.

Awka District

Annual Report, 1925–1926.

Calabar Province

Annual Report, 1942 and 1952.
    Report on Travelling and Agricultural instructional Work, 1913.
Report of the Agricultural Department, 1918.

Civil Secretary's Office, Enugu (CSO)

Correspondence, 1925, 1940, 1943.

Eket District

Annual Reports, 1935–1951.

Eastern States Interim Assets and Liability Agency (ESIALA)

ESIALA, 27/1/53, Matter relating to agricultural loans.
ESIALA, 63/1/70, Matters Relating to Biafra Development Corporation Crash food programme.
ESIALA, 63/1/70-SEC/217, vol. 1, "Emergency Food Production," Director, Food Production Directorate to the Chairman, BDC, 1968.
ESIALA, 64/1/1, Eastern Nigeria Development Corporation.
ESIALA, 64/1/3, Matters relating to Biafra Development Corporation.

Ministry of Local Government (MINLOC)

MINLOC, 6/1/175, "Intelligence Report on Ekwereazu and Ahiara Clan, Owerri Division."
MINLOC, 6/1/175. "Intelligence Report on Obowo and Ihitte Clan, Okigwe Division."

Obudu District

OBUDIST, 4/1/309, Produce Drive: Kernel and Rubber Return Prosecutions.

Ogoja Province

OGPROF, Annual Report, 1943.
OGPROF, OG 323, Food Situation in Obudu District.
ONDIST, 12/1/104, Economic Survey-Onitsha Province.
ONDIST, 12/1/1050, Agricultural Education Scheme.

Onitsha District

ONDIST, 12/1/578, Instruction for Farmers' Sons, 1933, 1934.

ONDIST, 12/1/578, Agricultural Instructions.

ONDIST, 13/1/2, Food Control, 1939–1942.

ONPROF, 7/12/92, Bands of Women Dancers Preaching Ideas of Desirable Reforms, 1925.

ONDIST, 12/1/90, Matters Relating to the Effects and Implications of War Conditions on Nigeria.

ONDIST, 12.1/104, Palm oil Production, 1939.

Owerri District

OWDIST, 4/13/70 File no. 91/27, Cultivation of Crops, Owerri District, 1928.

OWDIST, 4/13/70 File no. 91/27, Assessment Report, 1928.

Rivers Province

RIVPROF, 8/5/661, File No. OW 630/17, Trade Prices at Up Country Markets, 1917.

RIVPROF, 9/1/135 File no. Ow: 1636, Annual Report on the Social and Economic Progress of the People of Nigeria for the year 1933.

RIVPROF, 8/5/430, Policy of the Agricultural Department, 1939.

RIVPROF, 8/5/661, Cooperative Agricultural Settlements for Nigeria, 1940.

Umuahia Province

UMUPROF, 1/5/2, Women's Movement, 1929, 1930.

UMUPROF, 1/5/4, Women Movement, Aba Patrol Report 1930.

UMUPROF, 1/5/5, Women Movement, Bende Report, 1930.

Eastern Nigeria

Agricultural Division Annual Report, 1959–1961.

Annual Abstract of Statistics, 1964.

Annual Abstract of Statistics. Lagos: Acting Federal Government Press, 1961.

Annual Report and Statement of Accounts, 1966 (Lagos, 1967).

Annual Report of Agricultural Department, 1952–1953 (Enugu: Government Printer, 1953).

E.N.D.C. in the First Decade, 1955–1964. Enugu: ENPC (n.d).

Eastern Nigeria Development Plan 1962–1968: Official Document No. 8. Enugu: Government Printer, 1962.

Eastern Nigeria, Development Programme, 1958–1962: Eastern Region Official Document No. 2 of 1959. Enugu: Government Printer, 1959.

The Development of Agriculture in Eastern Nigeria, No. 1, 1965.

The First Three Years: A Report of the Eastern Nigeria Six Year Development Plan, 1965.

Imo State of Nigeria

Annual Report 1988. Owerri: Government Printer, 1989.

Government White paper on the Report of the Judicial Commission of Inquiry into the ADC, Owerri (Owerri: Government Printer, 1980).

Imo State Agricultural Development Project Law, 1992. Owerri: Government Printer, 1993.

ISADAP After Three Years. Information Bulletin No. 4 (nd).

Spotlight on Imo State Local Governments Areas. Owerri: Government Printer, 1982.

Nigeria: National Archive, Calabar (NAC)

CALPROF, 14/8/712, File no. E/1019/13, "Report on travelling and agricultural instructional work," 1913.

CALPROF, 14/7/1698: "Land Tenure-Orlu District."

CALPROF 14/8/711, File no. 1018/13, *The Quarterly Report of the Agricultural Department*, 1918.

CALPROF, 3/1/2329, Control of Locally Produced Foodstuffs, 1942.

United Kingdom. Public Record Office, London (PRO)

CO, 583/159/12, Introduction of Direct Taxation in Southern Provinces – Petition Regarding.

CO, 583/176/9, "Native Unrest," Correspondence arising out of the Report of the Aba Commission, 1930.

CO, 583/193/8, Palm oil Industry, 1933.

CO, 589/159/12, Native Revenue Amendment Ordinance, 1928.

FCO, 65/384, Intelligence Memorandum: CIA Food Crisis in Eastern Nigeria.

Oxford. Rhodes House, Boldhein, Library, Oxford University (RH)

Mss Afr. s. 1000, Falk, Edward Morris Papers as DO in Nigeria.

Mss Afr. s. 1000, Mrs Falk, Edward Morris Diary.

Mss Afr. s. 1073, Crowker Walter R, Papers.

Mss Afr. s. 1556, George Sandy Parker Papers.

Mss Afr. s. 1343, Belcher, Margaret Papers.

Mss Afr. s. 2127, Barwick N. Papers.

Mss Afr. s. 546, Carr F.B Papers as Resident, Owerri Province in Nigeria.

Mss Afr. s. 1779, Harington, G. N. Papers.

Mss Afr. s. 1520, Leith-Ross Slyvia, 1902–59 Papers.

Mss Afr. s. 823, Mackie J. R, Capt. Paper on Nigeria.

Mss Afr. s. 1505, Mayne, C. J. Papers.

Mss Afr. s. 2399, Nigeria-Biafra Association Papers.

II. ORAL HISTORY INTERVIEWS

Agu, Elija Ukaeme, Umunomo, Mbaise, 3 August 1999.

Anabalam, Comfort, Umuchieze, Mbaise, 13 December 1998.

Anabalam, Linus, Umuchieze, Mbaise, 13 and 14 December 1998.

Chidomere, Christopher, Umuchieze, Mbaise, 13 December 1998.

Chidomere, Grace, Umuchieze, Mbaise, 13 December 1998.

Chigbu, Mathais, Umunomo, Mbaise, 25 July 1999.

Eboh, James, Alike Obowo, Etiti, 2 January 2000.

Eke, Aplonsus, Umupko-Emeabiam, Owerri, 18 December 1999.

Emenike, Sarah, Owerri, 22 December 1998.

Enweremadu, Francis, Mbutu, Mbaise, 31 December 1999 & 2 January 2000.

Ibe, Annex, Umunomo, Mbaise, 17 December 1998.

Ibe, Robert, Umuchieze, Mbaise, 28 July 1999.

Ibekwe, Uzodinma, Nguru, Mbaise, 25 December 1999.

Iheaguta, Michael, Umuchieze, Mbaise, 2 August 1998.

Ihediwa, Eleazer, Owerrenta, Isiala Ngwa, 24 July 1999.

Iheonunekwu, Linus, Mbutu, Mbaise, 2 January 2000.

Ihuoma, Francis, Mbutu, Mbaise, 17 December 1998.

Iwu, Isaac, Umuchieze, Mbaise, 18 December 1999.

Iwuagwu, Chilaka, Umunomo, Mbaise, 31 July 1999.

Iwuagwu, Fidelis Ogu, Umunomo, Mbaise, 3 August 1999.

Iwuagwu, Susan, Umunomo, Mbaise, 31 July 1999

Iwuh, Alfred, Umuchieze, Mbaise, 16 December 1998.

Korie, Alphreda, Ihitteafoukwu, Mbaise, 23 December 1998.

Korieh, Amarahiaugwu,Umuchieze, Mbaise, 13 December 1998

Korieh, Isdore, Ihitteafoukwu, Umuchieze, Mbaise, 5 May 2008.

Korieh, Mbagwu, Umuchieze, Mbaise, 18 December 1998.

Korieh, Onyegbule, Ihitteafoukwu, Mbaise, 17 December 1998.

Marizu, Canice, Nguru, Mbaise, 25 December 1999.

Marizu, Christiana, Nguru, Mbaise, 25 December 1999.

Nkwocha, Gilbert, Umupko-Emeabiam, Owerri, 23 December 1999.

Nwoko, Paschal, Mbaise, 22 December 1998.
Nwosu, Ugwuanya, Owerri, 22 December 2000.
Nwosu, Lawrence, Umupko-Emeabiam, Owerri, 18 December 1999.
Nwosu, Sybilia, Umuchieze, Mbaise, 12 December 1998.
Obasi, Gabriel Nwokeke, Umunomo, Mbaise, 25 July 1999.
Obasi, Nwanyiafo, Umunomo, Mbaise, 25 July 1999.
Okafor, Ambrose, Umuchieze, Mbaise, 15 December 1998.
Okafor, Diala, Umuchieze, Mbaise, 17 December 1998.
Okere, Jonah, Umuekwune, Ngor Okpala, 12 December 1999.
Okere, Nwalozie, Emeabiam, Owerri, 22 December 1999.
Onyema, Theophilus, Umuorlu, Isu, 5 January 2000.
Onyesoh, Alban, Mbaise, 18 December 1999.
Osunwoke, Luke, Umuorlu, Isu, 7 January 2000.
Otuonye, Eugenia, Umuchieze, Mbaise, 23 December 1998.
Philip Simon, Njoku, Nguru, Mbaise, 12 January 2000.
Uche, Lucian, Umuevu, Mbaise, 20 January 2000.
Uzor, Gilbert, Umunomo, Mbaise, 22 July 2000.

III. THESES

Agiri, B. A. "Kola in Western Nigeria, 1850–1950." PhD dissertation, University of Wisconsin, 1972.

Akpalaba, Onyema. "The Role of Agriculture in the Economic Development of Ezinihitte since 1900." BA thesis, University of Nigeria, 1987.

Alua, Chioma N. "Cassava Industry and its Socio-Economic Impact on Umuokanne-Ohaji from 1900 to Present." BA thesis. University of Nigeria, 1966.

Anumihe, Ebere Chinyere. "The Role of Cassava in Ensuring Small farm Household Food Security in Imo State." PGD, Imo State University, 1998.

Anyawnu, S.E.N. "The Igbo Family Life and Cultural Change." PhD dissertation, Philipps-Universität, Marburg, 1976.

Chuku, Gloria I. "Changing Role of Women in Igbo Economy, 1929–1985." PhD thesis, Department of History, University of Nigeria, Nsukka, 1995.

Eland, Libuse Renata, "A Separate Spheres' Approach to the agricultural model." PhD thesis, University of Georgia, 1999.

Esse, Uwakwe O. A. "The Second World War and Resource management in Eastern Nigeria, 1939–1945." MA thesis, Department of History, University of Nigeria, 1997.

_____. "Road Transportation in Nigeria as a Private Enterprise Among the Igbo, 1920–1999." PhD thesis, Department of History, University of Nigeria, Nsukka, 2003.

Iwuagwu, Obichere Chilaka. "A Socio-economic History of Food Production in Igboland 1900–1980: A Study of Yam, Cocoyam and Cassava." PhD thesis. University of Lagos, 1998.

Jones, Elizabeth Bright. "Gender and Agricultural Change in Saxony, 1900–1930." PhD thesis, University of Minnesota, 2000.

Lee, Krug Karen. "Farm Women's Perspectives on the Agricultural Crisis, Ecological Issues, and United Church of Canada Social Teaching." PhD thesis, University of Victoria, Canada, 1995.

Lekwa, G. "The Characteristics and Classification of Genetic Sequences of Soil in the Coastal Plain Sands of Eastern Nigeria." PhD dissertation, Michigan State University, 1979.

Libuse, Elad Renata, "A 'Separate Spheres' Approach to the Agricultural Model." PhD thesis, University of Georgia, 1999.

Mah, Ahmed M. "The Colonial Discourse of Development in Africa: The Somalia Experience." MA thesis, University of Toronto, 1999.

Nakanyike, B. Musisi. "Transformations of Baganda Women: From the Earliest times to the demise of the Kingdom in 1966." PhD dissertation, University of Toronto, 1991.

Njoku, Ndubueze-Life. "Influence of International Economy on the Peasantry in Eastern Nigeria, 1900–1960." PhD thesis, University of Calabar, 1991.

Nwokeji, G. Ugo. "The Biafran Frontier: Trade, Slaves, and Aro Society, c. 1750–1905." PhD thesis, University of Toronto, 1999.

Obijuru, Ihuoma Adaure. "State Capitalism and the Peasantry: A Case Study of ADAPALM in Imo State." MA thesis, Imo State University, 1998.

Oham, Anthony. "Labor Migration from Southeastern Nigeria to Spanish Fernando Po, 1900–1968," MA thesis, Department of History, Central Michigan University, 2007.

Okolie, Andrew C. "Oil Revenue, International Credits and Food in Nigeria, 1970–1992." PhD thesis, University of Toronto, 1995.

Okoro, Caleb O. "The Uzouwani Farm Settlement and Socioeconomic Development in the Anambra Basis, 1961–1971." MA thesis, University of Nigeria, 1986.

Olsen, Margareta. "Patterns of Protest: Swedish Farmers in Times of Cereal Surplus Crisis." PhD thesis, University of Adelaide, 1993.

Onyewuenyi, M. A. "Railway Development and the Growth of Export Agriculture in Nigeria, 1900–1950." MA thesis, University of Ottawa, 1981.

Oyemakinde, O. "A History of Indigenous Labour on the Nigerian Railway, 1995–1945." PhD thesis, University of Ibadan, 1970.

Oyewumi, Oyeronke. "Mothers Not Women: Making an African Sense of Western Gender Discourse." PhD dissertation, University of California at Berkeley, 1993.

Puplampu, Peter K. "The State and Agriculture: The Social Dynamics of Agricultural Policy in Ghana, 1900–1994." PhD thesis, University of Alberta, 1998.

Soon, Kim C. "The Role of the Non-Western Anthropologist Reconsidered: Illusion versus Reality," *Current Anthropology* 31, no. 2 (1990): 196–201.

Stone, Magaret Priscilla. "Women, Work, and Marriage: A Restudy of the Nigeria Kofyar." PhD dissertation, University of Arizona, 1988.

Wilson, Myers Gregory. "This Is Not Your Land: An Analysis of the Impact of the Land Use Act in Southwest Nigeria." PhD thesis, University of Wisconsin-Madison, 1990.

IV. UNPUBLISHED PAPERS

Igwe, A. U. "The Role of Migrant Labour from Owerri Province in the Economy of Eastern Nigeria, 1915–1965." PhD Seminar, Department of History, University of Lagos, November 1998.

Ikpi, A. E., et al. "Cassava – A Crop for Household Food Security: A 1986 Situation Analysis from Oyo Local Government Area, Nigeria," 1986.

Imoudu, Peter, B. "Social Science Research and Public Policy Implications for Agricultural Development in Nigeria," 25–27 March 1996.

Labo, Adbullahi. "Social Research, Agricultural Policy and Rural Social Change in Nigeria." Paper presented at the Ninth General Assembly of Social Science Council of Nigeria, University of Lagos, 25–27 March 1996.

Nwokeji, G. Ugo. "African Conceptions of Gender and the Slave Traffic." Paper presented at Transatlantic Slavery and the African Diaspora; Using the W.E.B. DuBois Dataset of Slaving Voyages, Omohundro Institute of African American History and Culture, Williamsburg, Virginia, 11–13 September 1998.

Toyo, Eskor. "The Peasantry in Nigeria: Identity and Change," Unpublished paper, Department of Economics, University of Maiduguri, 1980.

V. NEWSPAPERS

New York Times, 12 April 1903.
Pall Mall Gazette, 12 April 1903.
Time, August 1968.
Toronto Globe and Mail, 2 October 1968.
Wall Street Journal, 25 November 1981.
West Africa, 1924.
West African Pilot, 1942–1945.

VI. SECONDARY SOURCES

Achebe, Chinua. *Things Fall Apart*. London: Heinemann, 1959.

Achebe, Nwando. *Farmers, Traders, Warriors, and Kings: Female Power and Authority in Northern Igboland, 1900–1960*. Portsmouth, NH: Heinemann, 2005.

Acker, Joan. "From Sex Roles to Gendered Institutions." *Contemporary Sociology* 21 (1992): 565–69.

Adams, R. "The Economic and Demographic Determinants for International migration in Rural Egypt." *Journal of Development Studies* 30, no. 1 (1993): 146–67.

Afigbo, Adiele E. *The Abolition of the Slave Trade in Southeastern Nigeria, 1885–1950*. Rochester, NY: University of Rochester Press, 2006.

———, ed. *Groundwork of Igbo History*. Lagos: Vista Books, 1992.

———. "Revolution and Reaction in Eastern Nigeria 1900–1929 (The Background to the Women's Riot of 1929)." *Journal of the Historical Society of Nigeria* 3, no. 3 (1966): 539–57.

———. "The Aro Expedition of 1901–1902: An Episode in the British Occupation of Iboland." *ODU: A Journal of West African Studies* 7 (1972): 3–25.

———. *The Warrant Chiefs: Indirect Rule in Eastern Nigeria 1891–1929*. Ibadan: Longman, 1972.

———. "The Calabar Mission and the Aro Expedition of 1901–1902." *Journal of Religion in Africa* 5, no. 2 (1973): 94–106.

———. "The Eclipse of the Aro Trading Oligarchy of Southeastern Nigeria, 1901–1927." *Journal of the Historical Society of Nigeria* 4, no. 1 (1971): 3–24.

———. *Ropes of Sand: Studies in Igbo History and Culture*. Nsukka: University Press, 1981.

Afonja, Simi. "Current Explanations of Sex Roles and Inequality: A Reconsideration." *Nigeria Journal of Social Studies* 22, no. 1 (1980): 85–105.

———. "Changing Modes of Production and the Sexual Division of Labour among the Yoruba." Signs 7, no. 2 (Winter 1981): 299–313.

———, and Bisi Aina. *Nigeria Women in Social Change*. Ile-Ife: Obafemi Awolowo University, 1995.

Ajaegbu, H. I. *Urban and Rural Development in Nigeria*. London: Heinemann, 1976.

Ajayi, S. I., M. A. Iyoha, A. Soyode, and E. C. Anusionwu. *The Nigerian Economy: A Political Economy Approach*. London: Longman, 1989.

Alford, Robert, and Roger Friedland, *Power of Theory: Capitalism, the State and Democracy*. New York: Cambridge University Press, 1985.

Allan, William. *The African Husbandman*. New York: Barnes & Noble, 1965.

Allman, Jean, Susan Geiger, and Nakanyike Musisi, eds. *Women in African Colonial Histories*. Bloomington and Indianapolis: University of Indiana Press, 2002.

Alpers, E. A. *Ivory and State in East Central Africa: Changing Patterns of International Trade in the Late Nineteenth Century*. London: Heinemann, 1975.

Amadiume Ifi. *Re-inventing Africa: Matriarchy, Religion and Culture*. London and New York: Zed Books, 1997.

_____. "Underdevelopment and Dependence in Black Africa." *Journal of Modern African Studies* 10, no. 4 (1974): 503–24.

Arghiri, Emmanuel, *Unequal Exchange: A Study of the Imperialism of Trade*. Translated from the French by Brian Pearce. London: Monthly Review Press, 1972.

Arrighi, Giovanni, and John S. Saul, *Essays on the Political Economy of Africa*. New York: Monthly Review Press, 1973.

Asante, S.K.B. "Food as a Focus of National and Regional Policies in Contemporary Africa." In *Food in Sub-Saharan Africa*, ed. A. Hansen and D. E. McMillan, 11–24. Boulder, CO: Lynne Rienner, 1986.

Ayadi, Olusegun Felix, and Caleb Ojo Falusi. "The Social and Financial Implications of Agricultural Farm Settlements in Nigeria." *Journal of Asian and African Studies* 31, no. 3–4 (1996): 191–216.

Ayittey, George. Africa in Chaos. New York: St. Martin's Press, 1998.

Babalola, Ademola. "Colonialism and Yoruba Women in Agriculture." In *Nigerian Women in Social Change*, ed. Simi Afonja and Bisi Aina, 47–55. Ile-Ife, Nigeria: Obafemi Awolowo University Press, 1995.

Baker, Jonathan. *Rural Communities Under Stress: Peasant Farmers and the State in Africa*. New York: Cambridge, 1989.

_____. *The Politics of Agriculture in Tropical Africa*. Berkley Hills: Sage, 1984.

_____. *Agricultural Change in Nigeria: Case Studies in the Developing World*. London: John Murray, 1989.

Baker, R. "Linking and Sinking: Economic Externalities and the Persistence of Destitution and Famine in Africa." In *Drought and Hunger in Africa*, ed. Michael H. Glantz, 149–170. Cambridge: Cambridge University Press, 1987.

Bamisaiye, E. A. "Solving the Food Crisis in Africa: The Role of Higher Education." *Journal of African Studies* 11, no. 4 (1985): 182–88.

Barber, W. J. "The Movement into the World Economy." In *Economic Transition in Africa*, ed. Melville J. Herskovits and Mitchell Harwitz, 299–329. Evanston, IL: Northwestern University Press, 1961.

Barry, Kathleen. "Biography and the Search for Women's Subjectivity," *Women's Studies International Forum* 12, 6 (1989): 561–77.

Basden, George Thomas. *Niger Ibos: A Description of the Primitive Life, Customs and Animistic Beliefs of the Ibo People of Nigeria by one who, for Thirty-five Years, Enjoyed the Privilege of Their Intimate Confidence and Friendship*. New York: Barnes & Noble, 1966.

Bates, Robert. *Essays on the Political Economy of Rural Africa*. Cambridge: Cambridge University Press, 1983.

_____. *Markets and States in Tropical Africa: The Political Basis of Agricultural Policies*. Los Angeles: University of California Press, 1981.

Baumann, Hermann. "The Division of Work According to Sex in African Hoe Culture." *Africa* 1, no. 5 (1928): 289–319.

Beer, C. *The Politics of Peasant Groups in Western Nigeria*. Ibadan: Ibadan University Press, 1975.

Beinart, W., and C. Bundy. *Hidden Struggles in Rural South Africa*. London: James Currey, 1987.

Beinart, W. "Soil Erosion, Conservation and Ideas about Development: A Southern Africa Exploration, 1900–1960." *Journal of Southern African Studies* 11, no. 1 (1984): 52–83.

Bell, Diane, Pat Caplan, and Wazir Johan Karim, eds. *Gendered Fields: Women, Men and Ethnography*. London and New York: Routledge, 1993.

Beneria, Lourdes, and Gita Sen. "Accumulation, Reproduction, and Women's Role in Economic Development: Boserup Revisited." In *Women's Work*, ed. H. Leacock and H. Safa, 42–51. South Hadley, MA: Bergin and Garvey, 1986.

Berkvens, R. "Backing Two Horses: Interaction of Agricultural and Non-Agricultural Activities in a Zimbabwean Communal Area." Working Paper no. 24. Leiden: African Studies Centre, 1997.

Bernstein, Henry. "African Peasantries: A Theoretical Framework." *Journal of Peasant Studies* 6, no. 4 (1979): 421–43.

Berry, Sara. "Macro-policy Implications of Research on Rural Households and Farming Systems." In *Understanding Africa's Rural Households and Farming Systems*, ed. Joyce Lewinger Moock, 199–216. Boulder, CO: Westview Press, 1986.

_____. "Oil and the Disappearing Peasantry: Accumulation, Differentiation and Underdevelopment in Western Nigeria." *African Economic History* 13 (1984):1–22.

_____. "Social Science Perspective on Food in Africa." In *Food in Sub-Saharan Africa*, ed. A Hansen and D. McMillan, 64–81. Boulder, CO: Lynne Rienner, 1986.

_____. "The Food Crisis and Agrarian Change in Africa: A Review Essay." *African Studies Review* 27, no. 2 (1984): 59–112.

_____. *Cocoa, Custom and Socio-economic Change in Western Nigeria*. Oxford: Clarendon Press, 1975.

_____. *Fathers Work for Their Sons: Accumulation, Mobility, and Class Formation in an Extended Yoruba Community*. Berkeley: University of California Press, 1985.

_____. *No Condition is Permanent: The Social Dynamics of Agrarian Change in Sub-Sahara Africa*. Madison: University of Wisconsin Press, 1993.

Boserup, Ester. *Women's Roles in Economic Development*. London: Allen & Unwin, 1970.

Bowlig, Simon. *Peasant Production and Market Relations: A Case Study of Western Ghana*. Copenhagen: Third World Observer, 1993.

Bradley, C. "Women, Weeding and Plow: A Comparative Test of Boserup's Hypothesis" *African Urban Quarterly* 5, nos. 3–4 (1990): 188–96.

Braidotti, Rosi. *Nomadic Subjects: Embodiment and Sexual Difference in Feminist Theory*. New York: Columbia University Press, 1994.

Breth, E. A. *Colonialism and Underdevelopment in East Africa: The Politics of Economic Change*. London: Heinemann, 1974.

Bryceson, Deborah Fahy, ed. *Women Wielding the Hoe: Lessons for Rural Africa from Feminist Theory and Development Practice*. Oxford and Washington, DC: Berg, 1995.

_____. "Sub-Saharan Africa Betwixt and Between: Rural Livelihood Practices and Policies." *African Studies Centre Working Paper* 43. Leiden: Afrika-Studiecentrum, 1999.

_____, and Vali Jamal, eds. *Farewell to Farms: De-agrarianisation and Employment in Africa*. Aldershot: Ashgate, 1997.

_____. "African Rural Labour, Income Diversification and Livelihood Approaches: Long-term Development Perspective." *African Studies Centre Working Paper* 35, Leiden: African Studies Center, 1999.

_____. "Household, Hoe and Nation: Development Policies of the Nyerere Era." In *Tanzania After Nyerere*, ed. Michael Hodd, 36–48. London and New York: Pinter, 1989.

_____, and Mooij, ed. *Disappearing Peasantries: Rural Labour in Africa, Asia and Latin America*. London. Intermediate Technology Publications, 2000.

Bryson, Judy C. "Women and Agriculture in Sub-Saharan Africa: Implications for Development (An Exploratory Study)." *Journal of Development Studies* 17, no. 3 (1981): 29–46.

Buchanan, K. M., and J. C. Pugh. *Land and People in Nigeria: The Human Geography of Nigeria and its Environmental Background*. London: University of London Press, 1955.

Bulbeck, Chilla, *Re-orienting Western Feminism: Women's Diversity in a Postcolonial World*. Cambridge: Cambridge University Press, 1998.

Bundy, Colin. *The Rise and Fall of the South African Peasantry*. London: Heinemann, 1979.

Burgess, Robert G. *Strategies of Educational Research: Qualitative Methods*. London: Falmer, 1985.

Bush, Ray. "The Politics of Food and Starvation," *Review of African Political Economy* 68 (1996): 169–95.

Cain, P. J., and A.G. Hopkins, *British Imperialism, 1688–2000*, 2nd ed. Harlow, England: Longman, 2001.

Carloni, A. *Integrating Women in Agricultural Projects: Case Studies of Ten FAO-Assisted Field Projects*. Rome: FAO, 1983.

Casinader, Rex A. Sepalika Fernando, and Karuna Gamage, "Women's Issues and Men's Roles: Sri Lankan Village Experience." In *Geography of Gender in the Third World*, ed. Janet Henshall and Janet Townsend, 309–322. Hutchinson: State University of New York Press, 1987.

Chaudhuri, Nupur, and Margaret Strobel, eds. *Western Women and Imperialism: Complicity and Resistance*. Bloomington: Indiana University Press, 1992.

Childe, V. G. *What Happened in History*. London: Routledge and Paul, 1942.

Chisholm, M. D. I. *Rural Settlement and Land Use*. London: Hutchison Press, 1962.

Chodorow, Nancy. *The Reproduction of Mothering: Psychoanalysis and Sociology of Gender*. Berkeley: University of California Press, 1978.

Chubb, L. T. *Ibo Land Tenure* (2nd ed.). Ibadan: Ibadan University Press, 1961.

Chuku, Gloria. "Igbo Women in the Economy of Igboland, 1900–1970." *African Economic History* 23 (1995): 37–50.

_____. *Igbo Women and Economic Transformation in Southeastern Nigeria, 1900–1960.* New York: Routledge. 2005.

Chukwuezi, Barth. "De-agrarianisation and Rural Employment in Igboland, Southeastern Nigeria." *ASC Working Paper 37/1999.* Leiden: Afrika-Studiecentrum, Centre for Research and Documentation (CRD), Kano, 1999.

Clark, F.A.S. "The Development of the West African Forces in the Second World War," *Army Quarterly* (1947): 58–72.

Clark, Peter Bentley. "Economic Planning for a Country in Transition: Nigeria." In *Planning Economic Development: A Study,* ed. Everett D. Hagen, 252–93. Homewood, IL: Richard D. Irwin, 1963.

Clarke, E. *My Mother who Fathered Me.* 2nd ed. London: George Allen and Unwin, 1966.

Clarke, J. D. "The Spread of Food Production in Sub-Saharan Africa." *Journal of African History* 3, no. 2 (1962): 211–28.

Cliffe, Lionel. "Nationalism and the Reaction to Enforced Agricultural Change in Tanganyika during the Colonial Period." In *Socialism in Tanzania: An Interdisciplinary Reader,* vol. 1, ed. L. Cliffe and John S. Saul, 17–24. Dar es Salaam: East Africa Publishing House, 1972.

Clifford, James, and George E. Marcus, eds. *Writing Cultures: The Poetics and Politics of Ethnography.* Berkeley: University of California Press, 1986.

Cloud, Kathlen. *Gender Issues in A.I.D.'s Agricultural Projects: How Efficient Are We?* Washington, DC: USAID, 1987.

Clute, Robert E. "The Role of Agriculture in African Development." *African Studies Review* 25, no. 4 (1982): 1–20.

Cohen, D. W. "Reconstructing a Conflict in Bunafu: Seeking Evidence Outside the Narrative Tradition." In *The African Past Speaks: Essays on Oral Tradition and History,* ed. Joseph C. Miller, 201–20. Hamden, CT: Archon, 1980.

Collins, Patricia Hill. *Black Feminist Thought: Knowledge, Consciousness, and the Politics of Empowerment.* New York: Routledge, 1990.

Commander, S. ed. *SAP and Agriculture: Theory and Practice in Africa and Latin America.* London: Overseas Development Institute, 1989.

Commins, S. K., M. F. Lofchie, and R. Payne. *Africa's Agrarian Crisis: The Roots of Famine.* Boulder, CO: Lynne Rienner, 1986.

Commonwealth Secretariat, *Engendering Adjustment for the 1990s: Report of a Commonwealth Expert Group on Women and Structural Adjustment.* London: Commonwealth Secretariat, 1989.

Connell, R. W. *Gender and Power: Society, the Person, and Sexual Politics.* Stanford, CA: Stanford University Press, 1987.

Cooper, Barbara M. "Oral Sources and the Challenge of African History." In *Writing African History,* ed. John Edward Philips, 191–215. Rochester, NY: University of Rochester Press, 2005.

Coquery-Vidrovitch, Catherine. *African Women: A Modern History.* Translated by Beth Gillian Raps. Boulder: Westview Press, 1997.

Cornell, Drucilla. *Transformations: Recollective Imagination and Sexual Difference.* New York: Routledge, 1994.

Creswell, John W. *Qualitative Inquiry and Research Design: Choosing among Five Traditions.* Thousand Oaks, CA: Sage, 1998.

Crowder, Michael. "World War II and Africa: Introduction." *Journal of African History* 26 (1985): 287–88.

Curry, John, et al., "Gender and Livestock in African Production Systems: An introduction." *Human Ecology: An Interdisciplinary Journal* 24, no. 2 (1996): 149–61.

Davison, Jean, ed. *Agriculture, Women and Land: The African Experience.* Boulder, CO: Westview Press, 1988.

Dike, Azuka A. "Urban Migrants and Rural Development," *African Studies Review* 25, no. 4 (1982): 85–94.

Dike, Onwuka K. *Trade and Politics in the Niger Delta,* 1830–1885. Oxford: Clarendon Press, 1956.

Dillon, Pattie. "Teaching the Past through Oral History." *Journal of American History* 87, no. 2 (2000): 602–07.

Duggan, W. R. *An Economic Analysis of Southern Africa Agriculture.* Westport, CT: Praeger, 1986.

Echeruo, Michael J. C. "Aro and Nri: Lessons of Nineteenth Century Igbo History." In *The Aftermath of Slavery: Transitions and Transformations in Southeastern Nigeria,* ed. Chima J. Korieh and Femi J. Kolapo, 228–47. Trenton: Africa World Press, 2007.

Edozien, E. C. "Poverty: Some Issues in Concept and Theory." In *Poverty in Nigeria,* ed. O. Teriba, 35–42. Ibadan: Nigerian Economic Society, 1975.

Eicher, Carl K. *Research on Agricultural Development in Five English Speaking Countries in West Africa.* New York: Agricultural Development Council, 1970.

_____. "The Dynamics of Long-term Agricultural Development in Nigeria." In *Growth and Development of the Nigerian Economy,* ed. C. K. Eicher and Carl Lieldholm, 1–15. East Lansing: Michigan State University Press, 1970.

_____. "Facing Up to Africa's Food Crisis." *Foreign Affairs* 61 (1982): 151–74.

Ekechi, Felix K. *Tradition and Transformation in Eastern Nigeria: A Sociopolitical history of Owerri and its Hinterland, 1902–1947.* Kent, OH: Kent State University Press, 1989.

_____. "Episodes of Igbo Resistance to European Imperialism." In *Olaudah Equiano and the Igbo World: History, Society and Atlantic Diaspora Connections,* ed. Chima J. Korieh, 219–43. Trenton: Africa World Press, 2009.

Ekejiuba, Felicia I. "Omu Okwei – The Merchant Queen of Ossomari: A Biographical Sketch." *Journal of the Historical Society of Nigeria* 3, no. 4 (1967): 663–46.

Elliot, Walter. "The Parliamentary Visit to Nigeria." *Journal of the Royal African Society* 27, no. 107 (1928): 205–18.

Elson, Diane. "Male Bias in the Development Process: An Overview." In *Male Bias in the Development Process*, ed. Diane Elson, 1–28. Manchester: Manchester University Press, 1995.

Enslin, Elizabeth. "Beyond Writing: Feminist Practice and the Limitations of Ethnography," *Cultural Anthropology* 9, 4 (1994): 537–68.

Ensminger, J. "Economic and Political Differentiation among Galole Women," *Ethnos* 52, nos. 1–2 (1987): 28–49.

Etienne, Mona, and Eleanor Leacock. *Women and Colonization: Anthropological Perspectives*. New York: Praeger, 1980.

Etienne, Mona. "Women and Men, Cloth and Colonization: The Transformation of Production-Distribution Relations among the Baule (Ivory Coast)." In *Women and Colonization: Anthropological Perspectives*, ed. Etienne Mona and Eleanor Leacock, 214–38. New York: Praeger, 1980.

Fairhead, James, and Melissa Leach. *Reframing Deforestation: Global Analysis and Local Realities: Studies in West Africa*. London and New York: Routledge, 1998.

Falcon, W. P., W. O. Jones, and S. R. Pearson. *The Cassava Economy of Java*. Stanford, CA: Stanford University Press, 1984.

Falconbridge, Alexander. *An Account of Slave Trade on the Coast of Africa*. London, 1788.

Fallers, L. A. "Are African Cultivators to be called 'Peasants'?" *Current Anthropology* 2, no. 2 (1961): 108–10.

Falola, Toyin. "Cassava Starch for Export in Nigeria during the Second World War." *Journal of African Economic History* 18 (1989): 73–98.

———. "'Salt is Gold': The Management of Salt Scarcity in Nigeria during World War II," *Canadian Journal of African Studies* 26, no. 3 (1992): 412–36.

———. "Mission and Colonial Documents." In *Writing African History*, ed. John Edward Philips, 266–83. Rochester, NY: University of Rochester Press, 2005.

———, and J. O. Ahazuem. "Production for the Metropolis: Agriculture and Forest Products." In *Britain and Nigeria: Exploitation or Development*, ed. Toyin Falola, 80–90. London: Zed Books, 1987.

———, and Christian Jennings, eds. *Sources and Methods in African History: Spoken, Written and Unearthed*. Rochester, NY: University of Rochester Press, 2003.

Faloyan, Adekunle. *Agriculture and Economic Development in Nigeria: A Prescription for the Nigeria Green Revolution*. New York: Vantage Press, 1983.

Fanon, Franz. *A Dying Colonialism*. New York: Grove Press, 1965.

Feldstein, H. S., and S. V. Poats. *Working Together: Gender Analysis in Agriculture*, vol. 1: *Case Studies*. Hartford: Kumarian Press, 1989.

Feldstein, Hilary S., and Jiggins Janice. *Tools for the Field: Methodologies Handbook for Gender Analysis in Agriculture*. West Hartford: Kumarian Press, 1994.

Finnegan, Ruth. *Oral Tradition and the Verbal Arts*. London and New York: Routledge, 1992.

Flinn, Juliana, Leslie Marshall, and Jocelyn Armstrong. *Fieldwork and Families: Constructing New Models for Ethnographic Research*. Honolulu: University of Hawai'i Press, 1998.

Flint, J. E. *Nigeria and Ghana*. Englewood Cliffs, NJ: Prentice-Hall, 1966.

Flores-Meiser, Enya. "Field Experience in Three Societies." In *Fieldwork: The Human Experience*, ed. Robert Lawless, Jr., Vinson Sutlive, and Mario Zamora, 1983.

Floyd, Barry. *Eastern Nigeria: A Geographical Review*. New York: Frederick C. Prager, 1969.

_____, and Monica Adinde. "Farm Settlements in Eastern Nigeria: A Geographical Appraisal." *Economic Geography* 43, no. 3 (1967): 189–230.

Food and Agricultural Organization, *Food and Agriculture Year Book, 1980*, 75–80. Rome: FAO, 1980.

Forestone, Shulamith, *The Dialectic of Sex*. New York: William Morrow, 1970.

Formes, Malia B. "Beyond complicity versus Resistance: Recent Work on Gender and European Imperialism." *Journal of Social History* (1995): 629–41.

Forsyth, Frederick, *The Making of an African Legend: The Biafra Story*. Harmondsworth: Penguin, 1977.

Fortmann, Louise. "The Plight of the Invisible Farmer: The Effects of National Agricultural Policy on Women in Africa." In *Women and Technological Change in Developing Countries*, ed. Cain Dauber, 204–14. Boulder, CO: Westview Press, 1981.

Foster, Johanna, "An Invitation to Dialogue: Clarifying the Position of Feminist Gender Theory in Relation to Sexual Difference Theory," *Gender and Society* 13, no. 4 (1999): 431–56.

Frank, A. G. *Latin America: Underdevelopment and Revolution*. New York: Monthly Review Press, 1969.

Freidl, Ernestine. *Men and Women: An Anthropologist's View*. New York: Holt, Rinehart and Winston, 1975.

Freund, Bill. *The Making of Contemporary Africa: The Development of African Societies since 1800* (Second edition). Boulder, CO: Lynne Rienner, 1998.

Freund, B. *Capital and Labour in the Nigerian Tin Mines*. Atlantic Highland, NJ: Humanities Press, 1981.

Friedmann, Harriet, "Household Production and the National economy: Concepts for the Analysis of Agrarian formations." *Journal of Peasant Studies* 7 (1980): 158–84.

Gailey, Harry. *The Road to Aba: A Study of British Administrative Policy in Eastern Nigeria*, New York: New York University Press, 1970.

Gakou, Mohamed. L. *The Crisis in African Agriculture*. London: Atlantic Highlands, 1987.

Gallagher, J., and R. E. Robinson, "The Imperialism of Free Trade," *Economic History Review* 6 (1953): 1–15.

Geiger, Susan, *Tanu Women: Gender and Culture in the Making of Tanganyika Nationalism, 1955-1965*. Portsmouth, NH: Heinemann, 1997.

Gertzel, Cherry. "Towards a Better Understanding of the Causes of Poverty in Africa in the late Twentieth Century," *Briefing Paper No. 42*. Canberra: Australian Development Studies Network, 1996.

Gladwin, Christina H., ed. *Structural Adjustment and African Women Farmers*. Center for African Studies. Gainesville: University of Florida Press, 1991.

Gleave, M. B., and A. L. Mabogunje. "Changing Agricultural Landscape in Southern Nigeria – The Example of Egba Division, 1850–1950." *Nigerian Geographical Journal* 7, no. 1 (1964): 1–15.

Goldman, A. "Population Growth and Agricultural Change in Imo State." In *Population and Agricultural Change in Africa*, ed. B. L. Turner II, Hyden Goran, and Robert W. Kates, 250–301. Gainesville: University of Florida Press, 1993.

Green, M. M. *Land Tenure in an Igbo Village in Southeastern Nigeria*. LSE Monograph on Social Anthropology no. 6. London: Berg, 1941.

_____. *Igbo Village Affairs: Chiefly with Reference to the Village of Umueke Agbaja*. London: Frank Cass, 1964.

Gubrium, Jaber F., and James A. Holstein, *The New Language of Qualitative Method*. New York: Oxford University Press, 1997.

_____. "Naturalism in Models of African Production," Man 19 (1984): 355–73.

_____. "Female Farming in Anthropology and African History." In *Gender at the Crossroads of Knowledge: Feminist Anthropology in the Postmodern Era*, ed. M. di Leonardo, 257–77. Berkeley, University of California Press, 1991.

_____. "Multiplication of Labour: Historical Methods in the Study of Gender and Agricultural Change in Modern Africa." *Current Anthropology* 29 (1988): 247–72.

Hafkin, Nancy, and Edna Bay. "Introduction." In *Women in Africa: Studies in Social and Economic Change*, ed. Nancy Hafkin and Edna Bay, 1–18. Stanford: Stanford University Press, 1976.

Haggblade, S., et al. "Non-Farm Linkages in Rural Sub-Saharan Africa." *World Development* 17, no. 8 (1989): 1173–201.

Hano-Sano, Otto. *The Political Economy of Food in Nigeria, 1960–1982: Discussion on Peasants, State and the World Economy*. Uppsala: Scandinavian Institute of African Studies, 1983.

Harneit-Sievers, Axel, Ahazuem Jones, and Sydney Emezue. *A Social History of the Nigerian Civil War: Perspectives from Below*. Enugu: Jemezie Associates, 1997.

Harris, J. "Paper on the Economic Aspects of the Ozuitem Ibo." *Africa* 14 (1943): 12–23.

_____. "Some Aspects of the Economics of 16 Ibo individuals." *Africa* 14 (1944): 302–35.

_____. "Human Relationships to the Land in Southern Nigeria." *Rural Sociology* 7 (1942): 89–92.

Hart Gillian, "Engendering Everyday Resistance: Politics, Gender and Class Formation in Rural Malaysia," *Journal of Peasant Studies* 19, no. 1 (1991): 93–121.

Hart, Keith. *The Political Economy of West African Agriculture*. New York: Cambridge University Press, 1982.

Helleiner, Gerald K. *Peasant Agriculture. Government and Economic Growth in Nigeria*. Homewood, IL: Richard D. Irwin, 1966.

_____. "The Fiscal Role of the Marketing Boards in Nigerian Economic Development, 1947–61." *Economic Journal* 74, no. 295 (1964): 582–610.

Henderson, Richard. *The King in Every Man: Evolutionary Trend in Onitsha Igbo Society.* New Haven, CT: Yale University Press, 1972.

Henige, David. "Oral Tradition as a Means of Reconstructing the Past." In *Writing African History*, ed. John Edward Philips, 169–90. Rochester, NY: University of Rochester Press, 2005.

Henrietta, Moore. *Space, Text and Gender: An Anthropological Study of Marakwet of Kenya.* Cambridge: Cambridge University Press, 1986.

Heyer, Judith, et al. *Rural Development in Tropical Africa.* New York: St. Martin's Press, 1981.

Hill, Polly. *Studies in Rural Capitalism in West Africa.* Cambridge: Cambridge University Press, 1970.

_____. *Rural Hausa: A Village and a Setting.* Cambridge: Cambridge University Press, 1972.

Hofmeier, R. *Transport and Economic Development in Tanzania.* Munich: Ifo-Institute für Wintschaftsforschung Weltforum Verlag, 1973.

Hogendorn, J. S. "Economic Initiative and African Cash-Farming: Pre-Colonial Origins and Early Colonial Developments." In *Colonialism in Africa 1870–1960, vol. 4: The Economics of Colonialism*, ed. L. H. Gann and Peter Duignan, 283–328. Cambridge: Cambridge University Press, 1975.

Hogendorn, J. S. "The Vent-for-Surplus Model and African Cash Agriculture to 1914." *Savanna* 5, no. 1 (1976): 15–28.

Hoopes, John. *Oral History: An Introduction.* Chapel Hill, NC: University of North Carolina Press, 1979.

Hoppenbrouwers, Peter, Jan Luiten van Zanden, and J. Luiten van Zanden, eds. *Peasants into Farmers? The Transformation of Rural Economy and Society in the Low Countries (Middle Ages – 19th Century) in Light of the Brenner Debate.* Turnhout: Brepols, 2001.

Hopkins, Anthony G. *An Economic History of West Africa.* New York: Columbia University Press, 1973.

Hunt, Nancy Rose. "Domesticity and Colonialism in Belgian Africa: Usumbura's Foyer and Social, 1946–1960." *Signs: Journal of Women in Culture and Society* 15 (1990): 447–74.

Hursh, Gerald D., Niels R. Roling, and Graham B. Kerr. *Innovation in Eastern Nigeria: Success and Failure of Agricultural Programs in 71 Villages of Eastern Nigeria: Diffusion of Innovations Research Report 8.* Michigan State University, 1968.

Huss-Ashmore, Rebecca, and Solomon H. Katz, eds. *African Food Systems in Crisis: Part One: Micro Perspectives.* Reading: Gordon and Breach Science, 1989.

Hyden, G. "The Anomaly of the African Peasantry." *Development and Change* 17, no. 4 (1986): 677–705.

_____. *No Shortcut to Progress: African Development Management in Perspective.* London: Heinemann, 1983.

Idachaba, F. S. *Rural Infrastructure in Nigeria*. Ibadan: Federal Department of Rural Development, 1985.

Ifemesia, Chieka. *Traditional Humane Living Among the Igbo: Historical Perspective*. Enugu: Fourth Dimension, 1979.

Ijere, M. O. "The Agricultural Economy of Igboland." In *Igbo Economics: 1989 Ahiajoku Lecture Colloquium*. Owerri: Ministry of Information and Culture, 1989.

_____. *The African Poor: A History*: Cambridge University Press, 1987.

Isichei, E. *A History of the Igbo People*. London, 1976.

_____. *Igbo Word: An Anthology of Oral Histories and Historical Descriptions*. Philadelphia: Institute for the Study of Human Issues, 1976.

Isichei, E. *A History of Nigeria*. London: Longman, 1983.

Iyegha, David. A. *Agricultural Crisis in Africa: The Nigerian Experience*. London: University Press of America, 1988.

Jewsiewicki, Bogumil, and D. Newbury, eds. *African Historiographies: What History for Which Africa?* Boulder, CO: Sage, 1986.

Jewsiewicki, Bogumil. "African Historical Studies, Academic Knowledge as 'Usable Past': A Radical Scholarship." *African Studies Review* 32, no. 3 (1989): 1–79.

Jiggins, Janice. "Gender-Related Impacts and the Work of the International Agricultural Research Centre." Washington, DC: *CGIAR Study Paper No. 17*, 1986.

Johnston, B. J. "Agricultural and Structural Transformation in Developing Countries: A Survey of Research." *Journal of Economic Literature* 8, no. 2 (1970): 369–401.

_____, and J. W. Mellor. "The Role of Agriculture in Economic Development." *American Economic Review* 51, no. 3 (1961): 566–93.

Jones, W. O. "Food and Agricultural Economy of Tropical Africa." *Food Research Institute Studies* 2 (1961): 3–20.

Kay, Hymer G. *The Political Economy of Colonialism in Ghana: A Collection of Documents and Statistics 1900–1960*. Cambridge: Cambridge University Press, 1972.

Kilby, Peter. "What Oil Wealth Did to Nigeria." *Wall Street Journal*, 25 November 1981, 26.

Killingray, David, and Richard Rathbone, eds. *Africa and the Second World War*. London: Macmillan, 1985.

King, Deborah. "Multiple Jeopardy, Multiple Consciousnesses: The Context of a Black Feminist Ideology." *Signs: Journal of Women in Culture and Society* 14 (1988): 265–95.

Kjaerby, F. *Problems and Contradictions in the Development of Ox-Cultivation in Tanzania. Research Report No. 66*. Copenhagen: Centre for Development Research, 1983.

Klein, Martin A. "Social and Economic Factors in the Muslim Revolution in Senegambia," *Journal of African History* 13 (1972): 414–41.

_____. *Peasants in Africa: Historical and Contemporary Perspectives*. London: Sage, 1980.

Kleis, Gerald K. "Confrontation and Incorporation: Igbo Ethnicity in Cameroon," *African Studies Review* 23, no. 3 (1980): 89–100.

Koikari, Mire. "Rethinking Gender and Power in the US Occupation of Japan, 1945–1952." *Gender and History* 11, no. 2 (1999): 313–35.

Kolawole, Are, Y. A. Ambi, and J. F Alamu. "Agricultural Transformation, Food Crises and the Accelerated Wheat Production Programme in Nigeria: The Case of Kano and Kaduna States." In *Agricultural Transformation and Social Change in Africa*, ed. Bernhand Nett, Wulf Volker, and Abdramare Diarra. Peter Lang, 1992.

Kopytoff, Igor. "Women's Roles and Existential Identities." In *African Gender Studies: A Reader*, ed. Oyeronke Oyewumi, 127–44. New York: Palgrave Macmillan, 2005.

Korieh, Chima J., and F. J. Kolapo. *The Aftermath of Slavery: Transitions and Transformation Southern Nigeria*. Trenton: Africa World Press, 2007.

_____. "Agriculture," in *Africa*, vol. 5: *Contemporary Africa*, ed. Toyin Falola, 417–36. Durham, NC: Carolina Academic Press, 2003.

_____. "Gender and Resistance in Colonial Eastern Nigeria: Recasting the Myth of the Invisible Woman, 1925–1945," in *Transformations of Nigeria: Essays in Honour of Toyin Falola*, ed. A. Adebayo, 623–46. Trenton, NJ: Africa World Press, 2003.

_____. "The Invisible Farmer? Women, Gender, and Colonial Agricultural Policy in the Igbo Region of Nigeria, c. 1913–1954," *African Economic History* 29 (2001): 117–62.

_____. "Urban Food Supply and Vulnerability during the Second World War," in *Nigerian Cities*, ed. Toyin Falola and Steven J. Salm, 127–52. Trenton: African World Press, 2004.

_____. "Food Production and the Food Crisis in Sub-Saharan Africa," in *Africa*, vol. 5: *Contemporary Africa*, ed. Toyin Falola, 391–416. Durham, NC: Carolina Academic Press, 2003.

_____. "The Nineteenth Century Commercial Transition in West Africa: The Case of the Biafra Hinterland." *Canadian Journal of African Studies* 34, no. 3 (2000): 588–615.

Labasse, J. "La Dakar – Niger et sa zone d'Action." *Revue de géographie de Lyon* 3 (1954): 183–204.

Laclau, C. "Feudalism and Capitalism in Latin America." *New Left Review* 67 (1971): 36–37.

Lagemann, J. *Traditional Farming Systems in Eastern Nigeria*. Munich: Weltforum, 1977.

Lamphere, L., and N. Z. Rosaldo. *Women, Culture, and Society*. Stanford: Stanford University Press, 1974.

Law, Robin. "The Historiography of the Commercial Transition in Nineteenth Century West Africa." In *African Historiography: Essays in Honour of Jacob Ade Ajayi*, ed. Toyin Falola, 91–115. London: Longman, 1993.

_____. *From Slavery to 'Legitimate' Commerce: The Commercial Transition in Nineteenth Century West Africa*. Cambridge: Cambridge University Press, 1995.

Lawrence, Peter. "The Political Economy of the 'Green Revolution' in Africa." *Review of African Political Economy* 15, no. 42 (1988): 59–75.

Leith-Ross, Sylvia. *African Women: A Study of the Igbo of Nigeria*. London, 1965.

Levine, Philippa, ed. *Gender and Empire*. Oxford: Oxford University Press, 2004.

Leys, Colin. "Confronting the African Tragedy." *New Left Review* 204 (1994): 33–47.

Likaka, Osumaka. "Rural Protest: The Mbole Against Belgian Rule, 1897–1959." *International Journal of African Historical Studies* 27, no. 3 (1994): 589–617.

Lofchie, Michael F., and Commins, S. K. "Food Deficit and Agricultural Policies in Tropical Africa." *Journal of Modern African Studies* 20, no. 1 (1982): 1–25.

Lofchie, Michael F. "Africa's Agricultural Crisis: An Overview." In *Africa Agrarian Crisis*, ed. S. K. Commins et al., 3–18. Boulder, CO: Lynne Rienner, 1986.

_____. "The Decline of African Agriculture: An Internalist Perspective." In *Drought and Hunger in Africa: Denying Famine a Future*, ed. Michael H. Glantz, 85–109. Cambridge: Cambridge University Press, 1987.

Lonsdale, John, and Bruce Berman, "Copping with the Contradictions: The Development of the Colonial State in Kenya, 1898–1914." *Journal of African History* 20 (1979): 487–505.

Lopata, Helen, and Barrie Thorne. "On the Term 'Sex Roles.'" *Sign: Journal of Women in Culture and Society* 3 (1978): 718–21.

Lorber, Judith. *Paradoxes of Gender*. New Haven, CT: Yale University Press, 1994.

Lugard, F. D. *Dual Mandate in Tropical Africa*. Edinburgh: Blackwood Press, 1922.

Mackintosh, Maureen. "Gender and Economics: The Sexual Division of Labour and the Subordination of Women." In *Of Women and Marriage: Women's Subordination Internationally and its Lessons*, ed. Kate Young et al., 1–15. London and New York: Routledge, 1984.

Macus, George E., and Michael M. J. Fischer. *Anthropology and Cultural Critique: An Experimental Movement in the Human Sciences*. Chicago: University of Chicago Press, 1986.

MacWillian, Scoth, France Desaubin, and Wendy Timms. *Domestic Food Production and Political Conflict in Kenya*. Western Australia: Indian Ocean Centre for Peace Studies, 1995.

Maday, Bela, ed. *Anthropology and Society*. Washington, DC: Anthropological Society of Washington, 1975.

Margot, Lovett. "Gender Relations, Class Formation and the Colonial State in Africa." In *Women and the State in Africa*, ed. Parpart Jane and Kathleen A. Staudt, 23–46. Boulder, CO: Lynne Rienner, 1989.

Martin Lynn. *Commerce and Economic Change in West Africa: The Palm Oil Trade in the Nineteenth Century*. Cambridge: Cambridge University Press, 1997.

Martin, Anne. *The Oil Palm Economy of the Ibibio Farmer*. Ibadan: Ibadan University Press, 1956.

Martin, Kay M., and Barbara Voorhies. *The Female of the Species*. New York: Columbia University Press, 1975.

Martin, Susan. "Slaves, Igbo Women and Oil Palm." In *From Slavery to Legitimate Commerce: The Commercial Transition in Nineteenth Century West Africa*, ed. Robin Law, 172–194. Cambridge: Cambridge University Press, 1995.

_____. *Palm Oil and Protest: An Economic History of the Ngwa Region, Southeastern Nigeria, 1800–1990*. Cambridge: Cambridge University Press, 1988.

_____. "Gender and Innovation: Farming, Cooking and Palm Processing Ngwa Region of Southeastern Nigeria, 1900–1930." *Journal of African History* 25 (1984): 411–27.

Mba, Nina. *Nigerian Women Mobilized: Women's Political Activity in Southern Nigeria, 1900–1965.* Berkeley: Institute of International Studies, University of California, 1982.

_____. "Heroines of the Women's War." In *Nigerian Women in Historical Perspective*, ed. Bolanle Awe, 73–88. Lagos: Sincere/Bookcraft, 1992.

McClintock, Ann. *Imperial Leather: Race Gender and Sexuality in the Colonial Contest.* New York and London: Routledge, 1995.

McKittrick, Meredith. "Capricious Tyrants and Persecuted Subjects: Reading between the Lines of Missionary Records in Precolonial Northern Namibia." In *Sources and Methods in African History: Spoken, Written and Unearthed*, ed. Toyin Falola and Christian Jennings, 219–36. Rochester, NY: University of Rochester Press, 2003.

Meillassoux, Claude. *L'Anthropologie économique des Guro de Cote d'Ivoire.* Paris: Mouton, 1964.

_____. *Maidens, Meal and Money.* Cambridge: Cambridge University Press, 1981.

Meredith, David. "Government and the Decline of the Nigeria Oil-Palm Export Industry, 1991–1939." *Journal of African History* 25 (1984): 311–29.

Merton, R. "Insiders and Outsiders: A Chapter in the Sociology of Knowledge," *American Journal of Sociology* 78 (1972): 9–47.

Mettrick H. *Oxenization in the Gambia: An Evolution.* London: Ministry of Overseas Development, 1973.

Miller, Joseph C. "Presidential Address: History and Africa/Africa and History," *American Historical Review* 104, no. 1 (1999): 1–32.

Mills, James, and Patricia Barton, eds. *Drugs and Empires: Essays in Modern Imperialism and Intoxication 1500–1930.* New York: Palgrave Macmillan, 2007.

Minnier, Y. *Poussière et la cendre: Paysages, dynamiques de formations vegetales et stratégies des sociétés en Afrique de l'ouest.* Paris: Agence de Coopération Culturelle et Technique, 1981.

Modleski, Tania. *Feminism without Women: Culture and Criticism in a "Postfeminist Age."* New York: Routledge, 1991.

Mok Chiu Yu, and Lynn Arnold, ed. *Nigeria-Biafra: A Reading into the Problems and Peculiarities of the Conflict.* Adelaide: Adelaide University Quaker Society, 1968.

Moniot, H. "Profile of a Historiography: Oral Tradition and Historical Research in Africa." In *Profile of a Historiography*, ed. B. Jewsiewicki and D. Newbury, 50–59. Boulder, CO: Sage, 1986.

Moore, Henrietta, and Vaughan Megan. *Cutting Down Trees: Gender, Nutrition, and Agricultural Change in the Northern Province of Zambia, 1890–1900.* Portsmouth: Heinemann, 1994.

Morgan, D. J. *The Official History of Colonial Development, vol. 1: The Origins of British Aid Policy, 1924–45.* Atlantic Highlands, NJ: Humanities Press, 1980.

Morgan, W. B. "The Influence of European Contacts on the Landscape of Southern Nigeria," *Geographical Journal* 125, no. 1 (1959): 48–64.

_____."*Farming Practice*, Settlement Pattern and Population Density in Southeastern Nigeria," Geographical Journal 121, no. 3 (1955): 320–33.

_____, and J. C. Pugh. *West Africa*. London: Methuen, 1969.

Morgen, S., et al. *Gender and Anthropology: Critical Reviews for Research and Teaching*. Washington, DC: American Anthropological Association, 1989.

_____. *Adapting to Drought: Farmers, Famines and Desertification in West Africa West Africa*. Cambridge: Cambridge University Press, 1985.

Moser, Caroline. "Gender Planning in the Third World: Meeting Practical and Strategic Gender Needs," *World Development* 17, no. 11 (1989): 1799–825.

Muntemba, Maud Shimwaayi. "Women and Agricultural Change in the Railway Region of Zambia: Dispossession and Counter Strategies, 1930–1970." In *Women and Work in Africa*, ed. Edna Bay, 83–103. Boulder, CO: Westview Press, 1982.

Mutsaers, H.J.W. *Peasants, Farmers and Scientists: A Chronicle of Tropical Agricultural Science in the Twentieth Century*. New York: Springer, 2007.

Murdock, George P. *Africa: Its Peoples and their Cultural History*. New York: McGraw-Hill, 1959.

Murray, C. *Families Divided: The Impact of Migrant Labour in Lesotho*. Johannesburg: Ravan, 1981.

Musisi, Nakanyike B. "A Personal Journey into Custom, Identity, Power, and Politics: Researching and Writing the Life and Times of Buganda's Queen Mother Irene Drusilla Namaganda (1896–1957)," *History in Africa* 23 (1996): 369–85.

Nafziger, E. Wayne. "The Economic Impact of the Nigerian Civil War," *Journal of Modern African Studies* 10, no. 2 (1972): 223–45.

Ndaywel, E. Nziem, "African Historians and Africanist Historian." In *Profile of a Historiography*, ed. Bogumil Jewsiewicki and David Newbury, 20–27. Boulder, CO: Sage, 1986.

Njoku, John E. *The Igbos of Nigeria: Ancient Rites, Change and Survival*. Lewiston, Queenston, Lamperter: Edwin Mellen Press, 1990.

Njoku, O. N. "Trading with the Metropolis: An Unequal Exchange." In *Britain and Nigeria: Exploitation or Development*, ed. Toyin Falola, 124–41. London: Zed Books, 1987.

Nnaemeka, Obioma. "Feminism, Rebellious Women, and Cultural Boundaries: Rereading Flora Nwapa and Her Compatriots." *Research in African Literature* 26, no. 2 (1995): 80–113.

_____. "Fighting on all Fronts: Gendered Space, Ethnic Boundaries, and the Nigeria Civil War." *Dialectical Anthropology* 22 (1998): 247–48.

_____. *Sisterhood, Feminisms and Power: From Africa to the Diaspora*. Trenton: Africa World Press, 1998.

_____. "Development, Cultural Forces, and Women's Achievement in Africa." *Law and Policy* 18, nos. 3 & 4 (1996): 251–79.

Norman, D. W. "Economic Analysis of Agricultural Production and Labor Utilization among the Hausa in the North of Nigeria." *African Rural Employment Paper*, no. 4, East Lansing: Michigan State University, 1973.

Northrup, David. *Trade without Rulers: Pre-colonial Economic Development in Southeastern Nigeria*. Oxford: Clarendon Press, 1978.

Nukunya, G. K. *Kinship and Marriage Among the Anlo Ewe*. New York: Humanities Press, 1969.

Nworah, K. Dike. "The Politics of Lever's West African Concessions, 1907–13." *Journal of African Historical Studies* 5 (1972): 248–64.

Nzegwu, Nkiru. "Confronting Racism: Toward the Formation of a Female-Identified Consciousness." *Canadian Journal for Women and the Law* 7, no. 1 (1994): 15–33.

Nziem, Ndaywel E. "African Historians and Africanist Historians." In *Profile of a Historiography*, ed. B. Jewswicki and D. Newbury, 20–27. Boulder, CO: Sage, 1986.

Nzimiro, Ikenna. "A Study of Mobility among the Ibos of Southern Nigeria." *International Journal of Comparative Sociology* 6, no. 1 (1965): 117–30.

_____. *Studies in Igbo Political Systems: Chieftaincy and Politics in Four Niger States*. London: Frank Cass, 1972.

_____. *The Green Revolution in Nigeria, or, The Modernization of Hunger*. Oguta, Nigeria: Zim Pan, 1985.

Oakley, A. *Sex, Gender and Society*. London: Temple Smith, 1972.

O'Connor, Anthony. *Railway and Development in Uganda: A Study in Economic Geography*. Nairobi: Oxford University Press, 1965.

_____. *Poverty in Africa: A Geographical Approach*. London: Pinter, 1991.

Oguntoyinbo, J. S. "The Changing Trends in Agricultural Production in Nigeria." *Nigerian Geographical Journal* 28 & 29, nos. 1 & 2 (1985/86): 1–14.

Ohadike, Don C., "The Influenza Pandemic of 1918–19 and the Spread of Cassava in the Lower Nigeria: A Case Study in Historical Linkages." *Journal of African History* 22 (1981): 379–91.

_____. *The Ekumeku Movement: Western Igbo Resistance to the British Conquest of Nigeria, 1883–1914*. Athens, OH: Ohio University Press, 1991.

_____. *Anioma: A Social History of the Western Igbo People*. Athens, OH: Ohio University Press, 1994.

_____. *Sacred Drums of Liberation: Religions and Music of Resistance in Africa and the Diaspora*. Trenton: Africa World Press, 2007.

Ofomata, G.E.K., ed. *A Survey of the Igbo Nation*. Onitsha, Nigeria: Africana First Publishers, 2002.

Okali, C. "Rural Sociology," in *Readings in Nigerian Rural Society and Rural Economy*, edited by Onigu Otite and Christine Okali. Ibadan: Heinemann, 1990.

Okediji, O. "Some Socio-Cultural Problems in the Western Nigeria Land Settlement Scheme: A Case Study." *Nigerian Journal of Economic and Social Studies* (November 1965): 301–10.

Okely, Judith. "'Anthropology and Autobiography' Participatory experience and Embodied Knowledge." In *Anthropology and Autobiography*, ed. Judith Okely and Helen Callaway, 1–29. London and New York: Routledge, 1992.

Okere, Linus C. *The Anthropology of Igbo Food in Rural Igboland, Nigeria: Socioeconomic and Cultural Aspects of Food and Food Habit in Rural Igboland*. Lanham, MD, London: University Press of America, 1983.

Okigbo, B. *Plants and Food in Igbo Culture and Civilization: 1980 Ahiajoku Lecture*. Owerri: Ministry of Information and Culture, 1980.

Okolie, Andrew C. "Oil Rents, International Loans and Agrarian Policies in Nigeria, 1970–1992." *Review of African Political Economy* 64 (1995): 199–212.

Olaleye, Oruene Taiwo. *Nation Builders: Women of Nigeria*. London: Change International Reports, 1985.

Olatunbosun, Dope. *Nigeria's Neglected Rural Majority*. Ibadan, Nigeria: Nigerian Institute of Social and Economic Research, 1975.

Oliver, Roland. *In the Realms of Gold: Pioneering in African History*. London: Frank Cass, 1997.

Olukoju, Ayodeji. "The Cost of Living in Lagos." In *African Urban Past*, ed. David M. Anderson and Richard Rathbone, 126–43. Oxford: James Currey, 2000.

Olusanya, G. O. *The Second World War and Politics in Nigeria, 1939–1945*. Lagos: University of Lagos, 1973.

Oluwasanmi, H. A. *Agriculture and Nigerian Economic Growth*. Ibadan: Oxford University Press, 1966.

Omari, C. K. *Persistent Principles Amidst Crisis*. Arusha, Tanzania: Evangelical Lutheran Church in Tanzania, 1989.

Onaiwu, W. Ogbomo, *When Men and Women Mattered: A History of Gender Relations Among the Owan of Nigeria*. Rochester, NY: University of Rochester Press, 1997.

Onimode, Bade. *Imperialism and Underdevelopment in Nigeria: The Dialectics of Mass Poverty*. London: Zed Books, 1982.

Onitiri, H.M.A., and D. Olatunbosun, eds. *The Marketing Board System: Proceedings of an International Conference*. Ibadan, Nigeria: Nigerian Institute of Social and Economic Research, 1974.

Onuoha, Enyeribe. *The Land and People of Umuchieze*. Owerri: Augustus, 2003.

Onwuka, Ralph I. "The Political Economy of the Igbo." In *Igbo Economics: Papers presented at the 1989 Ahiajoku Lecture (Onugaotu) Colloquium*, ed. G. M. Umezurike, et al., 7–17. Owerri: Ministry of Information and Culture, 1991.

Orewa, S. I. "Designing Agricultural Development Projects for the Small Scale Farmers: Some Lessons from the World Bank Assistance Small Holder Oil Palm Development Scheme in Nigeria." *Journal of Applied Sciences* 8, no. 2 (2008): 295–301.

Oriji, John. *Ngwa History: A Study of Social and Economic Change in Igbo Mini-State in Time Perspective*. New York: Peter Lang, 1991.

Ortner, S., and H. Whitehead, eds. *Sexual Meanings: The Cultural Construction of Gender and Sexuality*. Cambridge: Cambridge University Press, 1981.

Oshio, P. E. "Agricultural Policy and the Nigerian Land Use Decree: The Conflict," *Journal of African Law* 30, no. 2 (1986): 130–42.

Ottenberg, Phoebe V. "The Changing Economic Position of Women among the Afikpo Ibo." In *Continuity and Change in African Cultures,* ed. Bascon Russell William and Melvin J. Herskovits, 205–23. Chicago: University of Chicago Press, 1958.

Ottenberg, Simon. "Ibo Receptivity to Change." In *Continuity and Change in African Cultures,* ed. William Russell Bascom and Melvin J. Herskovits, 130–43. Chicago: University of Chicago Press, 1958.

———. *Farmers and Townspeople in a Changing Nigeria: Abakaliki during Colonial Times (1905–1960).* Ibadan: Spectrum, 2005.

Oyeronke, Oyewumi. *The Invention of Women: Making an African Sense of Western Gender Discourses.* Minneapolis: University of Minnesota Press, 1997.

———, ed. *African Gender Studies: A Reader.* New York: Palgrave Macmillan, 2005.

Phillips, Anne. *The Enigma of Colonialism: British Policy in West Africa.* London: James Currey, 1989.

Phillips, T. A. *An Agricultural Notebook.* Lagos: Longman, 1977.

Posnansky, M., ed. *Prelude to East African History.* London: Oxford University Press, 1966.

Pradham, A. "An Assessment of Rural Women's Existing and Potential Involvement in FAO-Assisted Field Projects in Nepal." Rome: FAO, 1983.

Purvis, Malcolm. *Report on the Survey of the Oil Palm Rehabilitation Scheme in Eastern Nigeria.* E. Lansing: Consortium for the Study of Nigerian Rural Development, 1968.

Ragin, Charles C. "The Logic of Qualitative Comparative Analysis." *International Review of Social History* 43 (1998): 165–84.

Ranger, Terence. "Growing from the Roots: Reflections on Peasants Research in Central and Southern Africa." *Journal of Southern African Studies* 5, no. 1 (1978): 99–133.

———. *Peasant Consciousness and Guerrilla Warfare in Zimbabwe.* London: James Curry, 1985.

Rathgeber, Eva M. "WID, WAD, GAD: Trends in Research and Practice." *Journal of Developing Areas* 24 (1990): 489–502.

Redfield, Robert. "Are African Cultivators to be Called 'Peasants'?" in Jack M. Potter, May N. Diaz, and George McClelland Foster, *Peasant Society: A Reader.* Boston: Little Brown, 1967.

Reskin, Barbara, and Irene Padadic, *Women and Men at Work,* Thousand Oaks, CA: Pine Forge Press, 1994.

Richard Jr., Mitchell, and Charmaz Kathy, "Telling Tales, Writing Stories: Postmodernist Visions and Realists Images in Ethnographic Writing." *Journal of Contemporary Ethnography* 25, no. 1 (1996): 144–66.

Riney-Kehrberg, Pamela. *Rooted in Dust: Surviving Drought and Depression in Southwestern Kansas.* Lawrence: University of Kansas Press, 1994.

Rodney, Walter. *A History of the Upper Guinea Coast 1545–1800* (Oxford: Clarendon Press, 1970).

———. *How Europe Underdeveloped Africa.* London: Bogle-L'Ouverture, 1972.

Roider, W. *Farm Settlements for Socio-Economic Development: The Western Nigeria Case.* Munchen: Weltforum, 1971.

Rosaldo, Michelle Z. "Women, Culture and Society: A Theoretical Overview." In *Women, Culture and Society* ed. M. Rosaldo and L. Lamphere, 17–42. Stanford, CA: Stanford University Press, 1974.

Rosen, David M. "The Peasant Context of Feminist Revolt in West Africa," *Anthropological Quarterly* 56, no. 1 (1983): 35–43.

Rosling, H. *Cassava Toxicity and Food Security: A Review of Health Effects of Cyanide Exposure from Cassava and of Ways to Prevent the Effects.* Ibadan: IITA-UNICEF, 1987.

Rotberg, Robert I., ed. *Imperialism, Colonialism and Hunger: East and Central Africa,* Toronto: Lexington, 1983.

Sachs, C. C. *The Invisible Farmers: Women in Agricultural Production.* Totowa, NJ: Rowman & Allanheld, 1983.

Sacks, K. *Sisters and Wives: The Past and Future of Sexual Equality.* Westport, CT: Greenwood, 1979.

Samuel, Raphael, *Village Life and Labour.* London and Boston: Routledge and Kegan Paul, 1975.

Sanday, P. *Female Power and Male Dominance.* Cambridge: Cambridge University Press, 1981.

Saul, John S., and Roger Woods. "African Peasantries." In *Peasants and Peasant Societies,* ed. T. Shanin, 401–16. London: Blackwell, 1987.

Schech, Susanne, and Jane Haggis, *Culture and Development: A Critical Introduction.* Oxford: Blackwell, 2000.

Scott, James C. *The Moral Economy of the Peasant: Rebellion and Subsistence in Southeast Asia.* New Haven, CT: Yale University Press, 1976.

_____. *Weapons of the Weak: Everyday forms of Resistance.* New Haven, CT: Yale University Press, 1985.

Sen, A. *Poverty and Famine: An Essay on Entitlement and Deprivation.* Oxford: Clarendon Press, 1981.

Shanin, T. *Peasants and Peasant Societies.* Harmondsworth: Penguin, 1976.

Shaxson, Nicholas. *Poisoned Wells: The Dirty Politics of African Oil.* New York: Palgrave Macmillan, 2007.

Shaw, T. M. "Towards a Political Economy of the African Crisis: Diplomacy, Debates, and Dialectics." In *Drought and Hunger in Africa,* ed. M. H. Glantz, 127–47. Cambridge: Cambridge University Press, 1987.

Shaw, T. M., and J. O. Ihonvbere. *Towards A Political Economy of Nigeria: Petroleum Policy at the (Semi) Periphery.* Brookfield, VT: Avebury, 1982.

Shepherd, Verene, et al., eds. *Engendering History: Caribbean Women in Historical Perspective.* Kingston: Ian Randle, 1995.

Siddle, David, and Kenneth Swindell. *Rural Change in Tropical Africa: From Colonies to Nation-States.* Oxford: Basil Blackwell, 1990.

Silverman, David. Review of *The New Language of Qualitative Method* by Jaber F. Gubrium and James A. Holstein. *American Journal of Sociology* 105, no. 2 (1999): 578.

Skinner, E. P. "West African Economic System." In *Economic Transition in Africa*, ed. M. J. Herskovits and M. Harwitz, 77–97. Evanston, IL: Northwestern University Press, 1961.

Smock, David. "Land Fragmentation and the Possibilities of Consolidation in Eastern Nigeria." *Bulletin of Rural Economics and Sociology* 11, no. 3 (1967): 194–210.

_____, and Audrey C. Smock. *Cultural and Political Aspects of Rural Transformation: A Case Study of Eastern Nigeria*. New York: Praeger, 1972.

Solarz, William, B. Morgan, and A. Jerzy. "Agricultural Crisis in Sub-Saharan African: Development Constraints and Policy Problems." *Geographical Journal* 160, no. 1 (1994): 57–73.

Spear, Thomas. *Mountain Farmers: Moral Economies of Land and Agricultural Development in Arusha and Meru*. Berkeley: University of California Press, 1997.

_____. "Section Introduction: New Approaches to Documentary Sources." In *Sources and Methods in African History: Spoken, Written and Unearthed*, ed. Toyin Falola and Christian Jennings, 169–172. Rochester, NY: University of Rochester Press, 2003.

Spring, Anita. *Agricultural Development and Gender Issues in Malawi*. Lanham, MD: University Press of America, 1995.

Stark, O. "On the Role of Urban to Rural Remittances." *Rural Development Journal of Development Studies* 16 (1980): 369–74.

Stoler, Ann Laura, and Karen Strassler. "Casting for the Colonial: Memory in 'New Order Java'." *Comparative Studies in Society and History* 42, no. 1 (2000): 4–48.

Swindell, K. "The Crisis in African Agriculture." *African Affairs* 88 (1989): 119–21.

Tariba, O. *Poverty in Nigeria: Proceeding of the 1975 Annual conference of the Nigeria Economic Society*. Ibadan: Nigeria Economic Society, 1976.

Tashakkori, Abbas, and Charles Teddlie. *Mixed Methodology: Combining Qualitative and Quantitative Approaches*. Thousand Oaks, CA: Sage, 1998.

Taylor, Verta. "Gender and Social Movements: Gender Processes in Women's Self-help Movements." *Gender and Society* 13, no. 1 (1999): 8–33.

Terray, Emmanuel. "Historical Materialism and Segmentary Lineage-Based Societies." In *Marxism and 'Primitive' Societies*, ed. Emmanuel Terray, 93–186. New York: Monthly Review Press, 1972.

Thomas, N. W. *Anthropological Report on the Igbo of Eastern Nigeria*. 6 vols. London, 1913–1914.

Thompson, Paul. *The Voice of the Past: Oral History*. Oxford: Oxford University Press, 1988.

Thomson, Alistair. *The Oral History Reader*. London and New York: Routledge, 1998.

Thornton, John. "European Documents and African History." In *Writing African History*, ed. John Edward Philips, 254–65. Rochester, NY: University of Rochester Press, 2005.

Tiffen, Mary, *The Enterprising Peasants: Economic Development in Gombe Emirate North Eastern States, Nigeria 1900–1968*. London, Overseas Research Publication No. 21, 1976.

Timberlake, Lloyd. *Africa in Crisis: The Causes, the Cures of Environmental Bankruptcy*. London: Earthscan, 1985.

Tong, Rosemarie. *Feminist Thought: A More Comprehensive Introduction*. 2nd ed. Boulder, CO: Westview Press, 1989.

Tosh, J. "The Cash Crop Revolution in Tropical Africa: An Agricultural Reappraisal." *African Affairs* 79 (1980): 79–94.

Uchendu, Victor. *The Igbo of Southeastern Nigeria*. New York: Holt, Rinehart and Winston, 1965.

_____. "Some Issues in African Land Tenure." *Tropical Agricultura* 44, no. 2 (1967): 91–101.

Udo, R. K. "The Migrant Tenant Farmer of Eastern Nigeria," *Africa: Journal of the International African Institute* 34, no. 4 (1964): 326–39.

Ukaegbu, B. N. *Production in the Nigerian Oil palm Industry, 1900–1954*. London: University of London Press, 1974.

Umar, Muhammad S. *Islam and Colonialism: Intellectual Responses of Muslims of Northern Nigeria to British Colonial Rule*. Leiden: Brill, 2006.

Umezurike, G. M., et al. *Igbo Economics: 1989 Ahiajoku Lecture (Onugaotu) Colloquium*. Owerri: Ministry of Information and Culture, 1989.

Usoro, E. J. *The Nigerian Palm Oil Industry: Government and Export Production 1906–1965*. Ibadan: Ibadan University Press, 1974.

Uzodinma, Nwala. "Some Reflections on British Conquest of Igbo Traditional Oracles, 1900–1924." *Nigeria Magazine* 142 (1982): 25–35.

Van-Allen, Judith. "Sitting on a Man: Colonialism and the Lost Political Institutions of Igbo Women." *Canadian Journal of African Studies* 6, no. 2 (1972): 168–81.

_____. "Aba Riot or Women's War?: Ideology, Stratification, and the Invisibility of Women." In *Women in Africa: Studies in Social and Economic Change*, ed. Nancy J. Hafkin and Edna G. Bay, 59–86. Stanford, CA: Stanford University Press, 1976.

Vansina, Jan. *Living with Africa*. Madison: University of Wisconsin Press, 1994.

_____. *Oral History*. London, 1965.

_____. *Oral Tradition as History*. London, 1985.

_____. *Oral Tradition: A Study in Historical Methodology*, 1961.

_____. "Knowledge and Perceptions of the African Past." In *Profile of Historiography*, ed. B. Jewswicki and D. Newbury, 28–41. Boulder, CO: Sage, 1986.

_____. "Recording the Oral History of the Bakuba – I: Methods," *Journal of African History* 1, no. 1 (January 1960): 43–53.

_____. "Recording the Oral History of the Bakuba – II: Results," *Journal of African History* 1, no. 2 (July 1960): 257–70.

Wallach Scott, Joan. *Gender and the Politics of History*. New York: Columbia University Press, 1988.

Wariboko, Waibinte E. "The Status, Role and Influence of Women in the Eastern Delta States of Nigeria, 1850–1900: Examples From New Calabar." In *Engendering History: Caribbean Women in Historical Perspective*, ed. Verene Shephard, Bridget Brereton, and Barbara Bailey, 369–83. Kingston: Ian Randle, 1995.

Warner-Lewis, Maureen. *Archibald Monteath: Igbo, Jamaican, Moravian*. Kingston, Jamaica: University of the West Indies Press, 2007.

Watts, Michael. *State, Oil and Agriculture in Nigeria*: Institute of International Studies, Berkeley: University of California, 1987.

_____. *Silent Violence: Food, Famine, and Peasantry in Northern Nigeria*. Berkeley: University of California Press, 1983.

Wells, Jerome C. *Agricultural Policy and Economic Growth in Nigeria, 1962–1968*. Ibadan: Oxford University Press, 1974.

Wilson, Charles. *The History of Unilever: A Study in Economic Growth and Social Change*, vol. 1. London, 1954.

Wolpe, H., ed. *Articulation of Modes of Production*. London: Routledge and Kegan Paul, 1980.

World Bank. "Poverty in Sub-Sahara Africa." *Report No. 11842-MLI*. Washington, DC, 1993.

_____. *Sub-Saharan Africa: From Crisis to Sustainable Development*. Washington, DC: World Bank, 1989.

_____. *Towards Sustainable Development in Sub-Saharan Africa: A Joint Program of Action*. Washington, DC: World Bank, 1984.

_____. *Accelerated Development in Sub-Saharan Africa: An Agenda for Action*. Washington, DC: World Bank, 1981.

Worster, Donald. *Dust Bowl: The Southern Plains in the 1930s*. New York: Oxford University Press, 1979.

Wright, M. "Technology, Marriage and Women's Work in the History of Maize Growers in Mazabuka, Zambia: A Reconnaissance," *Journal of Southern African Studies* 10, no. 1 (1983): 71–85.

Wright, Richard. "Blueprint for Negro Writing." In *The Black Aesthetics*, ed. Addison Gayle, Jr. New York: Doubleday, 1971.

Nwabughuoghu, Anthony I. "The Isusu: An Institution for Capital Formation among the Ngwo Igbo: Its Origins and Development." *Africa* 54 (1984): 46-58.

Yearwood, Peter J. "The Expatriate Firms and the Colonial Economy in Nigeria in the First World War." *Journal of Imperial and Commonwealth History* 26, no. 1 (1998): 49–77.

Zartman, I. W. *The Political Economy of Nigeria*. New York: Praeger, 1983.

INDEX

Bende, 48, 126, 147
    Division of, 179
Benin, 29, 39, 136
Benue, 257
Berry Sara, 7, 9,115, 232
Biafra, 38, 220, 221, 224
Biafra Development Corporation (BDC),
    225
Biakpan Rubber Estate, 203
Bight Negroes, 50
Bight of Biafra, 30, 41, 49, 50, 52, 89
    Hinterland of, 77, 292n40
biscuits, 173
Boje Cocoa Estate, 203
Boki, 208
Bonny, 31, 79, 89
    Coconut Estate in, 203
Boserup, Ester, 8
Braham, Alex J., 67
bread, 173
Bridges, A.F.B., 70, 113, 187
Britain, 168
    Empire of, 59
British and Continental African Company
        Limited, 298n14
British, 170, 172, 182, 259
Brockway, Lord Fenner, 221
Brown, J. B., 109
Butcher, H.L.M., 187
butter, 173

C

Calabar, 31, 61, 62, 75, 78, 89, 125, 147, 188,
        202, 261
    Catering Rest House in, 259
    District of, 259
    Province of, 137, 252, 259
Calaro Oil Palm Estate, 203
Cameroon, 40, 109, 259, 263
    River, 51
    Southern, 263
Campbell, Benjamin, 52
Cape of Good Hope, 29
Carew, W. E., 32
Caritas, 219

Carr, F. D., 86, 87, 157, 164, 166, 167, 300n
        52, 302n11
cassava, 114, 177, 214, 215, 245, 250, 252,
        254, 267
cattle, 30
Ceylon, 207
Chambers, Douglas, 49
charcoal fuel, 171
Chubb, L. T., 166, 191
Chuku, Gloria, 16
Chukuemeka Odumegwu Ojukwu, 217, 225
Church Missionary Society, 101
Church of Biafra, 107
Clifford, Hugh, 79, 81, 93
Clough, Raymond Gore, 57, 84, 132, 134
Clute, Robert, 8
cocoyams, 28, 30
Colonial Development and Welfare Act,
        190, 196
Committee for Peace in Nigeria, 221
confectionery, 173
Conference of Christian Rural Workers,
        102
Congo, 81
Cooks, A. E., 74, 84, 85, 126, 139, 147, 182
Coquery-Vidrovitch, Catherine, 39
Couper Johnstone and Company, 298n14
Coursey, D. G., 41
Cross River, 4, 29, 48, 127, 254

D

Dance Movement, 130
Davidson, Basil, 170
Davies, J. B., 258, 262
de Carli, Denis, 39
Delta Igbo, 283n7
Denman, Captain, Joseph, 52
Dependency theory, 288n50
Diamond, Stanley, 255
Dibia, Felix, x
Dike, Kenneth O., 42, 55, 253
Dikwa, 66
Dobinson, H. H., 41
Douglas Expedition, 298n15

U

U.S. Agency for International Development (USAID), 207

Ubahu, 104

Uburu, 49

Uchendu, Victor, xiii, 32, 270

Ude, Emmanuel, 254

Udi, 100, 256
    Division, 166, 176

Udo, R. K., 253

Ukaegbu, Basil, 115

Ukwu, Ukwu, I, 48

Ulakwo, 186

Ulonna North, 208, 240

Ulonna South, 208

Umon, 127

Umuagwo, 249

Umuahia, 100, 102, 110, 117, 134, 139, 187, 189
    Cocoa Estate in, 203
    District of, 134
    Division, 202

Umuakpo, 155

Umudike, 86, 110

Umudioka, 32

Umueri, 28

Umuhu, 153

Umuorlu, 40

Umuosi, 153

United Africa Company, 61, 87, 147, 158–59, 169, 191, 258, 262

United Church of Canada, 101

United Kingdom, 168, 207

United States, 79, 168

University of Biafra, 219

University of Nigeria, 219

Upper Guinea Coast, 39

USAID, 225

Usoro, Eno, 15, 113, 283n6

Uturu, 159
    Trade Centre in, 104

Uyo, 177, 198, 202

Uzere, 63

Uzoakoli, 48

Uzouwani, 208

V

Van Allen, Judith, 128

Victoria, 260

W

Walter, B. W., 100

Ward, C. H., 161

Wariboko, Waibite, 55, 61

Warrant Chief, 71–73, 146

Watson, McL. E., 174

Watts, Michael, 7, 237

Wells, Jerome C., 283n7

West Africa, 29, 55, 59, 91, 172

West African coast, 29, 30, 51, 53

West African Cocoa Control Board, 191

West African Motors, 147

West African Pilot, 163, 171, 185

West African Produce Control Board, 191

West Germany, 207

Western Nigeria, 9, 98

wild rubber, 171

wild silk, 171

Win the War Fund, 164

Windward Coast, 50

Women's Revolt, 75, 131

Women's War, xiv, 125, 157

World Bank, 6, 7, 245, 285n15

Worster, Donald, 243

Y

yam, xii, xiii, 11, 20, 28–32, 34, 38–39, 41–44, 46–47, 49, 50, 87, 174–78, 180–81, 183–85, 187, 208, 212, 214, 219, 220, 222–25, 227, 240–41, 245–46, 249, 250–53, 255, 266–68, 270, 273–74, 276, 279

Yoruba, 40, 293n66

Yorubaland, 9

9 781552 382684